The Biology of Crustacea
VOLUME 4

NEURAL INTEGRATION AND BEHAVIOR

The Biology of Crustacea

Editor-in-Chief

Dorothy E. Bliss

Department of Invertebrates
The American Museum of Natural History
New York, New York*

*Present address: Brook Farm Road, RR5, Wakefield, Rhode Island 02879

The Biology of Crustacea

VOLUME 4
Neural Integration and Behavior

Edited by

DAVID C. SANDEMAN
Department of Neurobiology
Research School of Biological Sciences
Australian National University
Canberra, Australia

HAROLD L. ATWOOD
Departments of Zoology and Physiology
University of Toronto
Toronto, Ontario
Canada

ACADEMIC PRESS 1982
A Subsidiary of Harcourt Brace Jovanovich, Publishers

New York   London
Paris   San Diego   San Francisco   São Paulo   Sydney   Tokyo   Toronto

COPYRIGHT © 1982, BY ACADEMIC PRESS, INC.
ALL RIGHTS RESERVED.
NO PART OF THIS PUBLICATION MAY BE REPRODUCED OR
TRANSMITTED IN ANY FORM OR BY ANY MEANS, ELECTRONIC
OR MECHANICAL, INCLUDING PHOTOCOPY, RECORDING, OR ANY
INFORMATION STORAGE AND RETRIEVAL SYSTEM, WITHOUT
PERMISSION IN WRITING FROM THE PUBLISHER.

ACADEMIC PRESS, INC.
111 Fifth Avenue, New York, New York 10003

United Kingdom Edition published by
ACADEMIC PRESS, INC. (LONDON) LTD.
24/28 Oval Road, London NW1 7DX

Library of Congress Cataloging in Publication Data
Main entry under title:

Neural integration and behavior.

(The Biology of crustacea ; v. 4)
Includes bibliographies and index.
1. Nervous system. 2. Sensory-motor integration.
3. Crustacea--Behavior. 4. Crustacea--Physiology.
I. Sandeman, D. C. (David C.) II. Atwood, H. L.
(Harold L.) III. Series.
QP356.N47   595.3'04188   81-22881
ISBN 0-12-106404-2   AACR2

PRINTED IN THE UNITED STATES OF AMERICA

82 83 84 85   9 8 7 6 5 4 3 2 1

*To C. A. G. Wiersma (1905-1979),
a pioneer in several fields of
crustacean neurobiology
and to
Fred Lang (1944-1978),
whose flourishing work on
crustacean neuromuscular systems
was prematurely terminated.*

Contents

List of Contributors   xi
General Preface   xiii
General Acknowledgments   xv
Preface to Volume 4   xvii
Contents of Volumes 1 – 3   xxi

1 Neural Integration in the Optic System

C. A. G. WIERSMA, JOAN L. M. ROACH, and RAYMON M. GLANTZ

I.	Interneurons of the Optic Peduncle in Brachyuran Crabs, Lobsters, and Crayfish	2
II.	Visual Interneurons of the Crayfish Circumesophageal Commissures	20
III.	The Caudal Photoreceptor	23
IV.	Summary and Hypothesis	25
	References	27

2 Control of Posture

CHARLES H. PAGE

I.	Introduction	33
II.	Principles of Postural Control	34
III.	Thoracic Leg Posture	35

	IV.	Abdominal Posture	44
	V.	Conclusion	52
		List of Abbreviations	53
		References	54

3 Locomotion and Control of Limb Movements

WILLIAM H. EVOY and JOSEPH AYERS

I.	Introduction	62
II.	Limb Movement	63
III.	Intrasegmental Motor Programs	64
IV.	Inter-Limb Coordination and Gaits	75
V.	Segmental Oscillators for Rhythmic Limb Movements	77
VI.	Intersegmental Control Systems	83
VII.	Sensory Modulation of Limb Motor Programs	83
VIII.	Summary and Conclusions	97
	References	98

4 Autotomy

A. McVEAN

I.	Introduction	107
II.	The Incidence and Consequences of Autotomy	109
III.	Physical Factors in Autotomy	112
IV.	Function of the B-l Levator Muscles in Normal Locomotion and Autotomy	121
V.	Modifying Sensory Influences on Autotomy	126
VI.	The Behavioral Roles of Autotomy	128
VII.	Perspectives	129
	References	130

5 Compensatory Eye Movements

DOUGLAS M. NEIL

I.	Introduction	133
II.	The Eye as a Motor System	134
III.	Visual Responses	138

IV.	Responses Induced by the Statocysts	144
V.	Responses to Substrate	152
VI.	Multimodal Interaction	154
	References	157

6 Control of Mouthparts and Gut

W. WALES

I.	Control of the Mouthparts	166
II.	Control of the Foregut	170
III.	Control of the Midgut	186
IV.	Movements of the Hindgut and Anus	186
V.	Perspectives and Conclusion	188
	References	189

7 Small Systems of Neurons: Control of Rhythmic and Reflex Activities

T. J. WIENS

I.	Introduction	193
II.	Several Small Systems: Motor Output and Its Mechanisms	194
III.	Principles of Operation	220
IV.	Trends and Prospects	230
	References	232

8 The Cellular Organization of Crayfish Escape Behavior

JEFFREY J. WINE and FRANKLIN B. KRASNE

I.	Introduction	242
II.	Multiple Systems for Escape: The Role of the Giant Axons	243
III.	Organization of the Afferent Pathways	246
IV.	Motor Control: Excitatory Pathways for Flexion	255
V.	Motor Control: The Extensor System	271
VI.	Command-Derived Inhibition	275

	VII.	Modulation of Escape Tendencies	283
	VIII.	Concluding Remarks	287
		Abbreviations	288
		References	289

9 Views on the Nervous Control of Complex Behavior

PETER J. FRASER

	I.	Introduction	293
	II.	Studying Interneurons	294
	III.	Multisegmental Interneurons and Behavior	309
	IV.	Perspectives	313
		References	314

Systematic Index — 321

Subject Index — 323

List of Contributors

Numbers in parentheses indicate the pages on which the authors' contributions begin.

Joseph Ayers (61), Department of Biology and Marine Science Institute, Northeastern University, Nahant, Massachusetts 01908

William H. Evoy (61), Laboratory for Quantitative Biology, Department of Biology, University of Miami, Coral Gables, Florida 33124

Peter J. Fraser (293), Department of Zoology, University of Aberdeen, Aberdeen AB9 2TN, Scotland, United Kingdom

Raymon M. Glantz (1), Department of Biology, Wiess School of Natural Sciences, Rice University, Houston, Texas 77001

Franklin B. Krasne (241), Department of Psychology, University of California at Los Angeles, Los Angeles, California 90024

A. McVean (107), Bedford College, University of London, London NW11 4NS, England

Douglas M. Neil (133), Department of Zoology, University of Glasgow, Glasgow G12 8QQ, Scotland, United Kingdom

Charles H. Page (33), Department of Physiology, Nelson Biological Laboratories, Rutgers University, Piscataway, New Jersey 08854

Joan L. M. Roach (1), Biology Division, California Institute of Technology, Pasadena, California 91125

W. Wales (165), Biology Department, University of Stirling, Stirling FK9 4LA, Scotland, United Kingdom

T. J. Wiens (193), Department of Zoology, University of Manitoba, Winnipeg, Manitoba, R3T 2N2 Canada

*C. A. G. Wiersma** (1), Biology Division, California Institute of Technology, Pasadena, California 91125

Jeffrey J. Wine (241), Department of Psychology, Stanford University, Stanford, California 94305

*Deceased.

General Preface

In 1960 and 1961, a two-volume work, "The Physiology of Crustacea," edited by Talbot H. Waterman, was published by Academic Press. Thirty-two biologists contributed to it. The appearance of these volumes constituted a milestone in the history of crustacean biology. It marked the first time that editor, contributors, and publisher had collaborated to bring forth in English a treatise on crustacean physiology. Today, research workers still regard this work as an important resource in comparative physiology.

By the latter part of the 1970s, need clearly existed for an up-to-date work on the whole range of crustacean studies. Major advances had occurred in crustacean systematics, phylogeny, biogeography, embryology, and genetics. Recent research in these fields and in those of ecology, behavior, physiology, pathobiology, comparative morphology, growth, and sex determination of crustaceans required critical evaluation and integration with earlier research. The same was true in areas of crustacean fisheries and culture.

Once more, a cooperative effort was initiated to meet the current need. This time its fulfillment required eight editors and almost 100 contributors. This new treatise, "The Biology of Crustacea," is for scientists doing basic or applied research on various aspects of crustacean biology. Containing vast background information and perspective, this treatise will be a valuable source for zoologists, paleontologists, ecologists, physiologists, endocrinologists, morphologists, pathologists, and fisheries biologists, and an essential reference work for institutional libraries.

In the preface to Volume 1, editor Lawrence G. Abele has commented on the excitement that currently pervades many areas of crustacean biology. One such area is that of systematics. The ferment in this field made it difficult for Bowman and Abele to prepare an arrangement of families of

Recent Crustacea. Their compilation (Chapter 1, Volume 1) is, as they have stated, "a compromise and should be until more evidence is in." Their arrangement is likely to satisfy some crustacean biologists, undoubtedly not all. Indeed, Schram (Chapter 4, Volume 1) has offered a somewhat different arrangement. As generally used in this treatise, the classification of Crustacea follows that outlined by Bowman and Abele.

Selection and usage of terms have been somewhat of a problem. Ideally, in a treatise, the same terms should be used throughout. Yet biologists do not agree on certain terms. For example, the term *ostracode* is favored by systematists and paleontologists, *ostracod* by many experimentalists. A different situation exists with regard to the term *midgut gland,* which is more acceptable to many crustacean biologists than are the terms *hepatopancreas* and *digestive gland.* Accordingly, authors were encouraged to use *midgut gland.* In general, however, the choice of terms was left to the editors and authors of each volume.

In nomenclature, consistency is necessary if confusion as to the identity of an animal is to be avoided. In this treatise, we have sought to use only valid scientific names. Wherever possible, synonyms of valid names appear in the taxonomic indexes. Thomas E. Bowman and Lawrence G. Abele were referees for all taxonomic citations.

Every manuscript was reviewed by at least one person before being accepted for publication. All authors were encouraged to submit new or revised material up to a short time prior to typesetting. Thus, very few months elapse between receipt of final changes and appearance of a volume in print. By these measures, we ensure that the treatise is accurate, readable, and up-to-date.

Dorothy E. Bliss

General Acknowledgments

In the preparation of this treatise, my indebtedness extends to many persons and has grown with each succeeding volume. First and foremost is the great debt owed to the authors. Due to their efforts to produce superior manuscripts, unique and exciting contributions lie within the covers of these volumes.

Deserving of special commendation are authors who also served as editors of individual volumes. These persons have conscientiously performed the demanding tasks associated with inviting and editing manuscripts and ensuring that the manuscripts were thoroughly reviewed. In addition, Dr. Linda H. Mantel has on innumerable occasions extended to me her advice and professional assistance well beyond the call of duty as volume editor. In large part because of the expertise and willing services of these persons, this treatise has become a reality.

Several biologists have provided valuable help of one sort or another during the preparation of these volumes. Worthy of special mention are Raymond B. Manning and John H. Welsh. Also deserving of thanks and praise are scientists who gave freely of their time and professional experience to review manuscripts. In the separate volumes, many of these persons are mentioned by name.

Thanks are due to all members of the staff of Academic Press involved in the preparation of this treatise. Their professionalism and encouragement have been indispensable.

No acknowledgments by me would be complete without mention of the help provided by employees of the American Museum of Natural History, especially those in the Department of Invertebrates and in the Museum's incomparable library.

Finally, the publication of the two volumes on neurobiology in this treatise would surely have been gratifying to Fred Lang and C. A. G. Wiersma. In memory of these contributors, both pre-eminent comparative physiologists, we dedicate these volumes.

Dorothy E. Bliss

Preface to Volume 4

The study of the nervous system and the behavior of animals has traditionally been divided between physiologists and ethologists. Physiologists may study problems surrounding the transduction of the external physical world into nerve impulses, mysteries of synaptic transmission, changing permeabilities of membranes that accompany the passage of the impulse along an axon, or events that occur when electrical changes in the motor nerve ending are followed by mechanical changes in muscle. Ethologists, on the other hand, hold that animals must be studied intact and viewed as a whole. Parameters in the environment are adjusted, and changes that are brought about in the overall behavior of an animal provide vital clues as to how the nervous system works.

Between these two extremes, there exists a vast middle ground—the study of the physiology of behavior, or neuroethology. The philosophy underlying the neuroethologist's approach, broadly stated, is that the complete anatomical and physiological description of the relationships between the central nervous elements will provide a satisfactory understanding of how the shifting pattern of excitation and inhibition in the individual neurons results in the behavior of the animal.

The physiological study of behavior is technically feasible only when the behavioral act chosen for analysis is simple and repeatable and involves a "system" comprising a set of appendages or a single body part. Nevertheless, the success of the neuroethologists in establishing their mechanistic theories of the generation and control of behavior has been impressive.

Neuroethological studies of crustaceans have been pursued for more than one-hundred years and have probably contributed more than those of any other group of animals in the description of basic neural mechanisms underlying behavior. This volume of "The Biology of Crustacea" is given over to reviews on the best neuroethologically researched "systems" in crustaceans. Research on these systems varies in the emphasis placed on physiological or behavioral aspects. Neural circuits associated with rapid flexion of the crayfish tail are well described (Chapter 8). More precise behavioral measurements of the orientations of crustacean eyes have been made (Chapter 5), but the underlying neural mechanisms are hardly known. Some neural systems have been studied because they can be isolated from the brain; examples include the ganglia that control the heart and those that control the stomatagastric apparatus (Chapter 7). Here, the emphasis is placed less on behavior and more on the interactions between neural elements. Thus, information gathered from each "system" contributes in its way a small piece to the overall picture we have of the nervous system.

The chapters in this volume summarize what has been an era in the application of this particular philosophy, and it is interesting to witness a re-evaluation at this point in time. To date, the emphasis has been on the simplicity and quantifiability of the responses of so-called "simple" systems. We now see, creeping quietly into the reports on these systems, an acceptance of the fact that they are not so simple after all and that in even the most direct reflex, in which the response is tightly coupled to the stimulus, variability is a very obvious and important part of the system (Chapters 2 and 3). It is fascinating to read here about some of the underlying neural mechanisms that could bring about this variability. Parallel integrative networks appear to be the most important of these mechanisms, and in the future the attention of researchers must surely be focused on this important aspect of neural processing (Chapter 1). Our comfortable ideas about neurons which we can identify as to morphology and function are no longer adequate. If multimodal neurons are really part of parallel interactive networks, their function may well change with the activity of their fellows. Can we then really assign a specific function to them and infer that they invariably "command" certain actions (Chapter 9)?

Variability in behavior may arise, not only from parallel processing, but also from changes in performance of various elements in a network, brought about by alteration in synaptic efficacy or by actions of neurohormones (Volume 3). Such effects are clearly seen in the cardiac and stomatogastric systems (Chapter 7) and in the circuits responsible for escape behavior in crayfish (Chapter 8). Ultimately, the physiology of the animal as a whole must be considered. The multiplicity of factors to be assessed in relation to control of behavior leaves us still far removed from any general "laws" of neural integration. Another decade or two of work may be necessary before general laws, if attainable, can be formulated.

This volume presents a review of the ground won by neuroethologists in their study of crustaceans. It also heralds a new and significant step in bridging the gap between the physiologists and the ethologists, namely, the search for neural mechanisms that underly variability—the essence of animal behavior.

We wish to acknowledge the following persons who read some of the manuscripts for us and offered helpful criticisms and suggestions: Dr. J. W. Bloom, Professor Peter Hallett, Dr. Stacie Moffett, Dr. C. Thompson, and Carole Breen. Additional readers were consulted by various authors during preparation of their chapters. Technical assistance during manuscript editing was provided by Mrs. Nina Murray, Mrs. Irene Kwan, Mrs. Marianne Hegstrom-Wojtowicz, Tess Falconer, and Susan Murray.

D. C. Sandeman [*]

H. L. Atwood [†]

[*]Present address: School of Zoology, University of New South Wales, Kensington, New South Wales 2033 Australia
[†]Present address: Department of Physiology, University of Toronto, Toronto, Ontario, Canada M55 1A8

Contents of Volumes 1–3

Volume 1: Systemics, the Fossil Record, and Biogeography

1. Classification of the Recent Crustacea
   Thomas E. Bowman and Lawrence G. Abele

2. Systematic Methods in Crustacean Research
   Patsy A. McLaughlin, George T. Taylor, and Martin L. Tracey

3. Origin of the Crustacea
   John L. Cisne

4. The Fossil Record and Evolution of Crustacea
   Frederick R. Schram

5. Evolution within the Crustacea
   Robert R. Hessler, Brian M. Marcotte, William A. Newman, and Rosalie F. Maddocks

6. Biogeography
   Lawrence G. Abele

Volume 2: Embryology, Morphology, and Genetics

1. Embryology
   D. T. Anderson

2. Larval Morphology and Diversity
   D. I. Williamson

3. Growth
   Richard G. Hartnoll

4. Comparative Morphology of Appendages
   Patsy A. McLaughlin

5. Sex Determination
   T. Ginsburger-Vogel and H. Charniaux-Cotton

6. Genetics
   Dennis Hedgecock, Martin L. Tracey, and Keith Nelson

Volume 3: Neurobiology: Structure and Function

1. Organization of the Central Nervous System
   David C. Sandeman

2. Organization of Neuromuscular Systems
   C. K. Govind and H. L. Atwood

3. Synapses and Neurotransmitters
   H. L. Atwood

4. Muscle
   William D. Chapple

5. Development of Nerve, Muscle, and Synapse
   C. K. Govind

6. Hormones and Neurosecretion
   Ian M. Cooke and Robert E. Sullivan

7. Photoreception
   Stephen R. Shaw and Sally Stowe

8. Chemoreception and Thermoreception
   Barry W. Ache

9. Mechanoreception
   B. M. H. Bush and M. S. Laverack

1

Neural Integration in the Optic System*

C. A. G. WIERSMA, JOAN L. M. ROACH, AND
RAYMON M. GLANTZ

I.	Interneurons of the Optic Peduncle in Brachyuran Crabs, Lobsters, and Crayfish	2
	A. Introduction	2
	B. Homolateral Visual Interneurons	3
	C. Heterolateral Visual Interneurons	17
	D. Bilateral Visual Interneurons	18
	E. Mechanoreceptive Interneurons	19
	F. Multimodal Interneurons	20
II.	Visual Interneurons of the Crayfish Circumesophageal Commissures	20
III.	The Caudal Photoreceptor	23
	A. Occurrence	23
	B. Anatomy	24
	C. Physiology	24
IV.	Summary and Hypothesis	25
	References	27

*Editors' note: This chapter was put into final form by Raymon Glantz after the death of Professor C. A. G. Wiersma in 1979.

I. INTERNEURONS OF THE OPTIC PEDUNCLE IN BRACHYURAN CRABS, LOBSTERS, AND CRAYFISH

A. Introduction

The optic system of higher decapods usually consists of four peripheral optic ganglia (contained in the eyestalk), which are connected to the supraesophageal ganglion (or brain) by the optic peduncle. Most of the investigations described here have been performed on the optic penducle, the brain, and the circumesophageal connectives.

The optic peduncle, or optic nerve, contains axons of afferent visual interneurons that originate in the peripheral optic ganglia, as well as efferent fibers that originate in the brain or even more caudally. In cross section the peduncle is observed to contain many thousands of axons (Nunnemacher et al., 1962; Nunnemacher, 1966). The crayfish optic nerve, for instance, contains about 17,000 axons. The largest are about 20 μm in diameter and the majority are less than 2 μm. The physiological results described here presumably reflect the characteristics of the largest members of this population. In addition to afferent visual fibers, the optic nerve contains efferent mechanoreceptive interneurons with extensive receptive fields, efferent axons from cephalic sense organs, particularly the statocysts (Wiersma and Yamaguchi, 1966), and efferent visual interneurons with receptive fields exclusively in the contralateral eye (Wiersma et al., 1964). These fibers presumably cross the midline of the brain in the tracts of the protocerebral commissures, which connect the two optic nerves (Hanström, 1928). Thus,

TABLE I

Known Visual Fibers in the Different Decapod Types Investigated[a]

Class of fiber	Crayfishes	Rock lobsters	Crabs
Sustaining	+	+	+
Dimming	+	+	+
Jittery movement	+	+	+
Light movement		+	
Medium movement		+	+
Fast movement	+?	+	+
Seeing		+	?
Slow movement			+
Space-constant	+	+	+
Unidirectional movement	+	+	+
Multimodal	+	+	+

[a] From Wiersma and Yanagisawa, 1971.

1. Visual Interneurons

it is possible to distinguish a large number of neuronal classes in the optic nerve. Those with visual input are shown in Table I for rock lobsters, crayfishes, and crabs.

B. Homolateral Visual Interneurons

These visual interneurons have been divided into classes on the basis of their responses to light or movement or both. Excitatory receptive fields are generally large (consisting of a few dozen to several hundred ommatidia) and overlap extensively. Furthermore, in any one species the receptive fields of the different functional classes of interneurons subdivide the visual space in the same manner (compare Figs. 1 and 2). Thus, a change in light intensity over any part of the retina elicits responses in several members of the same functional class and in members of several different classes with similar receptive fields. The integrative organization implies both substantial convergence (large receptive fields) and divergence (multiple representations of the same retinal locus). A similar pattern of convergence and divergence is also documented for the interneuronal representation of tactile hairs (Wiersma and Bush, 1963; Wiersma and Mill, 1965; Kennedy and Mellon, 1964; Kennedy, 1971).

In different species of decapod crustaceans, there are obvious similarities in the way the surface of the eye is subdivided into receptive fields (e.g., compare the subdivisions for *Panulirus interruptus*, Fig. 3, and *Carcinus*

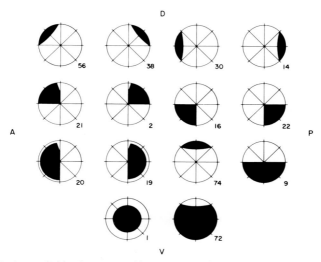

Fig. 1. Excitatory fields of sustaining fibers in *Procambarus clarkii*. Each established fiber's code number appears with its visual field. (From Wiersma and Yamaguchi, 1966.)

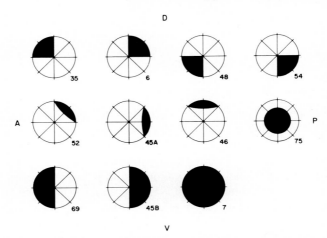

Fig. 2. Excitatory fields of jittery movement fibers in *Procambarus clarkii*. Each established fiber's coder number appears with its visual field. (From Wiersma and Yamaguchi, 1966.)

maenas, Fig. 4). Approximate outlines of the visual fields as determined (with manually controlled light probes) in crayfishes (*Procambarus clarkii*), rock lobsters (*Panulirus interruptus*), and two grapsoid crabs have been plotted on a Mercator projection of the retinas. In the crayfish (Figs. 1 and 2) and rock lobsters (Fig. 3), all areas are symmetrical with regard to horizontal and vertical axes. In *Carcinus* (Fig. 4), however, the horizontal axis is displaced ventrally, and in *Pachygrapsus crassipes* (Fig. 5), the frontal part of the eye is much larger, so that each frontal field is greater than the fields in the back half. Furthermore, whereas the crayfish has only one central area on the horizontal axis, *Panulirus* and *Pachygrapsus* have three areas in which the fields do not touch the rim at any point. As can be seen, the same fields obtain in most classes; but in those neurons with complex stimulus requirements, the receptive fields may lack sharp outlines at one or more boundaries. [Although this review is limited to decapods, most of the fiber classes discussed below have also been found in the optic nerve of the stomatopod *Oratosquilla oratoria* (Ochi and Yamaguchi, 1976).]

1. SUSTAINING FIBERS (SuF's)

Sustaining fibers (SuF's) respond to a step of illumination with a tonic "on" response that persists for the duration of the stimulus. The class has been most extensively studied in the crayfish (Aréchiga and Yanagisawa, 1973; Glantz, 1971, 1973b; Treviño and Larimer, 1970; Wiersma and Yamaguchi, 1966, 1967b; Woodcock and Goldsmith, 1970, 1973). In the rock lobsters, most work was performed on *Panulirus interruptus,* but considerable data were also obtained from *P. argus* and *P. gracilis.* In the rock

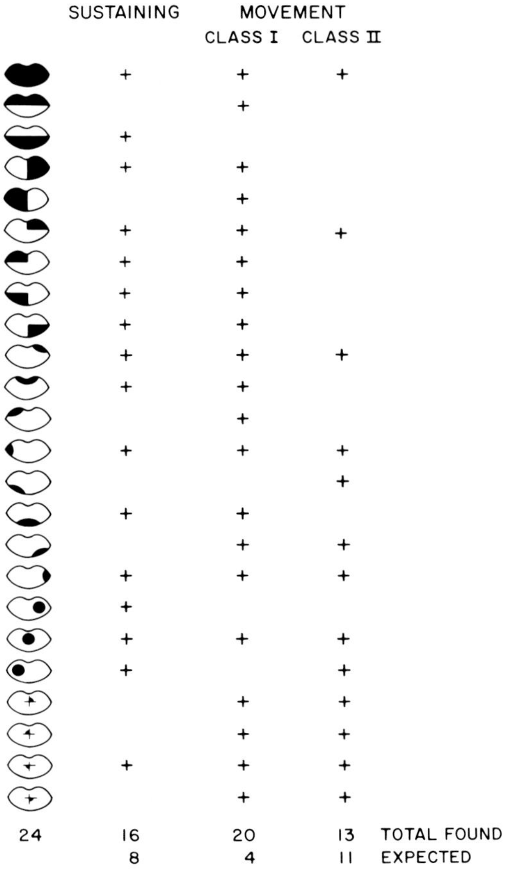

	SUSTAINING	MOVEMENT CLASS I	CLASS II
	+	+	+
		+	
	+		
	+	+	
		+	
	+	+	+
	+	+	
	+	+	
	+	+	
	+	+	+
	+	+	
		+	
	+	+	+
			+
	+	+	
		+	+
	+	+	+
	+		
	+	+	+
	+		+
		+	+
		+	+
	+	+	+
		+	+
24	16	20	13 TOTAL FOUND
	8	4	11 EXPECTED

Fig. 3. Excitatory receptive fields of sustaining and movement fibers of *Panulirus interruptus*. Crosses indicate the classes of fibers which have been found for each visual field. (From Wiersma and Yamaguchi, 1967a.)

lobster, SuF responsiveness to light is different from that of other decapods, and it is difficult to determine the outline of the fields even in light adapted eyes (Waterman and Wiersma, 1963; Wiersma and Yamaguchi, 1967a; Wiersma and Yanagisawa, 1971; York, 1972). In brachyuran crabs, the SuF's are more difficult to isolate, but they respond much like those of the crayfish (Aréchiga et al., 1974b; Waterman and Wiersma, 1963; Waterman et al., 1964; Wiersma, 1970; Wiersma et al., 1977).

In the crayfish, the most intensively investigated unit (coded 038) has an excitatory receptive field in the upper back rim. The location of this fiber in the nerve cross section is near the nerve sheath, in a bundle that also con-

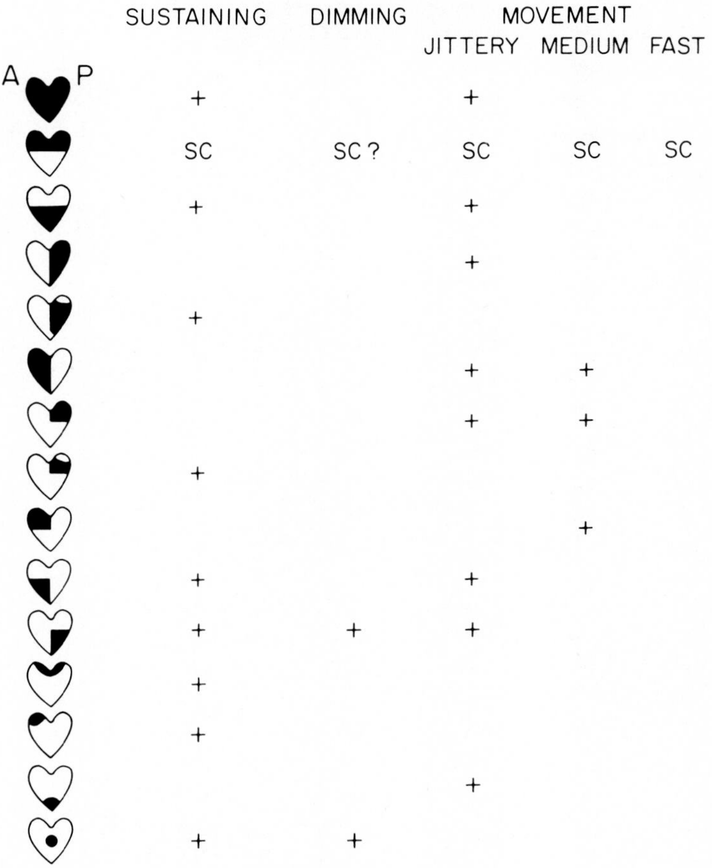

	SUSTAINING	DIMMING	MOVEMENT		
			JITTERY	MEDIUM	FAST
A P	+		+		
	SC	SC?	SC	SC	SC
	+		+		
			+		
	+				
			+	+	
			+	+	
	+				
				+	
	+		+		
	+	+	+		
	+				
	+				
			+		
	+	+			

Fig. 4. Excitatory receptive fields of sustaining, dimming, movement, and space-constant fibers in *Carcinus maenas*. Crosses indicate the classes of fibers that have been found for each visual field. (From Wiersma, 1970.)

tains the other SuF's. About fourteen different members of the class have been recognized and identified by their receptive fields. Intracellular iontophoresis of the fluorescent dye Lucifer Yellow-CH into crayfish SuF's reveals a large (20–30 μm), transversely oriented (with respect to the longitudinal axis of the eye) dendrite at the caudal margin of the medulla externa (Waldrop and Glantz, 1980). This major process gives rise to several parallel branches, which form a vertical uniplanar sheet within the medullary neuropil (Fig. 6A). Recently, Kirk *et al.* (1981) have shown that the SuF dendritic projections in the medulla correlate well with their corneal receptive fields (Fig. 6A3,B). The secondary and higher order branches of each SuF are restricted to that portion of the neuropil corresponding to its corneal

1. Visual Interneurons

VISUAL FIELD A　　P	Su F	JMF	MMF	FMF
		3	6	33
		2	4	49
		9	30	+
	+	10	13	
		15	27	
		34	47	
		19	18	
	43	8	17	+
	42	16	22	
		39	38	
		+	+	
	+	11	7	48
		31	41	+
		46		
		+	+	
		45		
		+	+	
		20	12	

Fig. 5. Excitatory receptive fields of sustaining and movement fibers in *Pachygrapsus crassipes*. Each established fiber's code number appears with its visual field; crosses indicate that a fiber has been found for the visual field but not enough to be established as an entity. (From Wiersma et al., 1977.)

receptive field following an 180° anterior-posterior reversal (due to the first optic chiasmata, Kirk and Glantz, 1981). These results provide a functional verification for the columnar and topological organization of the peripheral optic ganglia that is revealed with purely anatomical techniques (Nässel, 1977). An analogous finding is documented for dipteran visual neurons in the lobula (Eckert and Bishop, 1978). The SuF somata are located on the lateral surface of the medulla interna, and each SuF sends a thin neurite to join the expanding axonal process at the base of

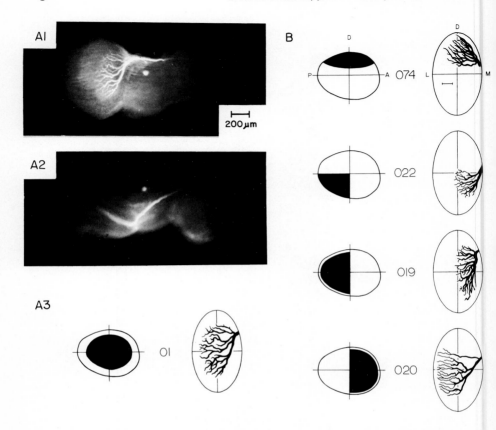

Fig. 6. Structure of crayfish SuF's and correlation with their corneal receptive fields. (A1) Medial view of a Lucifer Yellow injected 01. (A2) Dorsal view of 01. Note the uniplanar arrangement of the process sandwiched within the neuropil of the medulla externa. (A3) A tracing of 01 in A1 and A2 (made from photographic slides, X50, taken at successive focal planes) positioned within an outline of the medulla externa and paired with a sketch of its corneal receptive field. (B) Correspondence of SuF dendritic projections within the medulla externa and their associated corneal receptive fields. Note the 180° anterior–posterior reversal which arises from the fiber crossings at the first optic chiasmata. Calibration bar in B (applies also to A3): 100 μm. Abbreviations: A, anterior; P, poserior; D, dorsal; M, medial; L, lateral. (A and B modified from Kirk et al., in press.)

the medulla externa. Apparently, every SuF possesses an efferent process within the medulla terminalis as well as a tuft-like terminal arborization in the optic neuropil of the brain.

The usual response to illumination of its central field is a tonic discharge with a latency of 20–40 msec. The initial and steady-state firing rates are proportional to the log of the light intensity. The response starts with a burst at about 300 Hz. The discharge stops as soon as the light is turned off. The sensitivity diminishes by 10- to 100-fold between the center and the edge of the excitatory field (Glantz, 1973a). If a light spot is placed at the edge of the excitatory field, the response is weaker, often phasic and the stimulus may elicit a transient off-discharge.

Illumination anywhere on the retina outside of the SuF excitatory field inhibits the discharge. This is true for all SuF's in crayfish and crabs. In rock lobsters the inhibitory surround is more restricted. In addition, lateral inhibition is observed within the excitatory field but is normally masked by the simultaneous excitation (Glantz, 1973a). Sustaining fibers in the crayfish, rock lobsters, and crabs exhibit a synchronous bursting pattern when bright uniform illumination is applied to the receptive field. Glantz and Nudelman (1976) recorded from pairs of SuF's in the crayfish during such bursting. Cross-correlation histograms revealed that the several sustaining fibers burst in a fixed phase relationship with one another and with identical burst periods. The phase difference is at least partially controlled by the position of the stimulus. Repeated bursts with the same period are also elicited with brief flashes to the inhibitory surround. These data suggest that the bursting and synchronization are due to lateral inhibitory interactions.

Glantz (1971) examined the relationship between the receptor potentials in the retinular cells and the discharge rate of SuF 038. This neuron is particularly sensitive to the rate of change of the receptor potential. Its peak firing rate occurs during the transient rising phase of the receptor potential, and the peak of the receptor potential is often associated with a post-burst silent period. The increment threshold behavior and the time course of dark adaptation are entirely predicted from the properties of the retinular cells (Glantz, 1971). At low intensities, SuF discharge may start when the retinular cell depolarization is still in the noise level. This is presumably due to the fact that it sums the output of many retinular cells.

In dark adaptation, all SuF's show a low level discharge in crayfish and lobsters. In the crayfish, this discharge show a circadian rhythm. The firing rate is much greater at night than in daytime. A similar oscillation is observed for the light-evoked discharges (Aréchiga and Wiersma, 1969b). The excitatory field of the SuF's expands during dark adaptation, although the inhibitory effect of illuminating other parts of the eye remains (Aréchiga and Yanagisawa, 1973). Expansion of the excitatory field is most likely due to migration of retinal pigments, probably the shielding pigments. In dark-adapted eyes, expansion can be reversed by injection of sinus gland extracts (Aréchiga et al., 1974a).

Circadian rhythmicity in the SuF's was investigated in the crayfish using chronically implanted electrodes (Aréchiga and Wiersma, 1969b). The animals were maintained in darkness for up to 8 weeks and darkness interrupted by periodic test flashes of light. The number of spikes evoked by a test flash was about three to four times greater during subjective night than at dawn, with responsiveness gradually increasing during the day and evening and falling abruptly at dawn (Fig. 7). The free-running period of the circadian rhythm was less than 24 hr, so that by the sixth day of continuous darkness the cycle of responsiveness was phase shifted about 6 hr relative to an external clock. The level of spontaneous activity in the SuF's is very low in the dark during the subjective day, but it increases considerably during subjective night (Fig. 8). Circadian rhythmicity in the sensitivity of SuF's is also present in Carcinus (Aréchiga et al., 1974b).

Sustaining fiber responsiveness is also modulated in accordance with the general level of excitability of the animal. The "excited state" can be elicited by mechanical stimulation, or it can occur spontaneously; it is expressed

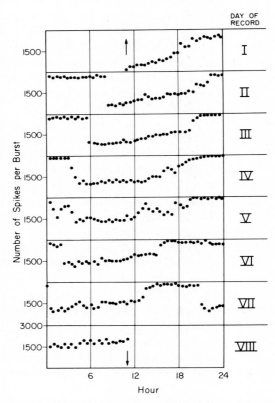

Fig. 7. *Procambarus clarkii.* Continuous recording of responses in SuF 038 to a 15-sec light pulse presented every 30 min for 1 week. (↑) Begin recording, (↓) stop recording. (From Aréchiga and Wiersma, 1969b.)

1. Visual Interneurons

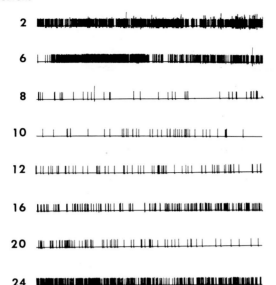

Fig. 8. *Procambarus clarkii.* Ten-minute samples of spontaneous activity in SuF 038 in continuous darkness. Numbers on left indicate the hour in which the sample was taken. (From Arechiga and Wiersma, 1969b.)

behaviorally by leg movements. Heightened reactivity in the SuF's occurs during leg movements that do not produce locomotion in freely moving crayfish. Mechanical stimulation that fails to elicit an excited state does not increase the SuF firing rate. Information on the excited state is presumably relayed to the SuF's by the activity fibers (see below).

a. The Shadowing Effect. A brief shadow over the excitatory field of an SuF results in a brief silent period and a renewed burst. By providing a shadow that lasts only milliseconds, but is repeated at a frequency of 4-10 Hz, the SuF's fire in repeated bursts at a frequency considerably higher than that of the sustained discharge. This phenomenon shows no sign of adaptation. The degree of enhancement is inversely related to the steady firing rate. Certain crab SuF's that are totally inhibited by constant illumination to the whole retina will discharge continuously in response to an oscillating shadow. Repeated shadowing elicits motor activity that is especially pronounced in the lobster. Similar results can be obtained by periodically modulating a stationary light source (Gordon, 1975). In crayfish, the SuF instantaneous firing rate varies linearly for sinusoidal modulation frequencies of up to 1.5 Hz. At higher stimulus modulation frequencies, the discharge is phase-locked to the input, with one or two bursts occurring on each stimulus cycle. As the modulation frequency increases from 2 to 15 Hz, the number of spikes per stimulus cycle declines, but the mean firing rate increases. These and other results can be described by a nonlinear model, which incorporates a high gain derivative-sensitive channel in parallel with a low gain DC channel.

   b. *Spectral Sensitivity.* Most SuF's are maximally sensitive to light at 560–575 nm, although a few are most sensitive near 460 nm, and two fibers were shown to have sensitivity peaks at both wavelengths (Woodcock and Goldsmith, 1970). Five SuF's whose receptive fields were mapped with chromatic stimuli showed slightly greater sensitivity to blue light (473 nm) in the dorsal part of the eye and to yellow light (573 nm) in the ventral part of the eye (Woodcock and Goldsmith, 1973). The spectral sensitivity is dependent on the state of dark adaptation (due to position and spectral absorbance of retinal screening pigment) (Woodcock and Goldsmith, 1970), as well as state of violet- or red-light adaptation (Treviño and Larimer, 1970).

   c. *Polarization Sensitivity.* Although behavioral and primary receptor responses to polarized light have been known in crustaceans for some time, it has only recently been possible to show them at the interneuronal level. Crayfish SuF's exhibited maximum responses to the horizontal and vertical vectors of plane polarized light when the e-vector was continuously rotated (T. Yamaguchi, personal communication; Yamaguchi, 1967b, cited in Yamaguchi *et al.*, 1976), and the sensitivity in either vector was enhanced by previous adaptation to a stationary presentation of the other vector. Since the same interneuron responds to both horizontal and vertical vectors, the response must be governed by a difference-detecting mechanism. Two types of interneurons sensitive to polarized light were distinguished in the medulla externa of the swimming crab *Scylla serrata* (Leggett, 1976); however, neither type appears to have the properties of SuF's.

2. DIMMING FIBERS (DiF's)

   These units are in most respects the exact opposite of the SuF's. They discharge continuously in the dark and are inhibited by the onset of illumination. In the crayfish as well as in other species, their course in the optic nerve is close to that of the SuF's, but it is uncertain whether they are all together in a single bundle or dispersed in several bundles between the SuF's. The visual fields of the DiF's examined are identical to those of the SuF's. In rock lobsters it is sometimes difficult to decide whether a unit is a DiF or a SuF modulated by a stimulus to its inhibitory receptive field. The ambiguity arises from the abnormal reactions of the rock lobster SuF's and the very rare finding of DiF's. When the rock lobster SuF's respond as in the crayfish, the identity of the DiF's can be established. In crayfish and crabs, the DiF's share with the SuF's an increase of sensitivity associated with excited states. In response to polarized light, DiF's—like SuF's—show a greater sensitivity for horizontal and vertical vectors (Yamaguchi et al., 1976). It is possible that their input comes from one of the two main secondary neurons in the lamina ganglionaris. The DiF's thus appear to form with the SuF's a perfect basis for a differential mechanism to determine light intensities in different parts of the retina, and such a model has been proposed by Yamaguchi and Ohtsuka (1973).

3. MOVEMENT FIBERS

The difference between the movement fibers as a class, and the SuF's and DiF's is not as sharp as the names might indicate. The movement fibers may also react to a change in light intensity, with a phasic light-on reaction and/or a light-off response. Most movement sensitive classes, however, show pronounced habituation to repeated stationary stimuli. The latter characteristic gives rise to their subdivision into a number of subclasses. The investigations of the swimming crab, *Podophthalmus,* which were of a relatively preliminary nature (Waterman et al., 1964), as well as earlier investigations in a number of crab species and in *Panulirus argus* in Bermuda (Waterman and Wiersma, 1963), indicated substantial differences among the functional subclasses. These experiments were followed by intensive studies in *Procambarus* (Wiersma and Yamaguchi, 1966, 1967b).

a. Jittery Movement Fibers. Jittery movement fibers (JMF's) are of general occurrence in *Procambarus clarkii, Panulirus interruptus* (Wiersma and Yamaguchi, 1967a; Wiersma and Yanagisawa, 1971), *Carcinus* (Wiersma, 1970), and *Pachygrapsus* (Wiersma et al., 1977). The most extensive receptive field determinations for JMF's have been made in crayfishes (Fig. 2) and crabs. The field geometries are in general similar to those of the crayfish SuF's and DiF's. They generally respond only to the termination of a pulse of light and may thus be considered phasic light-off fibers. This response exhibits pronounced habituation. They react vigorously to dark edges entering their fields if the movement is not too slow. The mean response frequency is a linear function of target velocity from 1°/sec to 35°/sec (Glantz, 1974b). The response to larger edges traversing the eye in a smooth, continuous fashion is accompanied by an inhibitory wave that precedes the moving edge. It is therefore necessary to determine the outlines of the receptive field with the aid of a small black target (Wiersma and Yamaguchi, 1966). The longest lasting discharges are obtained when this object moves in a jittery fashion, so that any newly entered area of the field is not inhibited.

In the crayfish, the most easily found JMF in the optic nerve has the whole retinal surface as its receptive field, and most experiments on JMF's include this fiber. If a projected black edge enters the visual field, it will cause a discharge, which stops or is substantially diminished when the edge moves across the eye (Fig. 9). However, when the movement is stopped and resumed somewhat later, a renewed burst occurs, which again adapts quickly. With a smaller edge, covering half of the field, the effect is almost the same, unless the direction of movement is changed, in which case a new burst will occur. Returning the stimulus to a field that has been stimulated earlier reveals a refractory period, which gradually diminishes with a time course of about 1 min. If the crayfish is exposed to a series of stimuli at the same locus, the habituation after 10 to 30 presentations can be profound. Recovery can require 15–20 min (Glantz, 1974a). Alternatively a 5° shift of target position can produce substantial recovery from habituation in the very next trial. These results imply that the locus of habituation is restricted to the site of stimulation.

Habituation has also been studied by Shimozawa et al. (Shimozawa et al., 1972, 1977; Shimozawa, 1975; Yamaguchi et al., 1973), using the light-off reaction rather than target movement. The JMF's do not respond to continuous flicker but will discharge in response to a pulse deletion. A deletion during a fast flicker (e.g., 10/sec) elicits a burst with shorter latency than that of the response to a slower flicker (e.g., 4/sec). The difference in response delay (150 msec) corresponds to the difference in stimulus periods. The system behaves as though it possessed a short-term memory, which, when aroused by the periodic light pulse, predicts the time of the subsequent pulse and suppresses the discharge unless it fails to occur.

In the crayfish, the excited state enhances the discharge and diminishes

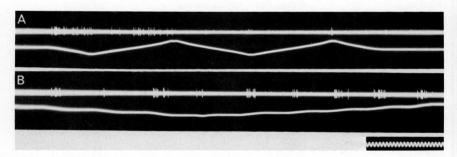

Fig. 9. *Procambarus clarkii.* Responses of JMF 07 to a small black target moving over a lighted dome. (A) Continuous movement back and forth results in habituation. (B) Habituation does not occur when the target is stopped frequently, and firing is resumed on each renewed movement. Lower trace indicates target movement. Time, 10/sec. (From Wiersma and Yamaguchi, 1966.)

1. Visual Interneurons

habituation (Aréchiga and Wiersma, 1969a; Glantz, 1974a). Spontaneous eye movements do not cause a discharge in the JMF's of the crayfish (Yamaguchi, 1967a). There is also evidence that the same occurs in other movement fibers. No inhibitory influence from other parts of the retina has been found.

A property that seems to be characteristic of all movement fibers, including those of vertebrates, is that the movement-elicited response is relatively independent of the background intensity and of the contrast between target and background. For the animal's behavior, this is quite understandable, since the movement of an object, be it of prey or predator, is of equal importance under a variety of light levels.

b. Slow Movement Fibers. This interesting group has been found regularly in *Podophthalmus,* and there is some evidence for their occurrence in grapsoid crabs. The units respond with a sustained discharge as long as an object moves across the field, but once exposed, a receptive area becomes refractory for a long time. In contrast to the JMF's, there is no inhibitory influence traveling in front of the stimulus. Unfortunately, JMF's have not been found in *Podophthalmus;* hence, it is not certain that both fiber types are present in the same species. It is also not known if there is more than one fiber of this type in each optic nerve.

c. Medium Movement Fibers (MMF's). These fibers have been studied in different crabs and in rock lobsters, but they have not yet been found in the crayfish optic nerve. They respond to light on- and off-stimuli and to approaching light and dark objects. Their adaptation rate is much less pronounced than that of the JMF's. They are under strong influence of the CNS in that, during a regular series of responses to an approaching object, one or more approaches will cause little or no reaction, whereas the next stimulus in the series can elicit a strong reaction. These fibers run, like the JMF's, in a bundle, and the members of the class have receptive fields that are indistinguishable in their outlines from those of the JMF's. Their reactions to objects moving parallel to the eye surface are much less pronounced than those of the JMF's, and they habituate to such stimuli more clearly than to approaching objects.

d. Fast Movement Fibers. These fibers have been found in crabs and rock lobsters. They react to rapidly approaching or receding black objects. Although large, they are very difficult to investigate, because after an initial response to the stimulus they habituate and cannot be excited again for some time.

4. SPACE CONSTANT FIBERS (SCF's)

All classes of ipsilateral visual interneurons described above contain at least one member that has the unique property of reacting exclusively to a stimulus that is above the horizon, regardless of the position of the animal. The visual fields of the SCF's change when the animal is rotated so that only those ommatidia that are above the horizon act as inputs (Fig. 10). Space constant fibers have been found in crabs, crayfish, and rock lobsters; in the crayfish (*Procambarus*), the SCF is the only fast movement fiber found. Removal of the statocysts in the crayfish causes a loss of the space constant property at least temporarily; in other species, the leg joints may also provide information to the SCF's.

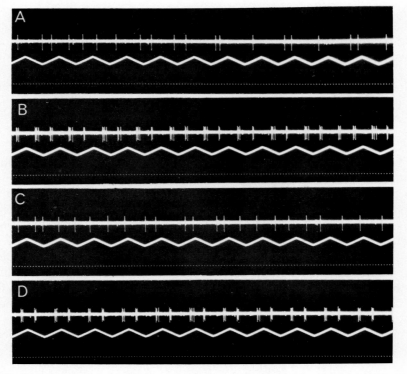

Fig. 10. *Procambarus clarkii*. Responses in SC SuF 023 to a shadow moving at 5 times/sec through its potential excitatory receptive field. (A) Animal in its normal position, shadow moved over ventral half of the eye. (B) Animal in its normal position, shadow moved over dorsal half of the eye. (C) Animal in upside down position, shadow moved over dorsal half of the eye. (D) Animal in upside down position, shadow moved over ventral half of the eye. Time, 1/60 sec. (From Wiersma and Yamaguchi, 1967b.)

5. LIGHT MOVEMENT FIBERS

These fibers have been found only in the rock lobster; they react to a light source moving at a medium rate through their visual field and also to light onset. Similar units have been found in the anomuran, *Pagurus* (Glantz, 1973b), and grapsoid crabs, where they also react to other moving objects.

6. UNIDIRECTIONAL MOVEMENT FIBERS

Unidirectional movement fibers have been found in crabs and rock lobsters, and there is some indication of their presence in crayfish. They are triggered by black or white objects moving at a medium speed either in a horizontal or a vertical direction. In the rock lobster, the subset preferring vertical movement is also space constant.

7. SEEING FIBERS (SeF's)

The SeF's are complex units whose excitatory visual fields are composed of two areas, a central field that reacts both to moving and stationary objects and a peripheral one that reacts only to moving objects. These interneurons have been studied in rock lobsters (Labhart and Wiersma, 1976; Wiersma and Yanagisawa, 1971; Wiersma and York, 1972), but they may also occur in crabs (Waterman and Wiersma, 1963). The boundaries of their visual fields (Fig. 11) are much less distinct than those of the simpler visual fibers, although most receptive fields correspond to the fields of the other units. One fiber (a multimodal unit) represents the anteroventral half of the eye, i.e., its boundary runs along a diagonal line of facets. Some of the SeF's are more strongly excited by movement in one direction, and in these units the visual field is expanded in the direction in which movement is preferred.

Seeing fibers habituate, although not as profoundly as the JMF's. Only that part of the field which has been stimulated is affected in some fibers, whereas in others, there is a general habituation. Habituation is more likely to occur in the peripheral part of the visual field than in the central area, but the areas most susceptible to habituation seem to be unique for each SeF. Seeing fibers are inhibited by objects outside their excitatory field. They react to a moving stripe, but the response is inversely proportional to the length of the stripe (Fig. 12). Inhibition can also occur in response to other visual and nonvisual stimuli, e.g., mechanical stimulation of the abdomen.

C. Heterolateral Visual Interneurons

Heterolateral counterparts of the homolateral fibers (Wiersma et al., 1964) of all visual classes have been found; they run in approximately the same location in the optic nerve as do the homolateral units. They have been less

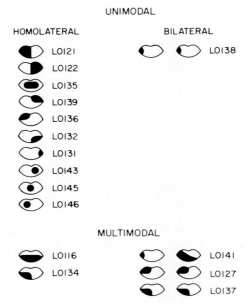

Fig. 11. Excitatory receptive fields of the unimodal and multimodal seeing fibers in *Panulirus interruptus*. The eye shown on the left for the bilateral fibers is the contralateral one. (From Wiersma and York, 1972.)

extensively studied because of the difficulty in recording from them. The only differences between the properties of the afferent and efferent fibers are longer latency and fewer spikes in the latter. The heterolateral fibers could be branches of the homolateral fibers.

D. Bilateral Visual Interneurons

A few bilateral interneurons with only visual input have been found in crayfish (*Procambarus*), rock lobster (*Parulirus*), and *Pachygrapsus*. Only movement-sensitive units have been recorded, so it appears that the large bilateral dimming fiber found in the crayfish commissure (see below) must originate in the brain. All three such interneurons established in the crayfish reacted to fast movements, but there was some doubt whether these were purely visual units rather than multimodal fibers in which the mechanoreceptive input was inhibited. Multimodal interneurons with bilateral visual input have been impaled in the crayfish brain and filled with the fluorescent dye Lucifer Yellow (Glantz et al., 1981). The somata are located in the protocerebrum and a major process projects to each optic nerve and to one of the circumesophageal connectives. More confidence was placed

1. Visual Interneurons

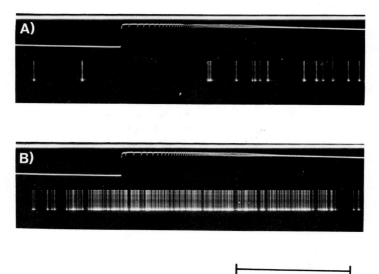

Fig. 12. *Panulirus interruptus* SeF 141. (A) Inhibition of background discharge due to movement of a vertical black strip composed of 5° square black targets. (B) Response to movement of the middle black target after the targets above and below it were removed. Drum speed, 8°/sec. (From Wiersma and York, 1972.)

in the optic origin of the units found in *Pachygrapsus* and the rock lobster. In the crab, habituation to approaching objects occurred when the stimulus was presented to either eye. Presentation of the stimulus to the contralateral eye restored the responsiveness of the habituated ipsilateral eye. In the bilateral seeing fiber of the lobster, stimulation of the contralateral eye was somewhat less effective than stimulation of the ipsilateral one. The importance of bilateral units is unknown; however, it was noted by Wiersma et al. (1977) that the stimulus requirements of such units were similar to that of the defensive reactions of the antennae directed to the site of the stimulus.

E. Mechanoreceptive Interneurons

Ipsilateral, contralateral, and bilateral mechanoreceptive neurons responsive to stimulation anywhere on the animal have been found in crabs, rock lobsters and crayfish and have been most thoroughly studied in the crayfish, *Procambarus* (Wiersma and Yamaguchi, 1966; Aréchiga et al., 1975). Units with receptive fields on the head may have more restricted fields, while those responsive to body stimulation generally have relatively large sensory fields; for example, one fiber responds to hair stimulation anywhere on the body. A few small fields (e.g., the fiber for the basal joint of the homolateral uropod) are also represented. As with the visual interneurons, there is considerable overlap in the receptive fields, and stimulation of any one area causes activity in more than one fiber.

F. Multimodal Interneurons

Ispilateral, contralateral, and bilateral interneurons that react to visual stimuli and to mechanoreceptive stimuli have been found in crabs, crayfish, and rock lobster. Movement is the effective visual stimulus in all cases, and in most units the whole eye is the receptive field. However, in *Carcinus* there is a fiber that reacts visually to the back half of the eye only. The mechanoreceptive sensory fields are similarly widespread, covering the whole animal or one side.

II. VISUAL INTERNEURONS OF THE CRAYFISH CIRCUMESOPHEGEAL COMMISSURES

Unimodal visual interneurons and multimodal interneurons with visual input have been recorded in the circumesophageal commissures of the crayfish, *Procambarus clarkii* (Wiersma and Mill, 1965; Aréchiga and Wiersma, 1969a) and *Carcinus* (Fraser, 1974). Sustaining, dimming, and movement sensitive types of visual input are all represented. Some of the fibers increase their firing rate during an excited state, while others are inhibited.

In crayfish, at least twelve entities have been established that respond to light onset. These neurons can be divided into four groups, including tonic neurons that closely resemble the optic nerve SuF's (SuF's are not found in the connective) and movement-sensitive phasic neurons. The receptive fields are large (90-180°) and often binocular. Synaptic interactions among these neurons were examined by Wood and Glantz (1980a) with intracellular recordings in the brain. When a tonic "on" unit is depolarized with outward current from the electrode, both the impaled cell and other descending visual neurons are observed to discharge in synchrony. Furthermore, EPSP's in the the impaled neuron can be observed in 1:1 association with the axon spikes of other descending neurons. Both the efficacy and temporal features of these interactions indicate that they are monosynaptic. One effect of these dendrodendritic interactions is that the light-elicited impulse activity of small groups of cells is highly coordinated. In a few instances, it has been shown that nearly all of the visual input of a descending neuron is derived from other descending visual cells.

With multiple extracellular electrodes, up to six descending units have been simultaneoulsy monitored (Wood and Glantz, 1980b). Cross-correlograms between the units reveal that they are interconnected in a cascaded hierarchically organized network (Fig. 13). A single cell may receive input from up to five descending neurons and project to a similar number. The networks can be subdivided into three functional layers. The first layer consists of at least two cells with optic nerve input. They exhibit the strongest visual responses and they connect monosynaptically to the second and third layers. The second layer consists of at least four neurons interconnected by excitatory reciprocal connections. These cells also project to the third layer. The third layer contains multimodal neurons with labile visual responses. The cells project exclusively out of the brain via the circumesophageal connectives. The cascade organization of the network is similar to the network of mechanoreceptive interneurons in the abdominal ganglia, which provides excitatory synaptic input to the lateral giant fiber (Kennedy, 1971; Zucker, 1972).

The network behaves as a sensory filter in two regards. For powerful, novel inputs, the cascade organization ensures a massive activation of the third layer cells by first and second layer neurons. The EPSP's in the third layer that originate from activity in the second layer will be delayed com-

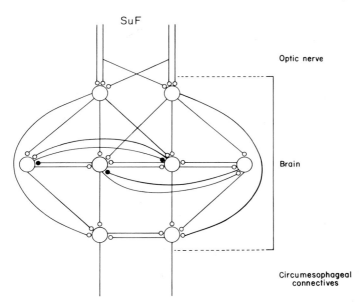

Fig. 13. Synaptic interactions among a population of descending "light-on" interneurons. The diagram is a composite based upon 15 preparations in which five or six cells were monitored simultaneously. The synaptic interactions are inferred from cross-correlation analyses of the spike trains. (After Wood and Glantz, 1980b.)

pared to the first layer input. This delay can behave as a temporal bandpass filter. The second source of feature selection arises from the many sites for spatiotemporal summation inherent in the network structure. Thus, spatially appropriate stimulus configurations will co-activate a particular subset of neurons whose outputs will sum on a selected subset of postsynaptic neurons. Since summation of EPSP's is required for postsynaptic spike activity, the combined effects of the stimulus configuration and the pattern of presynaptic convergence will determine the subset of excited postsynaptic neurons.

The motion-sensitive neurons of the crayfish circumesophageal connectives can be separated morphologically into two groups: (1) jittery motion fibers that originate in the optic ganglia; and (2) motion detectors that originate in the brain (Fig. 14) (R. M. Glantz, unpublished observations). The trigger features of the two groups are similar, but the units originating in the brain have larger and binocular receptive fields (Wiersma and Mill, 1965; Glantz, 1977).

The descending motion-sensitive neurons are excited by visual stimuli identical to those that elicit the defense reflex (Wiersma and Mill, 1965; Glantz, 1977). Direct stimulation of three such neurons individually can elicit the entire defense posture (Glantz, 1977). It should be noted however,

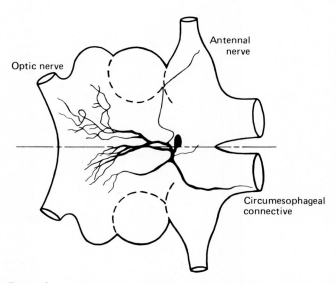

Fig. 14. Descending motion-sensitive interneuron in the crayfish brain. Drawing of brain revealing the soma and major neurites of the impaled cell. The cell was filled with Lucifer Yellow-CH. The figure was drawn from a whole mount fluorescent image. The cell body is at the posterior margin of the brain near the origin of the circumesophageal connectives. The axon exits via the left connective (lower right). Two sets of dendrites terminate in the left and right optic neuropils of the protocerebrum (figure left). (R. M. Glantz, unpublished observations.)

that stimulation of individual motion detectors at physiological firing rates elicited defense postures with a substantially longer latency than observed with natural stimuli. The more rapid natural response is probably controlled by a small co-activated population of motion-sensitive interneurons. Descending motion detectors have also been monitored with chronically implanted electrodes in freely moving animals (R. M. Glantz, unpublished observations). When the defense reflex is elicited with visual stimuli, the neurons commence firing 50-100 msec prior to the onset of cheliped muscle activity. The behavior is only weakly correlated to the firing of individual neurons, but it is highly correlated to the population firing rate.

There is also evidence to suggest that units of the connectives that respond to light onset may promote walking behavior. Thus, Larimer and Gordon (1977) have shown bouts of walking behavior following the onset of illumination. The appropriate stimulus for walking also excites many of the descending "on" units (Wiersma and Mill, 1965; Larimer and Gordon, 1977). Direct stimulation of "light-on" interneurons in the connectives promotes walking leg motor activity (Glantz, 1977).

III. THE CAUDAL PHOTORECEPTOR

A. Occurrence

Extraretinal light reception has been shown behaviorally in a number of decapod crustaceans, but the only interneurons that have been demonstrated to be involved are the bilaterally paired caudal photoreceptors (CPR) of the sixth abdominal ganglion. The electrophysiological responses of the CPR were first described by Prosser (1934) in crayfish (*Orconectes virilis, Procambarus clarkii, O. limosus*). He recorded impulses in the ventral nerve cord and found that illumination of the sixth abdominal ganglion caused an increased discharge frequency as far forward as the circumesophageal connectives. He noted the high threshold (20 meter-candles) and long latency (2-2½ sec), as well as the attempted behavioral responses of the animal (active leg movements occurred even though the crayfish was pinned to the substrate). Welsh (1934) investigated the photonegativity of the intact animal and demonstrated that leg movements occur upon illumination of the CPR when the eyes had been removed; such leg movements were not seen when the CPR's were also removed. He used *Procambarus clarkii* and *Orconectes virilis*, and stated that the CPR's were present in *Panulirus argus* and *P. guttatus* but not in certain Bermuda prawns.

Hess (1940) has reported that in freshly molted *Alpheus armillatus, Panulirus argus, Homarus americanus,* and *Orconectes limosus*, light over many regions of the body caused walking leg movements: the threshold to light increased as the exoskeleton pigmentation darkened. *Crangon* larvae are photopositive and orient toward light only when the eyes are intact, but "they did show evidence of being sensitive to light even though the caudal segment of the abdomen had been removed." Therefore, it appears that at least in larval stages and immediately after molting, photosensitivity is not limited to the sixth ganglion. Hess has claimed that in the tail of *Alpheus armillatus,* the uropods are sensitive to very locally applied light beams, and he described sensory cell bodies; he suggested that the photosensitive elements are cell bodies of neurons that serve other functions and that little or no pigment must be present in the exoskeleton for them to function as photoreceptors. Wells (1959) studied the behavior of two cavernicolous species of crayfish; he found that light on the cephalic region, but not the tail, caused leg movements in *Cambarus setosus* whether the eyes were present or not; however, there was no response to light by adult *C. hubrichti,* although the one juvenile specimen tested responded to light in the cephalic region (no other regions were discussed). A systematic electrophysiological study of the CPR in decapods was performed by Wilkens and Larimer (1976).

B. Anatomy

Kennedy (1963) showed physiologically that there are only two CPR's (one on each side of the sixth abdominal ganglion) in *Procambarus clarkii* Girard, and that they run in the cord as far as the brain. Wilkens and Larimer (1972) have shown, by injecting Procion Yellow, that the axons of the CPR branch extensively in the sixth abdominal ganglion; however, they traced the axons only as far as the fifth abdominal ganglion, where they also branch. The somata are in the contralateral half of the sixth ganglion, and although the two CPR axons seem to be near each other when they decussate, there does not appear to be synaptic contact between them. Physiological responses were recorded in the axons only when illumination was over the dendritic branching in the sixth abdominal ganglion; thus, the somata do not appear to be involved in the sensory response. The ultrastructure of the CPR's is unknown, although organelles that might be light-sensitive have been found in the cord and in the sixth abdominal ganglion (Hama, 1961; Uchizono, 1962; de Lorenzo and Zonana, 1963; Novikoff and Holtzman, 1976).

C. Physiology

The crayfish CPR's, which are spontaneously active in light or darkness, show an increased firing frequency in response to illumination of the sixth abdominal ganglion. They have a high threshold, react with a long latency, adapt slowly, and show a pronounced after-discharge when the light is turned off. Although these properties have been known for some time (Prosser, 1934; Kennedy, 1958a,b), it was not until the advent of single unit analysis that Kennedy (1963) proved that the paired CPR's are not only primary sensory receptors for light but also sensory interneurons integrating information from proprioceptors, or tactile receptors, or both, on the telson and uropods. Stimulation of ipsilateral tactile receptors caused excitation followed by inhibition of light-induced or spontaneous discharges in the interneurons, whereas contralateral stimulation causes only inhibition. The site of impulse initiation is unknown; since no potential changes can be recorded in the soma, it is assumed that it is far removed from the site of impulse initiation and that it does not contribute to the recorded electrical activity in the axon (Wilkens and Larimer, 1972).

Spectral sensitivity of the CPR has been measured in several species of epigeal crayfish (Bruno and Kennedy, 1962; Larimer et al., 1966) and one species of blind cavernicolous crayfish (Larimer et al., 1966). The visual pigment is probably a carotenoid protein, as in the compound eye; however, the peak sensitivity is 500-505 nm in epigeal forms and about 497 nm in the cave species. Why the sensitivity of the CPR should be shifted more toward the blue end of the spectrum than the compound eye is enigmatic although, as Bruno and Kennedy (1962) suggest, longer wavelengths would be screened out by the blue hypodermis which runs as a strip along the cord in some crayfish.

IV. SUMMARY AND HYPOTHESIS

Properties of the receptive fields of known decapod visual interneurons can be summarized as follows: (1) the visual environment is subdivided into 14-20 overlapping segments, each consisting of a few dozen to several hundred ommatidia; (2) every ommatidium is contained in the receptive fields of two or more (typically four to five) members of each interneuronal class; (3) the receptive field divisions are repeated in each of three to eight different classes of units; (4) each class of visual interneurons is particularly sensitive to one or two parameters of visual stimulation; and (5) the selectivity of a given class is generally either complementary or antagonistic to that of another class(e.g., dimming versus sustaining fibers and slow movement versus fast movement).

The significance of these divisions can be better appreciated if we examine the consequences, in the optic nerve, of a simple visual stimulus. Imagine that a bright object on a grey field assumes a stationary position in a decapod visual field. If the object occupies 20° in any direction, then, depending on its position, three or more SuF's will be excited to various degrees, and the remaining SuF's will be partially or totally inhibited. The dimming fibers will presumably exhibit the reverse behavior. One can regard the representation of the stimulus in terms of the activity of the one or two most excited units or in terms of the distribution of excitation and inhibition in the entire SuF and DiF array. If the representation is determined by the activity in a single SuF, then the organism will confuse variations in stimulus intensity and size. Furthermore, spatial resolution can be no better than the dimensions of individual receptive fields, which are typically 20-90° in any direction.

Although representation of the stimulus by single cells is possible, the coding of information will be ambiguous. In the example described above, it is clear that any natural visual stimulus must co-activate several members of any one class and may also modulate the activity of members of two or more classes. The ensemble output thus carries a spatial representation within each class and a quantitative analysis of one or several visual parameters through the selection of the activated classes. The output of the population exhibits less ambiguity regarding stimulus intensity and size, and the spatial resolution substantially exceeds that of any one cell. All of this information is delivered to the brain in a parallel array and at approximately the same time. A central issue for further study is the nature of the subsequent integration. Is the decoding achieved by independent weighting of every cell, by population interactions within a class, or by population analysis across several classes at a given retinal locus? The visual units in the circumesophageal connectives generally exhibit larger receptive fields than do optic nerve fibers, but the class distinctions are generally preserved. This would imply that the brain integrates information within each of the separate classes but does not perform an interclass analysis. It should be pointed out, however, that if such an interclass analysis were performed, the consequences could be quite subtle and easily missed with the techniques applied to date. Furthermore, there are several examples of units in the crayfish connectives (Wiersma and Mill, 1965) that appear to integrate the output of more than one class of optic nerve fiber.

The ultimate goal of the studies described is to understand the physiological basis of crustacean visual perception. We are severely handicapped, however, by the absence of adequately developed behavioral criteria for determining what these organisms see and precisely how visual information is used. This defect is particularly apparent in any attempt to interpret the characteristics of the descending visual interneurons that arise in the brain. The activity of these cells is almost certainly involved in the control or release of visually modulated behavior. Descending neurons differ from optic nerve fibers in several important respects: (1) receptive fields are larger and more often multimodal; (2) trigger features are more selective; (3) the neurons habituate more rapidly; and (4) they are interconnected in extensive networks that coordinate their impulses.

In the case of the motion detectors, binocular neurons originating in the brain are found adjacent to their monocular input elements, which originate in the optic ganglia. Similarly, higher order descending "on" units can be found in the connective adjacent to their presynaptic drivers. These parallel pathways suggest that the brain acts as a cascade multiplier to amplify the response to appropriate stimuli. Lower order interneurons are activated by broad classes of stimuli, and they propagate their activity to more posterior ganglia. When the stimulus contingencies are appropriate, the lower order cells may recruit a number of higher order interneurons, which convey their output on parallel pathways. Since the higher order elements in the brain are often multimodal, their responsiveness and thus the amplification will be contingent upon nonvisual inputs as well as the efficacy of the visual stimulus.

How is this message deciphered? Our best guess is derived by analogy from the input pathway of the lateral giant-escape system. If the cascaded elements converge upon common target neurons (i.e., motoneurons or premotor interneurons), and if these target cells exhibit relatively high thresholds, then the cascade multiplier could mediate the behavioral efficacy of the stimulus. Since the higher order elements are multimodal and highly sensitive to the general level of excitation in the animal, this mechanism could also serve as a switch, which could selectively reinforce a given premotor pathway in accordance with the strength of the visual as well as non-visual contingencies.

Although the above hypothesis is consistent with the available evidence, it is by no means a necessary conclusion from this evidence. A critical but untested requirement is that the cascaded ensemble elements converge upon common target sites. Furthermore, it is essential to demonstrate that the target sites (i.e., premotor or motoneurons) are sensitive to variations in the distribution of activity within the convergent population of interneurons. The sensory information transmitted by this scheme would thus depend on individual neuronal firing rates, the geometrical and dynamic rules for convergence of the excited elements, and the relative efficacy of the synapses to the target neurons.

REFERENCES

Aréchiga, H., and Wiersma, C. A. G. (1969a). The effect of motor activity on the reactivity of single visual units in the crayfish. *J. Neurobiol.* **1,** 53–69.

Aréchiga, H., and Wiersma, C. A. G. (1969b). Circadian rhythm of responsiveness in crayfish visual units. *J. Neurobiol.* **1,** 71–85.

Aréchiga, H., and Yanagisawa, K. (1973). Inhibition of visual units in the crayfish. *Vision Res.* **13,** 731–744.

Aréchiga, H., Fuentes-Pardo, B., and Barrera-Mera, B. (1974a). Influence of retinal shielding pigments on light sensitivity in the crayfish. *Acta Physiol. Lat. Am.* **24,** 601–611.

Aréchiga, H., Huberman, A., and Naylor, E. (1974b). Hormonal modulation of circadian neural activity in *Carcinus maenas* (L.). *Proc. R. Soc. London, Ser. B* **187,** 229–313.

Aréchiga, H., Barrera-Mera, B., and Fuentes-Pardo, B. (1975). Habituation of mechanoreceptive interneurons in the crayfish. *J. Neurobiol.* **6,** 131–144.

Bruno, M. S., and Kennedy, D. (1962). Spectral sensitivity of photoreceptor neurons in the sixth ganglion of the crayfish. *Comp. Biochem. Physiol.* **6,** 41–46.

de Lorenzo, A. J., and Zonana, H. (1963). Ultrastructure of photoreceptor neurons following exposure to light and darkness. *Anat. Rec.* **145,** 221–222.

Eckert, H., and Bishop, L. G. (1978). Anatomical and physiological properties of the vertical cells in the third optic ganglion of the *Phaehicia sericata* (Diptera, Calliphoridae). *J. Comp. Physiol.* **126,** 57–86.

Fraser, P. J. (1974). Interneurons in crab connectives (*Carcinus maenas* (L)): Giant fibres. *J. Exp. Biol.* **61,** 593–613.

Glantz, R. M. (1971). Peripheral versus central adaptation in the crustacean visual system. *J. Neurophysiol.* **34,** 485–492.

Glantz, R. M. (1973a). Spatial integration in the crustacean visual system: Peripheral and central sources of nonlinear summation. *Vision Res.* **13,** 1801–1814.

Glantz, R. M. (1973b). Five classes of visual interneurons in the optic nerve of the hermit crab. *J. Neurobiol.* **4,** 301–319.

Glantz, R. M. (1974a). Habituation of the motion detectors of the crayfish optic nerve. *J. Neurobiol.* **5,** 489–510.

Glantz, R. M. (1974b). Defense reflex and motion detector responsiveness to approaching targets: The motion detector trigger to the defense reflex pathway. *J. Comp. Physiol.* **95,** 297–314.

Glantz, R. M. (1977). Visual input and motor output of command interneurons of the defense reflex pathway of the crayfish. In "Identified Neurons and Behavior of Arthropods" (G. Hoyle, ed.), pp. 259–274. Plenum, New York.

Glantz, R. M., Kirk, M., and Viancour, T. (1981). Interneurons of the crayfish brain: The relationship between dendritic location and afferent input. *J. Neurobiol.* **12,** 311–328.

Glantz, R. M., and Nudelman, H. B. (1976). Sustained, synchronous oscillations in discharge of sustaining fibers of crayfish optic nerve. *J. Neurophysiol.* **39,** 1257–1271.

Gordon, W. H. (1975). A nonlinear model of the sustaining visual interneurons of the crayfish optic nerve. Doctoral Thesis, Rice University, Houston, Texas.

Hama, K. (1961). A photoreceptor-like structure in the ventral nerve cord of the crayfish, *Cambarus virilus. Anat. Rec.* **140,** 329–336.

Hanström, B. (1928). "Vergleichende Anatomie des Nervensystems der wirbellosen Tiere unter Berücksichtigung seiner Funktion." Springer-Verlag, Berlin and New York.

Hess, W. N. (1940). Regional photosensitivity and photoreceptors of *Crangon armillatus* and the spiny lobster, *Panulirus argus. Carnegie Inst. Washington Publ.* **517,** 155–161.

Kennedy, D. (1958a). Responses from the crayfish caudal photoreceptor. *Am. J. Ophthalmol.* **46,** 19–26.

Kennedy, D. (1958b). Electrical activity of a "primitive" photoreceptor. *Ann. N.Y. Acad. Sci.* **74,** 329–336.

Kennedy, D. (1963). Physiology of photoreceptor neurons in the abdominal nerve cord of the crayfish. *J. Gen. Physiol.* **46,** 551–572.

Kennedy, D. (1971). Crayfish interneurons. *Physiologist* **14,** 5–30.

1. Visual Interneurons

Kennedy, D., and Mellon, DeF. (1964). Synaptic activation and receptive fields of crayfish interneurons. *Comp. Biochem. Physiol.* **13,** 275-300.

Kirk, M. D., and Glantz, R. M. (1981). Morphological representation of crayfish sustaining fiber visual receptive fields. *Neurosci. Abstr.* **7** (In Press).

Kirk, M. D., Waldrop, B., and Glantz, R. M. (1981). The crayfish sustaining fibers. I. Morphological representation of visual space in the second optic ganglion. *J. Comp. Physiol.* (In press).

Labhart, T., and Wiersma, C. A. G. (1976). Habituation and inhibition in a class of visual interneurons on the rock lobster, *Panulirus interruptus. Comp. Biochem. Physiol.* A **55,** 219-224.

Larimer, J. L., and Gordon, W. H. (1977). Circumesophageal interneurons and behavior in crayfish. In "Identified Neurons and Behavior of Arthropods" (G. Hoyle, ed.), pp. 243-258. Plenum, New York.

Larimer, J. L., Treviño, D. L., and Ashby, E. A. (1966). A comparison of spectral sensitivities of caudal photoreceptors of epigeal and cavernicolous crayfish. *Comp. Biochem. Physiol.* **19,** 409-415.

Leggett, L. M. W. (1976). Polarised light-sensitive interneurons in a swimming crab. *Nature (London)* **262,** 709-711.

Nässel, D. R. (1977). Types and arrangements of neurons in the crayfish optic lamina. *Cell Tiss. Res.* **179,** 45-75.

Novikoff, A. B., and Holtzman, E. (1976). "Cells and Organelles," 2nd ed. Holt, New York.

Nunnemacher, R. F. (1966). The fine structure of optic tracts of Decapoda. In "The Functional Organization of the Compound Eye" (C. G. Bernhard, ed.), pp. 363-375. Pergamon, Oxford.

Nunnemacher, R. F., Camougis, G., and McAlear, J. H. (1962). The fine structure of the crayfish nervous system. *Electron Microsc., Proc. Int. Congr., 5th, 1962* Vol. 2, Artic. N-11.

Ochi, K., and Yamaguchi, T. (1976). Single unit analysis of visual interneurons in the optic nerve of the mantis shrimp, *Oratosquilla oratoria. Biol. J. Okayama Univ.* **17,** 47-60.

Prosser, C. L. (1934). Action potentials in the nervous system of the crayfish. II. Responses to illumination of the eye and caudal ganglion. *J. Cell. Comp. Physiol.* **4,** 363-377.

Shimozawa, T. (1975). Response entrainment of movement fibers in the optic tract of crayfish. *Biol. Cybernet.* **20,** 213-222.

Shimozawa, T., Takenda, T., and Yamaguchi, T. (1972). Movement perception by the movement fibre in the optic tract of the crayfish-Analysis of temporal factors in movement perception. *Jpn. J. Bio-Med. Eng.* **10,** 186-195.

Shimozawa, T., Takeda, T., and Yamaguchi, T. (1977). Response entrainment and memory of temporal pattern by movement fibers in the crayfish visual system. *J. Comp. Physiol.* **114,** 267-287.

Treviño, D. L., and Larimer, J. L. (1970). The responses of one class of neurons in the optic tract of crayfish *(Procambarus)* to monochromatic light. *Z. Vergl. Physiol.* **69,** 139-149.

Uchizono, K. (1962). The structure of possible photoreceptive elements in the sixth abdominal ganglion of the crayfish. *J. Cell Biol.* **15,** 151-154.

Waldrop, B., and Glantz, R. M. (1980). Localization of dendritic fields and terminal arborizations in sustaining fibers of the crayfish visual system. *Neurosci Abstr.* **6,** 221.

Waterman, T. H., and Wiersma, C. A. G. (1963). Electrical responses in decapod crustacean visual systems. *J. Cell. Comp. Physiol.* **61,** 1-16.

Waterman, T. H., Wiersma, C. A. G., and Bush, B. M. H. (1964). Afferent visual responses in the optic nerve of the crab, *Podophthalmus. J. Cell. Comp. Physiol.* **63,** 135-155.

Wells, P. H. (1959). Responses to light by cave crayfishes. *Occas. Pap. Natl. Speleol. Soc.* No. 4, pp. 4-15.

Welsh, J. N. (1934). The caudal photoreceptor and responses of the crayfish to light. *J. Cell. Comp. Physiol.* **4,** 379-388.

Wiersma, C. A. G. (1970). Neuronal components of the optic nerve of the crab, *Carcinus maenas. Proc. K. Ned. Akad. Wet., Ser. C* **73,** 25-34.

Wiersma, C. A. G., and Bush, B. M. H. (1963). Functional neuronal connections between the thoracic and abdominal cords of the crayfish, *Procambarus clarkii* (Girard). *J. Comp. Neurol.* **121,** 207-235.

Wiersma, C. A. G., and Mill, P. J. (1965). "Descending" neuronal units in the commissure of the crayfish central nervous system; and their integration of visual, tactile and proprioceptive stimuli. *J. Comp. Neurol.* **125,** 67-94.

Wiersma, C. A. G., and Yamaguchi, T. (1966). The neuronal components of the optic nerve of the crayfish as studied by single unit analysis. *J. Comp. Neurol.* **128,** 333-358.

Wiersma, C. A. G., and Yamaguchi, T. (1967a). The integration of visual stimuli in the rock lobster. *Vision Res.* **7,** 197-204.

Wiersma, C. A. G., and Yamaguchi, T. (1967b). Integration of visual stimuli by the crayfish central nervous system. *J. Exp. Biol.* **47,** 409-431.

Wiersma, C. A. G., and Yanagisawa, K. (1971). On types of interneurons responding to visual stimulation present in the optic nerve of the rock lobster, *Panulirus interruptus. J. Neurobiol.* **2,** 291-309.

Wiersma, C. A. G., and York, B. (1972). Properties of the seeing fibers in the rock lobster: Field structure, habituation, attention and distraction. *Vision Res.* **12,** 627-640.

Wiersma, C. A. G., Bush, B. M. H., and Waterman, T. M. (1964). Efferent visual responses of contralateral origin in the optic nerve of the crab, *Podophthalmus. J. Cell. Comp. Physiol.* **64,** 309-326.

Wiersma, C. A. G., Hou, L.-H.R., and Martini, E. M. (1977). Visually reacting neuronal units in the optic nerve of the crab *Pachygrapsus crassipes. Proc. K. Ned. Akad. Wet., Ser. C* **80,** 135-143.

Wilkens, L. A., and Larimer, J. L. (1972). The CNS photoreceptor of crayfish: Morphology and synaptic activity. *J. Comp. Physiol.* **80,** 389-407.

Wilkens, L. A., and Larimer, J. L. (1976). Photosensitivity in the sixth abdominal ganglion of decapod crustaceans: A comparative study. *J. Comp. Physiol.* **106,** 69-75.

Wood, H. L., and Glantz, R. M. (1980a). Distributed processing by visual interneurons of the crayfish brain. I. Response characteristics and synaptic interactions. *J. Neurophysiol.* **43,** 729-740.

Wood, H. L., and Glantz, R. M. (1980b). Distributed processing by visual interneurons of the crayfish brain. II. Network organization and stimulus modulation of synaptic efficacy. *J. Neurophysiol.* **43,** 741-753.

Woodcock, A. E. R., and Goldsmith, T. H. (1970). Spectral responses of sustaining fibers in the optic tracts of crayfish (*Procambarus*). *Z. Vergl. Physiol.* **69,** 117-133.

Woodcock, A. E. R., and Goldsmith, T. H. (1973). Differential wavelength sensitivity in the receptive fields of sustaining fibers in the optic tract of the crayfish *Procambarus. J. Comp. Physiol.* **87,** 247-257.

Yamaguchi, T. (1967a). Effects of eye motions and body position on crayfish movement fibers. *In* "Invertebrate Nervous Systems" (C. A. G. Wiersma, ed.), pp. 285-288. Univ. of Chicago Press, Chicago, Illinois.

Yamaguchi, T. (1967b). (No title given.) *Zool. Mag.* **76,** 443 (in Japanese); cited by Yamaguchi et al. (1976).

Yamaguchi, T., and Ohtsuka, T. (1973). Dual channelling mechanism of brightness and dimness information in the crayfish visual system. *J. Fac. Sci., Hokkaido Univ., Ser. 6* **19,** 15-30.

Yamaguchi, T., Ohtsuka, T., Katagiri, Y., and Shimozawa, T. (1973). Some spatial and temporal properties of movement fibres in the optic tract of the crayfish. *J. Fac. Sci., Hokkaido Univ., Ser. 6* **19,** 31–49.

Yamaguchi, T., Katagiri, Y., and Ochi, K. (1976). Polarized light responses from retinula cells and sustaining fibers of the mantis shrimp. *Biol. J. Okayama Univ.* **17,** 61–66.

York, B. (1972). Sustaining fibers in the rock lobster. *J. Neurobiol.* **3,** 303–309.

Zucker, R. (1972). Crayfish escape behavior and central synapses. I. Neural circuit exciting the lateral giant fiber. *J. Neurophysiol.* **35,** 599–620.

2

Control of Posture

CHARLES H. PAGE

I.	Introduction	33
II.	Principles of Postural Control	34
III.	Thoracic Leg Posture	35
	A. Motor Organization	35
	B. Sensory Receptors and Leg Reflexes	36
	C. Control of Thoracic Leg Posture	41
IV.	Abdominal Posture	44
	A. Motor Organization	44
	B. Abdominal Proprioceptors and Reflexes	47
	C. Control of Abdominal Posture	49
V.	Conclusion	52
	List of Abbreviations	53
	References	54

I. INTRODUCTION

The posture adopted by an animal—be it a crayfish brooding its eggs, a crab defending itself from a predator, or a hermit crab withdrawn into its shell—reflects the ongoing level of tonic activity in sensory, interneuronal, and motor systems that control the movements of the appendages and body segments. Each behavior pattern—standing, walking, swimming, defense, grooming, and others—requires a specific positioning of the cephalothorax and abdomen. The positions that various appendages adopt contribute to the behavior pattern. The postural stances of the thoracic legs are particularly important, because they determine the degree of elevation and orientation of the cephalothorax. In crustaceans that possess a large multisegmented abdomen (e.g., crayfish, lobsters, and hermit crabs), abdominal posture affects body orientation, since changes in abdominal position shift the center of gravity of the animal.

Examination of postural control systems has been confined for the most part to large decapods: crayfish (e.g., *Procambarus* and *Astacus*), crabs (e.g., *Cancer, Carcinus,* and *Callinectes*), lobsters (*Homarus*), rock lobsters (*Palinurus*), and hermit crabs (*Pagurus*). The analysis of postural control has been primarily directed at the sensory, central, and motor mechanisms that control leg stance and abdominal position.

II. PRINCIPLES OF POSTURAL CONTROL

The components to be considered in control of posture are reflexes evoked by sensory stimulation, sensory receptor systems, central control by interneurons, and tonic musculature and its efferent innervation.

A broad range of sensory receptor systems contributes to the regulation of posture. These include proprioceptors and contact receptors, both in the abdomen and in the legs. The simplest neural responses that contribute to the control of posture are resistance reflexes—i.e., negative feedback reflexes that resist movement imposed on a joint (Bush, 1965). The mechanical sensitivity of some proprioceptors is under efferent control, and the strength of the resistance reflexes evoked by stimulation of these sensory systems depends on the level of activity in the efferent nerves to them.

While most resistance reflexes are localized within the stimulated (and perhaps neighboring) segments, other reflexes have widespread effects on the posture of the entire animal; these generally involve interneurons. They include responses evoked by a variety of complex stimuli. Visual stimulation of the eye with an approaching object elicits a "defense posture" (claws raised, abdomen extended; Glantz, 1977); mechanical stimulation resulting from loss of contact between the legs and substrate produces an extension of the abdomen (Larimer and Eggleston, 1971); and stimulation of the legs and statocysts by a change in body position produces compensatory movements of the thoracic limbs and abdomen (Schöne, 1951, 1954; Schöne et al., 1978).

Although many postural stances can be adjusted by sensory reflex modulation of ongoing motor neuron activity, each posture that is adopted by the animal is thought to result from the excitation of a different set of premotor interneurons (command fibers). Some command fibers control the movements of a single appendage, while others evoke complex coordinated movements of all body parts. Examples of the latter include command fibers for standing, swimming, walking, and feeding (Atwood and Wiersma, 1967; Bowerman and Larimer, 1974a, b).

2. Control of Posture

Long-term maintenance of postural positions is achieved by slow, sustained tonic contractions of the body muscles. The muscles important for postural control are innervated by tonic motor neurons. When compared with phasic motor neurons, tonic motor neurons are characterized by their small size, high levels of spontaneous discharge, and resistance to synaptic fatigue (Atwood, 1973; Davis, 1971; Kennedy and Takeda, 1965a, b).

Posture in crustaceans is determined by the activity of motor and peripheral inhibitory neurons that innervate the muscles controlling the legs and abdomen. Most muscles in the leg are mixed and contain tonic, intermediate, and phasic muscle fibers (Atwood, 1973). Although the majority of these muscle fibers are innervated by both phasic and tonic motor neurons, there is a tendency for tonic fibers to be innervated by tonic motor neurons and phasic fibers by phasic motor neurons (Atwood, 1963, 1965). In the abdomen, postural muscles are composed solely of tonic muscle fibers and are innervated by tonic motor neurons (Kennedy and Takeda, 1965b; Parnas and Atwood, 1966).

III. THORACIC LEG POSTURE

All decapod crustaceans possess five pairs of legs. In most species (the palinurids, or spiny lobsters, are an exception) the first pair of legs support large chelae. The second through fifth pairs of legs (pereiopods) are smaller and thinner; in most decapods, they are nonchelate but in astacurans, the second and third pairs of legs possess small chelae.

A. Motor Organization

Each leg is composed of seven segments, which form a complex multijointed appendage (Fig. 1). Movement of these joints occurs primarily in two planes: the thoracicocoxal (T-C) and carpopropal (C-P) joints move the leg in the yaw (frontal) plane, while movement of the coxobasal (C-B), merocarpal (M-C), and propodactyl (P-D) joints occurs in the roll (transverse) plane (Ayers and Davis, 1977a; Clarac, 1977). In most brachyuran decapods, the basoischial (B-I) joint is fused, but in lobsters this joint is mobile. Movement of the ischiomeral (I-M) joint is also very restricted, but more mobile in lobsters than in crabs (Wales et al., 1970).

Movements of the leg of a typical decapod result from the contraction of 16 different muscles. The pattern of motor neuron innervation of the muscles in the distal leg segments is well known (Wales et al., 1970; Wiersma and Ripley, 1952); however, little is known concerning the innervation of the most proximal leg muscles. A minimal estimate of the number of motor neurons that innervate the muscles of one leg would be 30–35.

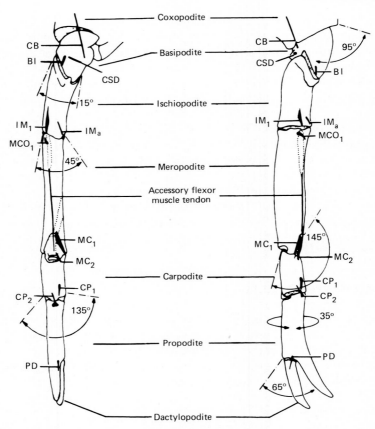

Fig. 1. The left leg of *Homarus gammarus*. Positions of the chordotonal organs (CB, BI, IM_1, IM_a, MC_1, MC_2, CP_1, CP_2, and PD), the myochordotonal organs (MCO_1), and the cuticular stress detectors (CSD) are indicated. The range of movement across each joint is also indicated. (From Mill, 1976.)

B. Sensory Receptors and Leg Reflexes

The crustacean leg contains a large number of proprioceptors that respond to a variety of mechanical conditions important in the reflex control of leg posture (Fig. 1). They include receptors sensitive to joint movement and position, muscle length, muscle tension, and cuticular stress. Details of these sensory structures are described in Chapter 9 of Volume 3; here, their involvement in reflexive control is outlined.

1. RESISTANCE REFLEXES

Passive movement of a leg joint invariably produces a resistance reflex response in the muscles moving the joint (Fig. 2). For example, the opening

2. Control of Posture

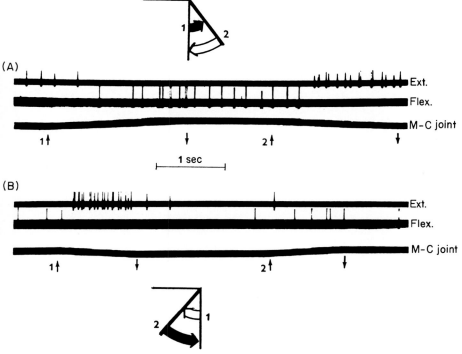

Fig. 2. Resistance reflex responses elicited by extension (A) or flexion (B) of the M-C joint in the crab, *Cancer magister*. Upper trace-extensor efferent; middle trace-flexor efferent; lower trace-M-C joint movement. Upward arrows mark initiation of movement; downward arrows indicate termination of the movement. (From Evoy and Cohen, 1969.)

of the P-D joint elicits reflex responses that excite both the tonic motor neuron to the closer muscle and the inhibitor to the opener muscle (Bush, 1962, 1963). Resistance reflexes that modulate tonic motor activity have been described in response to passive movement of each of the leg joints (Ayers and Davis, 1977b; Bush, 1965).

2. CHORDOTONAL ORGANS

Resistance reflexes are mediated by the chordotonal organs, which are formed by thin innervated elastic strands located near each leg joint (Fig. 1; see Chapter 9 of Volume 3, and Mill, 1976; Wales et al., 1970; Whitear, 1962). A resistance reflex can be elicited by direct stimulation of each of the chordotonal organs (Bush, 1965; Clarac, 1970). The resistance reflex produced by passive movement of a joint can be eliminated (C-B, I-M, C-P, and P-D) or weakened (M-C) either by removal or by denervation of the chordotonal organs associated with that joint (Bush, 1965). To eliminate the resistance reflex of the M-C joint, both the chordotonal organs (MC_1 and MC_2) and the myochordotonal organ must be surgically inactivated (Bush, 1965). Elimination of the resistance reflex of the T-C joint in brachyurans requires the denervation of the T-C muscle receptor organ (Bush and Roberts, 1968). In astacurans, the T-C chordotonal organ also contributes to the T-C resistance reflex (Bush, 1977).

3. THORACICOCOXAL MUSCLE RECEPTOR

The principal T-C joint receptor is a muscle receptor organ situated in parallel with the promotor muscle (Bush, 1976; Whitear, 1965). The muscle receptor consists of a thin bundle of tonic muscle fibers innervated by two motor neurons, one of which is shared with the promotor muscle (Bush and Cannone, 1974). Stimulation of the muscle receptor either by stretch (remotion of the T-C joint) or by efferent discharge (contraction of the receptor muscle) elicits the reflex excitation of motor neurons to both receptor and promotor muscles (Bush and Cannone, 1974; Bush and Roberts, 1968). The co-activation of motor neurons to promotor and receptor muscles prevents the slackening of tension in the muscle receptor (and the consequent loss of stretch sensitivity) during active promotion of the T-C joint (Bush and Cannone, 1974).

4. MYOCHORDOTONAL ORGANS

The I-M joint contains one (in astacurans) or two (in brachyurans, anomurans, and palinurids) specialized chordotonal organs which are connected to the proximal head of the accessory flexor muscle (AFM) in the meropodite (Alexandrowicz, 1972; Clarac and Masson, 1969; Mill, 1976). These form the myochordotonal organs (MCO's) which, as a result of their mechanical connections to the accessory flexor, are situated in parallel with the flexor muscle of the meropodite.

The accessory flexor is innervated by two efferent neurons: a single motor neuron (AFMN) and an inhibitory neuron shared with the M-C extensor muscle (Angaut-Petit et al., 1974). Contraction of the accessory flexor stimulates MCO units sensitive to M-C joint extension. Excitation of the extension-sensitive MCO units elicits a resistance reflex, in which the M-C flexor excitors and the extensor inhibitor are activated, while the flexor inhibitor and the extensor excitors are centrally suppressed (Fig. 3) (Evoy and Cohen, 1969; Vedel et al., 1975). Flexion-sensitive MCO units, responding both to flexion of the joint and to relaxation of the accessory flexor, elicit a resistance reflex in which the extensor muscles are excited and the flexor muscles are inhibited.

In brachyurans, the AFMN can be excited by stimulation of either flexion-sensitive or extension-sensitive MCO units (Fig. 3) (Evoy and Cohen,

2. Control of Posture

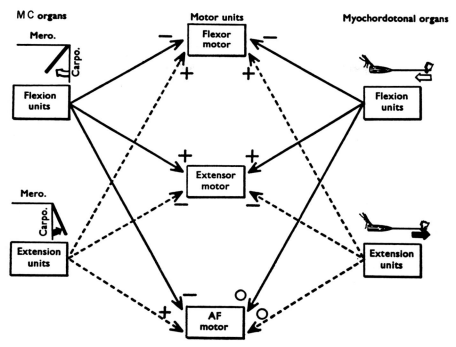

Fig. 3. Summary of the organization of the proprioceptive reflexes that control movement of the M-C joint in *Cancer*. Solid lines indicate extension; broken lines indicate flexion. Excitatory connections are (+); inhibitory connections are (−); ○, excitatory when moved from rest position in appropriate direction. (From Evoy and Cohen, 1969.)

1969). The strength of the AFMN response depends on the differences in the length of the resting muscle (determined by the level of ongoing efferent activity) and of the stimulated muscle. In astacurans, the reflex control of AFMN differs from that described above for brachyurans, in that the AFMN is excited by M-C extension and centrally inhibited by M-C flexion (Vedel et al., 1975).

Excitation of the flexors is usually accompanied by co-activation of the AFMN (Bush et al., 1978; Clarac and Ayers, 1977; Vedel et al., 1975). This prevents relaxation of the accessory flexor muscle during flexion of the M-C joint.

5. CUTICULAR STRESS DETECTORS

Cuticular receptors are located on both sides of the B-I joint (Clarac 1976; Wales et al., 1971). Each is innervated by 40 to 60 sensory neurons that respond to a variety of stimuli: water vibration, pressure applied to the B-I cuticle, movement of the B-I joint, and contraction of the depressor and levator muscles (Clarac et al., 1971).

6. APODEME TENSION RECEPTORS

The apodemes of several limb muscles of brachyurans are innervated by sensory fibers that are excited by an increase in tension in the muscle apodeme (MacMillan, 1976; MacMillan and Dando, 1972). The resistance reflex responses at the M-C joint are inhibited by stimulation of the apodeme tension receptors in the extensor and flexor apodemes (Clarac and Dando, 1973). The importance of these receptors in crustacean motor control is not known, and apodeme tension receptors have not been found in the limb muscles of other decapods (Clarac, 1977; Field, 1974; MacMillan, 1976; Vedel et al., 1975).

7. DISTRIBUTED REFLEXES

Joint movement evokes a number of distributed reflexes (Ayers and Davis, 1977b) in other regions of the stimulated leg. For example, in the American lobster both the T-C and C-B muscles are excited by movement of the T-C, C-B, M-C, and C-P joints, while the M-C flexor muscle is activated in response to movement of the T-C, C-B and M-C joints (Ayers and Davis, 1977b). In rock lobsters (*Palinurus*), movement of the C-B chordotonal organ modulates tonic motor activity throughout the entire leg (Fig. 4) (Clarac et al., 1978). Elevation of the C-B joint excites both the flexor and accessory flexors of the M-C joint while depression of the C-B joint excites the M-C extensor (Bush et al., 1978; Clarac et al., 1978). Stimulation of the cuticular stress detectors modulates the motor neurons that control leg movement in the roll (transverse) plane (Clarac, 1976); i.e., the depressors and elevators at the C-B joint, the flexor, extensor, and accessory flexor at the M-C joint, and the opener and closer at the P-D joint (Clarac, 1976; Clarac and Wales, 1970; Vedel et al., 1975). Other distributed reflexes include excitation of the bender muscle by closing of the P-D joint (Muramoto and Shimozawa, 1970), and excitation of the efferent to the T-C receptor muscle by bending of the C-P joint (Moody, 1970). Since most of the thoracic weight is supported by contractions of the depressor and flexor muscles when the animal is standing, distributed reflexes that affect the muscles at the C-B and M-C joints are particularly important in the maintenance of a stable leg posture.

Some reflexes spread to other legs (intersegmental reflexes: see Chapter 3, this volume). Stimulation of either the M-C myochordotonal or chordotonal organs produces reflex responses in the flexor and extensor muscles of other ipsilateral legs (Evoy and Cohen, 1969). Passive movement of the dactyl evokes a resistance reflex in the contralateral legs (Muramoto, 1965). The

2. Control of Posture

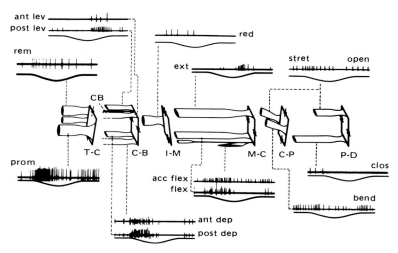

Fig. 4. Leg muscle activity elicited by stimulation of the CB chordotonal organ in rock lobster, *Palinurus*. CB organ stretch (limb depression) indicated by upward deflection; CB relaxation by downward deflection of lower trace in each set of recordings. Potentials recorded from the following muscles: rem, remotor; prom, promotor; ant and post dep, anterior and posterior depressors; red, reductor; ext, extensor; acc flex, accessory flexor; flex, flexor; stret, stretcher; bend, bender; open, opener; clos, closer; ant lev, anterior levator; and post lev, posterior levator. (From Clarac et al., 1978.)

flexor and accessory flexor muscles can be excited by stimulation of the contralateral legs and of the cephalic and abdominal appendages (Vedel et al., 1975).

C. Control of Thoracic Leg Posture

1. COMMAND FIBERS

A large number of premotor interneurons, which produce characteristic thoracic postures when electrically stimulated, have been located in the circumesophageal connectives of the crayfish, *Procambarus clarkii* (Atwood and Wiersma, 1967; Bowerman and Larimer, 1974a). On the basis of the postural responses that are generated when the fibers are stimulated electrically, they have been identified as command fibers.

Perhaps the best known is the single interneuron that elicits the defense reflex, during which the chelipeds and anterior walking legs are lifted or flexed and the abdomen is strongly extended (Fig. 5) (Atwood and Wiersma, 1967; Bowerman and Larimer, 1974a; Glantz, 1977). Other fibers that affect thoracic leg posture include interneurons evoking promotion of the legs, flexion of the legs, crossing of the chelipeds, and turning. At least one

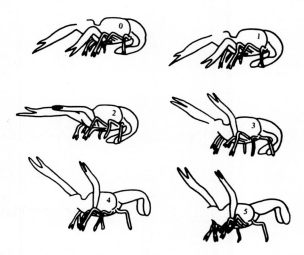

Fig. 5. Responses of a crayfish (*Procambarus clarkii*) to stimulation of a defense command fiber. Numbers indicate time (sec) from the beginning of stimulation. (From Bowerman and Larimer, 1974a.)

"statue fiber" has been described; this suppresses all ongoing movement and freezes the animal in whatever position it held immediately before stimulation (Bowerman and Larimer, 1974a). In addition to the postural fibers noted above, numerous interneurons have been described that evoke phasic motor patterns such as walking and swimming. These actions require that the thoracic appendages adopt a particular posture to serve as a base for their successful execution (Atwood and Wiersma, 1967; Bowerman and Larimer, 1974b; Evoy and Cohen, 1971).

2. EFFERENT SENSORY CONTROL

The postural positions of the T-C and M-C leg joints can be modulated by varying the neural discharge along two different efferent pathways: (1) efferent neurons that innervate the principle muscles of these joints (promotor, remotor, extensor, and flexor muscles); and (2) the efferent neurons to the T-C receptor muscle and the accessory flexor muscle of the MCO.

Increased activity in the efferents to the T-C receptor muscle increases the strength of the resistance reflex (excitation of promotor motor neurons) generated in response to a given remotion of the T-C joint (Bush, 1976; Bush and Roberts, 1968). The net result is to promote the T-C joint.

Increased discharge of the AFMN can produce an increase in the flexion of the M-C joint, since contraction of the accessory flexor raises the extension sensitivity of the MCO, thereby increasing the strength of the resistance reflex evoked by extension of the M-C joint. The resting position of the M-C joint may be set by the level of activity in the AFM efferents; the more intense the AFMN discharge, the greater the flexion of the M-C joint (Evoy and Cohen, 1969; Vedel et al., 1975).

3. CENTRAL VERSUS REFLEX CONTROL

Numerous attempts have been made to obtain qualitative estimates of the relative importance of reflex or centrally generated activity in maintenance of leg stance. Present evidence suggests that resistance reflexes do not contribute to the generation of many motor programs that originate centrally, especially those which involve active leg movement (walking—Ayers and Davis, 1977a; Barnes, 1977; Barnes et al., 1972; Vedel et al., 1975; cheliped cleaning—Field, 1974). In some instances, inactivation or elimination of sensory receptors has altered leg motor programs: removal of the MC (Cohen, 1965) or IM (Clarac, 1968) chordotonal organs alters the resting positions of the M-C and I-M joints, respectively; blocking C-B joints eliminates responses to substrate tilt (Schöne et al., 1976). Other observations are more ambivalent. If the M-C joint is fixed in an extended position, it will not affect the leg stance in either crabs (Carcinus maenus) (Clarac and Beaubaton, 1969) or lobsters (Homarus americanus) (MacMillan, 1975). Removal of the MCO from a single crab leg has no effect on leg posture (Cohen, 1965; Evoy and Fourtner, 1973), while the effects of eliminating the MCO's in all legs depend on the type of crab that is used in the experiments: the resting posture of Carcinus is disrupted (Evoy and Cohen, 1971), while the legs of Cardisoma guanhumi adopt a strong hyperflexion (Fourtner and Evoy, 1973). The results of these experiments may reflect the multiple reflex interactions that are produced between different leg segments (Clarac, 1977). For example, contraction of the flexor muscle is affected by stimulation of at least seven different receptors in the crab leg—three chordotonal organs (MC_1, MC_2, and CB), two myochordotonal organs (MCO_1 and MCO_2), an apodeme tension receptor, and the cuticular stress detectors. As a result of these many reflex interactions, elimination of any one sensory receptor system might produce only a slight alteration in motor activity.

A primary function of resistance reflexes in the legs may be to alter the motor program and compensate for the effects of those external forces opposing a centrally directed active leg movement. The imposition of an external force or load on the leg may evoke a resistance reflex that resists any alteration in leg posture from that which is specified by the ongoing central command (Ayers and Davis, 1978). Resistance reflexes can be readily evoked by imposing a variety of passive movements on the legs (Bush, 1965; Ayers and Davis, 1977b) or, in a more natural setting, by movement over a rough or slippery surface (Barnes et al., 1972; Barnes, 1977). The intensity of the resistance and distributed reflexes that are generated in response to the imposition of a load across the T-C or M-C joints is controlled by the level of ongoing activity in the efferents to the T-C muscle receptor and the MCO's (Barnes, 1977; Bush, 1977; Clarac, 1977; Evoy and Cohen, 1971; Vedel et al., 1975).

IV. ABDOMINAL POSTURE

The last six body segments form the abdomen in decapod crustaceans. In contrast to the thoracic body segments, which are immobilized as a result of their fusion with the overlying carapace, the abdominal segments are unfused and mobile. There are marked differences in the size of the abdomen in different decapods. Astacurans (crayfish, lobsters), palinurids (rock lobsters), and carids (shrimp) possess a large, well-developed abdomen, which is important in postural control, locomotion, and reproduction. In some anomurans (hermit crabs), the abdomen has both morphological and functional asymmetries which enable the crab to secure itself in its shell. The abdomen in brachyurans (true crabs) is reduced in size and held in a tightly flexed position against the ventral thorax, except in brooding females where it is extended to accommodate the egg mass.

Most of the information that is available concerning the sensory, motor, and neuronal elements controlling abdominal posture has been obtained from astacurans (especially crayfish). Data are also available concerning the organization of many elements that contribute to the control of abdominal posture in anomurans (primarily for the hermit crab).

A. Motor Organization

The abdominal musculature of decapods is divided into tonic and phasic systems. Contractions of the deep extensor and flexor muscles, which occupy most of the central core of the abdomen, generate a broad range of rapid, phasic abdominal movements—swimming, escape tail flips, etc. (Abbott and Parnas, 1965; Kennedy and Takeda, 1965a; see Chapter 8, this volume). Postural movements of the abdomen result from the tonic contractions of the superficial extensor (SEM) and superficial flexor (SFM) muscles, which span the dorsal and ventral aspects of the first through fifth abdominal segments, respectively, (Fig. 6) (Fields, 1966; Kennedy and Takeda, 1965b). These muscles are composed entirely of thin sheets of tonic muscle fibers (Kennedy and Takeda, 1965b; Parnas and Atwood, 1966; Jahromi and Atwood, 1967).

The tonic extensors and flexors are each innervated by six efferent neurons in astacurans; five are excitatory motor neurons (SEMN's and

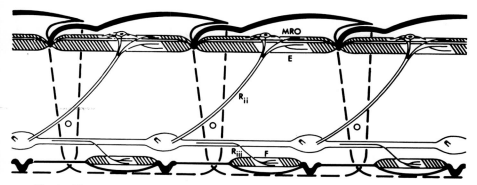

Fig. 6. The organization of the postural muscles and muscle receptor organ in the crayfish abdomen. E, superficial extensors; F, superficial flexors; MRO, tonic muscle receptor organ; R_{iii}, superficial branch of the third root; R_{ii}, second root. Circles mark the pivot points between abdominal segments. (From Fields et al., 1967.)

SFMN's) and one is a peripheral inhibitor (Table I). Each neuron can be identified on the basis of its morphological characteristics—e.g., size, position in the ganglion, and distribution of its terminals over the sheet of muscle fibers (Kennedy et al., 1966; Velez and Wyman, 1978; Wine and Hagiwara, 1977; Wine et al., 1974), and its physiological characteristics—e.g., amplitude of extracellular action potential, conduction velocity, pattern of spike discharge, and junctional potential evoked in muscle fibers (Fields, 1966; Kennedy and Takeda, 1965b; Kennedy et al., 1966; Sokolove and Tatton, 1975). By convention, each efferent is identified by the relative diameter of its axon, determined by either the amplitude or the conduction velocity of the extracellularly recorded action potentials (Fields, 1966; Kennedy and Takeda, 1965b; Sokolove and Tatton, 1975). The smallest (and most tonic) motor neurons are SEMN #1 and SFMN #1; the two peripheral inhibitors are #5 and the largest (and most phasic) motor neurons are SEMN #6 and SFMN #6 (Table I).

Although each of these efferents has "tonic" characteristics (Atwood, 1973), they differ in the degree of their tonicity (Fields, 1966; Kennedy and Takeda, 1965b). The smaller motor neurons are more tonic; when excited, they respond with a prolonged train of action potentials, and they have a high rate of spontaneous discharge. The largest motor neurons have more phasic characteristics; when excited they produce a brief burst of action potentials, and their rate of spontaneous discharge is very low. Their effects on the tonic muscle fibers are more variable. In the crayfish (*Procambarus*), the characteristics of the postsynaptic potential (PSP), which is evoked by discharge of a motor neuron, are related more closely to the location of the SFM fiber in the muscle sheet than to the identity of the active motor neuron

TABLE I

Efferents to the Superficial Abdominal Muscles in Crayfish[a,b]

Efferent	Muscle fibers innervated (percent)	Motor effect	Sensory inputs
SEM			
1	30–60 (also RM_1)	E	$MRO_1(1)$
2	100	E	$MRO_1(E)$
3	—	E	
4	30–60 (also RM_1)	E	CSR (E)
5	30	I	
6	40	E	
SFM			
1	45	E	VMR (1)
2	27	E	VMR (1)
3	86	E	VMR (1)
4	55	E	VMR (1)
5	100	I	
6	53	E	VMR (1)

[a] Data from Evoy and Beranek, 1972; Fields et al., 1967; Kennedy et al., 1966.
[b] E, excitatory; I, inhibitory; SEM, superficial extensor muscle; SFM, superficial flexor muscle; MRO_1, tonic muscle receptor organ; CSR, cord stretch receptors; VMR, ventral mechanoreceptor; RM_1, receptor muscle of MRO_1.

(Velez and Wyman, 1978). The smaller motor neurons tend to elicit small excitatory postsynaptic potentials (EPSP's) in the lobster SFM's with strong facilitation, while the largest motor neurons produce large EPSP's with minimal facilitation (C. S. Thompson and C. H. Page, unpublished data).

The organization of the abdominal musculature of the hermit crab (*Pagurus pollicaris*) differs from the astacuran pattern as a result of the decalcification of the abdominal cuticle of the hermit crab and the importance of the abdomen for securing the animal in its shell. Major differences include (1) the absence of the deep extensor muscle in the hermit crab; (2) the assymmetry of its deep flexors (the right muscle is much larger) (Chapple, 1966); (3) the division of its dorsal superficial muscles (homologous with the SEM's of other decapods) into three layers—tonic (lateral), intermediate (superficial central), and phasic (deep central) (Chapple, 1977a); and (4) the separation of its superficial ventral muscles (homologous with SFM's) into circular and longitudinal muscle layers (Chapple, 1969). The presence of circular superficial ventral fibers not present in other decapods (Kahan, 1971; Pilgrim and Wiersma, 1963) supports the theory that hydrostatic pressure is used as a mechanism for developing rigidity in the decalcified abdomen of the hermit crab (Chapple, 1969).

2. Control of Posture

The efferent innervation of the dorsal and ventral superficial muscles in hermit crabs differs slightly from the astacuran pattern; five motor neurons and two peripheral inhibitors innervate the dorsal superficial muscles (Chapple, 1977b).

B. Abdominal Proprioceptors and Reflexes

1. MUSCLE RECEPTOR ORGAN

A pair of muscle receptor organs (MRO's) spans the dorsal margin of each half abdominal segment (Fig. 6) (Alexandrowicz, 1951, 1967; Fields, 1976). They are found in close association with the superficial extensor muscles in all decapods except brachyurans and some hermit crabs (Chapple, 1977b; Fields, 1976; Pilgrim, 1960). Similar receptors also occur in the last two thoracic segments of most decapods (Pilgrim, 1974; Wiersma and Pilgrim, 1961). Physiological analysis of the MRO as a stretch receptor and as part of a reflex system has been confined to astacurans (crayfish and lobsters) and palinurids (rock lobsters).

Each MRO consists of a small receptor muscle (RM), which is situated adjacent to and in parallel with the SEM's, and a sensory neuron (SR), whose dendrites are embedded among the fibers of the RM (Figs. 6 and 7) (See Chapter 9 of Volume 3 for more detail). The sensory neuron is excited by an increase in receptor muscle tension produced either by its being stretched or by its own contraction (Wiersma et al., 1953).

Responses of the two MRO's differ (Wiersma et al., 1953). The lateral, or tonic, MRO readily responds to abdominal movements and postures that are

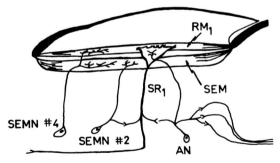

Fig. 7. Diagram of the SR_1-AN reflex in crayfish. Suppression of the SR_1-SEMN #2 resistance reflex during postural flexion of the crayfish abdomen. SR_1 activity excites both SEMN #2 and AN in the same segment and the AN's in the more anterior segments. Axons from the right which innervate the AN originate from SR_1's in the more posterior segments. SR_1 is the sensory neuron of the tonic MRO, AN is the accessory nerve, SEMN #2 is the second smallest superficial extensor motoneuron. (From Page and Sokolove, 1972.)

produced by contractions of the tonic musculature (SEM's and SFM's) (Fields, 1966). The sensory neuron of the lateral MRO (SR_1) is characterized by its high sensitivity to stretch and its low rate of sensory adaptation (Wiersma et al., 1953; Nakajima and Onodera, 1969). The lateral RM (RM_1), which possesses the histological and physiological characteristics of a tonic muscle, is innervated by at least two SEMN's (Fields and Kennedy, 1965; Kuffler, 1954).

In contrast, the medial, or phasic, MRO is characterized by its unresponsiveness during postural abdominal movements (Kennedy et al., 1966). The sensory neuron of the medial MRO (SR_2) responds only to stimuli such as those generated by contractions of the phasic musculature during swimming and escape responses (Wiersma et al., 1953).

The lateral (tonic) MRO receives several efferent neurons that modulate its stretch sensitivity. Although there is morphological evidence that RM_1 is innervated by three SEMN's (Alexandrowicz, 1967), electrophysiological analysis has identified only two SEMN's [a medium sized SEMN (#4) and the smallest SEMN (#1)] innervating RM_1 (Table I) (Fields et al., 1967). Excitation of these SEMN's increases the SR_1 discharge (Fields and Kennedy, 1965; Fields et al., 1967). The SR_1 is innervated by at least three inhibitory efferents—the thick and thin accessory nerves (AN's) and Fiber X (Alexandrowicz, 1951, 1967; Jansen et al., 1971; Wine and Hagiwara, 1977).

The lateral (tonic) MRO is the sensory element in two reflexes regulating abdominal postural movements. Discharge of the SR_1 excites both the thick and thin AN's (Jansen et al., 1970a) and one of the small SEMN's (#2) innervating most of the SEM fibers (Fig. 7) (Fields, 1966; Fields and Kennedy, 1965). The SR_1-AN reflex is a self-inhibitory reflex, which suppresses SR_1 discharge during abdominal flexion (Jansen et al., 1971; Kuffler and Eyzaguirre, 1955; Page and Sokolove, 1972). The SR_1-SEMN #2 reflex contributes to the maintenance of a constant abdominal posture (Fields, 1966; Fields et al., 1967).

The two reflexes (SR_1-AN and SR_1-SEMN #2) differ in the segmental distribution of their effects. Excitation of an SR_1 by flexion of an abdominal segment elicits accessory nerve discharge in both stimulated and adjacent segments (Fields et al., 1967; Jansen et al., 1970b). In contrast, the SR_1-SEMN #2 reflex is confined to the stimulated segment (Fields, 1966).

2. NERVE CORD STRETCH RECEPTORS

Stretching the sheath of the abdominal nerve cord in the crayfish (*Procambarus*) excites about twenty cord stretch receptors (Grobstein, 1973a; Hughes and Wiersma, 1960). These abdominal sensory neurons can be divided into tonic and phasic types.

Stretching of the nerve cord (as during abdominal extension) increases

SEMN discharge and suppresses ongoing SFMN activity (Grobstein, 1973b; Kennedy et al., 1966). Stretching of the nerve cord also excites the SR_1 of the tonic MRO; one of the SEMN's that responds to nerve cord stretch is shared (SEMN #4, which innervates both RM_1 and SEM) (Table I) (Fields et al., 1967). This positive feedback reflex pathway provides a mechanism for increasing SEMN activity during abdominal extension, as follows: (1) SEMN discharge evokes contraction of the superficial extensor muscles; (2) the resulting abdominal extension stretches the nerve cord, thus exciting the cord stretch receptors; (3) discharge of the cord stretch receptors excites the SEMN's (Grobstein, 1973b).

3. VENTRAL MECHANORECEPTORS

The hypodermis of the ventral abdominal cuticle is innervated by sensory endings of pressure-sensitive mechanoreceptors (VMR's) in crayfish (Pabst and Kennedy, 1967; Laverack, 1976). These receptors occur near the origins and insertions of the SFM and are stimulated by SFM contractions. Stimulation of the VMR's (as during abdominal flexion) inhibits SFMN activity (Table I).

C. Control of Abdominal Posture

1. COMMAND FIBERS

The command fibers controlling abdominal postural movement are divided into four groups on the basis of their motor effects: (1) those eliciting abdominal flexion; (2) those eliciting abdominal extension; (3) those controlling the generation of complex sequences of abdominal movements, such as turning or swimming; and (4) those evoking a tonic central inhibition of all efferent motor activity (Bowerman and Larimer, 1974a; Evoy and Kennedy, 1967; Kovac, 1974).

Each command fiber elicits a specific pattern of activity in the efferents to the superficial extensor and flexor muscles. Flexion command fibers drive one or more of the excitatory SEMN's and the inhibitor to the SEM, while centrally inhibiting the excitatory SEMN's and the SFM peripheral inhibitor (Evoy and Kennedy, 1967; Page, 1975a). All extension command fibers centrally inhibit both the SFMN's and the SEM inhibitor. Extension command fibers are divided into two groups on the basis of whether or not they activate SEMN #4, which innervates RM_1 (Fields et al., 1967; Page, 1975a). Approximately half of the extension command fibers drive only those SEMN's not innervating RM_1 (SEMN's #2, #3, and #6). Stimulation of the second group of command fibers excites only SEMN #4, which innervates RM_1.

Both extension and flexion command fibers produce their motor effects by activating reciprocal sets of SEMN's and SFMN's. Reciprocity in the activities of the SEM and SFM efferents is also observed in recordings of "spontaneous" efferent activity from the isolated abdominal nerve cord (Evoy and Kennedy, 1967; Kennedy et al., 1966). Analysis of simultaneous efferent spike trains has demonstrated that premotor neuron connections are primarily responsible for the reciprocity that is observed in the discharge of the antagonistic sets of SEM and SFM efferents. Inhibitory and excitatory cross-connections between many of these efferents also contribute to the generation of a reciprocal motor output (Evoy et al., 1967; Sokolove and Tatton, 1975; Tatton and Sokolove, 1975).

The hypothesis that command fibers control the abdominal postural movements of an intact animal is supported by several experimental results. First, the efferent activities recorded during abdominal movements evoked by sensory stimulation of the intact crayfish are similar to recordings of efferent activity obtained in response to the stimulation of single command fibers (Larimer and Eggleston, 1971; Sokolove, 1973; Page, 1975a). Second, the motor program evoked by stimulation of the backward-walking command fiber closely resembles the pattern of SEM and SFM efferent activity recorded during backward walking of the intact animal (Kovac, 1974). Furthermore, support for the hypothesis that command fibers control postural motor activity has been obtained from the observation that the presence or absence of extension command fibers in the circumesophageal connectives is correlated with generic differences in the sensitivity of the extension reflex (Page 1975a, b). One species of crayfish, *Procambarus clarkii*, has extension command fibers and a well-developed extension reflex, while another species, *Orconectes virilis*, lacks these features.

2. REFLEX CONTROL

Several proprioceptive reflexes contribute to the capacity of the abdominal motor system to adjust the activity of the SEM and SFM efferents and to resist the effects of any external forces (gravity, load, etc.) that are imposed upon the abdomen. These reflexes include the inhibitory SR_1-AN reflex, which facilitates flexion, and the SR_1-SEMN #2 and cord stretch receptor reflexes, both of which increase extension.

A major function of the SR_1-AN reflex is the suppression of the intrasegmental SR_1-SEMN #2 resistance reflex during voluntary abdominal flexion movements (Fig. 7) (Fields et al., 1967; Page and Sokolove, 1972). Postural flexion usually begins in the most posterior segment and proceeds anteriorly. As a result, excitation of the SR_1's (in response to flexion-induced stretch of RM_1) occurs first in the most posterior segment. Discharge of these SR_1's excites the AN's in the more anterior segments through intersegmental reflex connections. Premotor inputs common to both AN's and SFMN's also excite AN's during flexion (Sokolove and Tatton, 1975). Accessory nerve discharge

2. Control of Posture

inhibits the SR_1's, thus suppressing the activation of the SR_1-SEMN #2 resistance reflex during flexion movement.

The SR_1-SEMN #2 resistance reflex may be activated either through flexion of the abdomen by an external force or through the excitation of SEMN #4, which innervates RM_1 (Fig. 8) (Fields, 1966; Fields et al., 1967). On the basis of the known pathways that lead to the excitation of SEMN #4 during extension and the observation that the AN is inhibited during extension by inhibitory cross connections with one of the medium-sized SEMN's (Tatton and Sokolove, 1975), one would anticipate that excitation of the SR_1-SEMN #2 resistance reflex would contribute to the generation of postural extension movements. However, recordings of the SEM efferent activity during extensions of the intact animal are characterized by the absence of SR_1 activity (C. H. Page, unpublished data; Sokolove, 1973). These voluntary extensions are generated primarily as a result of the activity of the SEMN's that do not innervate RM_1 (SEMN's #2, #3, and #6). Thus, the absence of SR_1 activity probably reflects a decrease in RM_1 tension (i.e., unloading), resulting from contraction of the SEM's (Page, 1975a).

3. LOAD COMPENSATION

It has been proposed by Fields (1966) that the SEMN #4-SR_1-SEMN #2 reflex loop provides a mechanism for load compensation during postural abdominal movements (Fields, 1966; Fields et al., 1967). A set posture for the abdomen is established by the level of SEMN #4 activity, which determines the degree of RM_1 contraction. Imposition of a load that flexes the abdomen stretches the RM_1, exciting SR_1 to discharge. The intensity of the resulting SR_1 discharge is a function of the difference between the stretched and the set lengths of RM_1. The excitation of SEMN #2 by activation of the

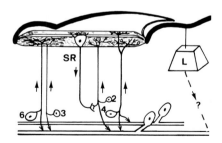

Fig. 8. Diagram of the reflex systems that contribute to load compensation of the crayfish abdomen. Load compensation results form the activation of two reflex systems: (A) SEMN #4-SR-SEMN #2 reflex activity; (B) reflex excitation of SEMN's #2, #3, #4, and #6. SR, stretch receptor neuron; (?) indicates unidentified load sensitive receptor; arrows indicate directions of impulse traffic.

SR_1-SEMN #2 reflex arc elicits an increase in SEM contraction, which permits the relaxation of RM_1 to its set length.

The above hypothesis that the SEMN #4-SR_1-SEMN #2 reflex loop provides a mechanism for load compensation has been examined in two preparations. (1) Recordings from intact crayfish while they extended their abdomens against a load in response to mechanical stimulation of their legs demonstrated that extension against a load results in increased SEMN discharge (especially in the largest neuron, SEMN #6), which is in turn accompanied by SR_1 activity (Sokolove, 1973). (2) Analysis of SR_1 and SEMN activity during command fiber-evoked extensions of the loaded abdomen suggest that there are at least two reflex systems that contribute to the increased SEMN activity observed during abdominal extension against a load (Fig. 8) (Page, 1978).

While the SEMN #4-SR_1-SEMN #2 reflex loop does contribute to the load-compensating increases in SEMN activity observed when the abdomen extends against a load, load-compensating increases in SEMN activity can also be observed in the absence of SR_1 activity. These increases must result from the excitation of a second load-compensating reflex. This reflex is responsible for the large increases in the activity of several SEMN's (especially SEMN's #3 and #6) observed when the abdomen extends against a load (Page, 1978).

In the hermit crab (*Pagurus*), abdominal flexion and elevation of the right fifth leg raises the shell as the crab moves across the substrate (Chapple, 1973). Although the abdomen contributes only 10% of the total lift that supports the shell, abdominal flexion varies with the size of the load.

V. CONCLUSION

The control of the postural stances of the legs and abdomen of crustaceans is an essential prerequisite for the successful execution of many patterns of behavior. Each posture that is adopted by an animal requires that a specific pattern of efferent activity be generated in the motor and inhibitory neurons innervating the muscles of the legs and abdomen. These patterns of activity result both from central motor programs and from the effects of reflex responses to sensory stimulation.

Understanding the neuronal processes that generate a particular postural stance requires detailed information concerning the activities and interconnections of the sensory, central, and motor elements that are involved in the production of the posture. At the present time, this information is not available for any crustacean postural control system. Although detailed descriptions of physiological activity are available for many of the sensory and motor elements in the abdomen and legs of several decapods, little is known concerning the central elements that generate and control the motor programs producing specific postures. More information is required concerning command fibers and other interneurons that control postural efferent activity and the central pathways forming the neural substrate for the postural reflexes. In most instances, this information can be obtained only by the use of intracellular recording and dye-filling techniques to analyze the synaptic interactions and interconnections of these neurons. Analysis of these problems is impeded by the difficulty of recording the activity of specific central neurons in intact, normally behaving animals. This is especially true when one wishes to monitor simultaneously the activities generated in several different neurons. In many instances, microelectrodes may have to be implanted at multiple sites in the central nervous system in order to examine the activity of specific central neurons during the generation of a postural motor program.

LIST OF ABBREVIATIONS

Chordotonal organs in decapod legs:
- BI — Basoischial chordotonal organ
- CB — Coxobasal chordotonal organ
- CP_1, CP_2 — Carpopropal chordotonal organs
- IM_1, IM_2 — Ischiomeral chordotonal organs
- MC_1, MC_2 — Merocarpal chordotonal organs
- PD — Propodactal chordotonal organ
- TC — Thoracicocoxal chordotonal organ

Leg joints in decapods:
- B-I — Basoischial joint
- C-B — Coxabasal joint
- C-P — Carpopropal joint
- I-M — Ischiomeral joint
- M-C — Merocarpal joint
- P-D — Propodactal joint
- T-C — Thoracicocoxal joint
- AFM — Accessory flexor muscle (in the meropodite segment of the decapod leg; component of a myochordotonal organ)
- AFMN — Accessory flexor motoneuron
- AN — "Accessory nerve" (inhibitory neuron to MRO)
- CSD — Cuticular stress detector

MCO	Myochordotonal organ
MCO_1	(= Barth's organ) myochordotonal organ associated with the proximal accessory flexor muscle in the decapod leg
MCO_2	(=IM_2) myochordotonal organ associated with the proximal accessory flexor muscle in the decapod leg
MRO	Muscle receptor organ
MRO_1	"Tonic" muscle receptor organ in the abdominal segments of long-bodied decapods
MRO_2	"Phasic" muscle receptor organ in the abdominal segments of long-bodied decapods
RM	Receptor muscle (muscular component of MRO)
RM_1	"Tonic" receptor muscle in the abdominal segments of long-bodied decapods
RM_2	"Phasic" receptor muscle in the abdominal segments of long-bodied decapods
SEM	Superficial extensor muscle in the abdominal segments of long-bodied decapods
SEMN	Motoneuron of the superficial abdominal extensor muscle
SFM	Superficial flexor muscle in abdominal segments of long-bodied decapods
SFMN	Motoneuron of the superficial abdominal flexor muscle
SR	Sensory neuron component of the muscle receptor organ
SR_1	"Tonic" sensory neuron of abdominal MRO in long-bodied decapods
SR_2	"Phasic" sensory neuron of abdominal MRO in long-bodied decapods
VMR	Ventral mechanoreceptor

ACKNOWLEDGMENT

The author was supported by NIH Grant NINDS 12262 while writing this chapter.

REFERENCES

Abbott, B. C., and Parnas, I. (1965). Electrical and mechanical responses in deep abdominal extensor muscles of crayfish and lobster. *J. Gen. Physiol.* **48,** 919-931.

2. Control of Posture

Alexandrowicz, J. S. (1951). Muscle receptor organs in the abdomen of *Homarus vulgaris* and *Palinurus vulgaris*. *Q. J. Microsc. Sci.* **92,** 163–199.

Alexandrowicz, J. S. (1967). Receptor organs in thoracic and abdominal muscles of crustacea. *Bio. Rev. Cambridge Philos. Soc.* **42,** 288–326.

Alexandrowicz, J. S. (1972). The comparative anatomy of leg proprioceptors in some decapod crustacea. *J. Mar. Biol. Assoc. U.K.* **52,** 605–634.

Angaut-Petit, D., Clarac, F., and Vedel, J. P. (1974). Excitatory and inhibitory innervation of a crustacean muscle associated with a sensory organ. *Brain Res.* **70,** 148–152.

Atwood, H. L. (1963). Differences in muscle fiber properties as a factor in "fast" and "slow" contraction in *Carcinus*. *Comp. Biochem. Physiol.* **10,** 17–31.

Atwood, H. L. (1965). Excitation and inhibition in crab muscle fibers. *Comp. Biochem. Physiol.* **16,** 409–426.

Atwood, H. L. (1973). An attempt to account for the diversity of Crustacean muscles. *Am. Zool.* **13,** 357–378.

Atwood, H. L., and Wiersma, C. A. G. (1967). Command interneurons in the crayfish central nervous system. *J. Exp. Biol.* **46,** 249–261.

Ayers, J. L., and Davis, W. J. (1977a). Neuronal crontrol of locomotion in the lobster, *Homarus americanus*. I. Motor programs for forward and backward walking. *J. Comp. Physiol.* **115,** 1–27.

Ayers, J. L., and Davis, W. J. (1977b). Neuronal control of locomotion in the lobster, *Homarus americanus*. II. Types of walking leg reflexes. *J. Comp. Physiol.* **115,** 29–46.

Ayers, J. L., and Davis, W. J. (1978). Neuronal control of locomotion in the lobster, *Homarus americanus*. III. Dynamic organization of walking leg reflexes. *J. Comp. Physiol.* **123,** 289–298.

Barnes, W. J. P. (1977). Proprioceptive influences on motor output during walking in the crayfish. *J. Physiol. (Paris)* **73,** 543–564.

Barnes, W. J. P., Spirito, C. P., and Evoy, W. H. (1972). Nervous control of walking in the crab, *Cardisoma guanhumi*. II. Role of resistance reflexes in walking. *Z. Vergl. Physiol.* **76,** 16–31.

Bowerman, R. F., and Larimer, J. L. (1974a). Command fibers in the circumoesophageal connectives of crayfish. I. Tonic fibers. *J. Exp. Biol.* **60,** 95–117.

Bowerman, R. F., and Larimer, J. L. (1974b). Command fibres in the circumoesophageal connectives of crayfish. II. Phasic fibers. *J. Exp. Biol.* **60,** 119–134.

Bush, B. M. H. (1962). Proprioceptive reflexes in the legs of *Carcinus maenas (L.)*. *J. Exp. Biol.* **39,** 89–105.

Bush, B. M. H. (1963). A comparative study of certain limb reflexes in decapod crustaceans. *Comp. Biochem. Physiol.* **10,** 273–290.

Bush, B. M. H. (1965). Leg reflexes from chordotonal organs in the crab, *Carcinus maenas*. *Comp. Biochem. Physiol.* **15,** 567–587.

Bush, B. M. H. (1976). Non-impulsive thoracic-coxal receptors in crustaceans. In "Structure and Function of Proprioceptors in the Invertebrates" (P. J. Mill, ed.), pp. 115–151. Chapman & Hill, London.

Bush, B. M. H. (1977). Non-impulsive afferent coding and stretch reflexes in crabs. In "Identified Neurons and Behavior of Arthropods" (G. Hoyle, ed.), pp. 439–460. Plenum, New York.

Bush, B. M. H., and Cannone, A. J. (1974). A positive feed-back reflex to a crustacean muscle receptor. *J. Physiol. (London)* **236,** 37p–39p.

Bush, B. M. H., and Roberts, A. (1968). Resistance reflexes from a crab muscle receptor without impulses. *Nature (London)* **218,** 1171–1173.

Bush, B. M. H., Vedel, J. P., and Clarac, F. (1978). Intersegmental reflex actions from a joint sensory organ (CB) to a muscle receptor (MCO) in decapod crustacean limbs. *J. Exp. Biol.* **73**, 47–63.

Chapple, W. D. (1966). Asymmetry of the motor system in the hermit crab *Pagurus granosimanus*. Stimpson. *J. Exp. Biol.* **45**, 65–81.

Chapple, W. D. (1969). Postural control of shell position by the abdomen of the hermit crab, *Pagurus pollicarus*. I. Morphology of the superficial muscles and their nerves. *J. Exp. Zool.* **171**, 397–408.

Chapple, W. D. (1973). Role of the abdomen in the regulation of shell position in the hermit crab *Pagurus pollicarus*. *J. Comp. Physiol.* **82**, 317–332.

Chapple, W. D. (1977a). Diversity of muscle fibers in the abdominal dorsal superficial muscles of the hermit crab, *Pagurus pollicarus*. *J. Comp. Physiol.* **121**, 395–412.

Chapple, W. D. (1977b). Motoneurons innervating the dorsal superficial muscles of the hermit crab *Pagurus pollicarus*, and their reflex asymmetry. *J. Comp. Physiol* **121**, 413–431.

Clarac, F. (1968). Proprioception by the ischio-meropodite region in legs of the crab, *Carcinus mediterranus* C. *Z. Vergl. Physiol.* **61**, 224–245.

Clarac, F. (1970). Fonctions proprioceptives an niveau de la région basi-ischio-méropodite chez *Astacus leptodactylus*. *Z. Vergl. Physiol.* **68**, 1–24.

Clarac, F. (1976). Crustacean cuticular stress detectors. *In* "Structure and Function of Proprioceptors in the Invertebrates" (P. J. Mill, ed.), pp. 299–321. Chapman & Hill, London.

Clarac, F. (1977). Motor coordination in crustacean limbs. *In* "Identified Neurons and Behavior of Arthropods" (G. Hoyle, ed.), pp. 167–186. Plenum, New York.

Clarac, F., and Ayers, J. (1977). La marche chez les Crustacés: Activité motrice programme et régulation périphérique. *J. Physiol. (Paris)* **73**, 523–544.

Clarac, F., and Beaubaton, D. (1969). Perturbations reversibles des programmes locomoteurs induites par blocage articulaire chez le crabe *Carcinus*. *C. R. Seances Soc. Biol. Sis Fil.* **163**, 2646–2649.

Clarac, F., and Dando, M. R. (1973). Tension receptor reflexes in the walking legs of the crab, *Cancer pagurus*. *Nature (London)* **243**, 94–95.

Clarac, F., and Masson, C. (1969). Anatomie comparée des propriocepteurs de la région basi-ischio-meropodite chez certains Crustacés décapodes. *Z. Vergl. Physiol.* **65**, 242–273.

Clarac, F., and Wales, W. (1970). Contrôle sensoriel des muscles élévateurs au cours de la marche et de l'automie chez certain Crustacés decapodes. *C.R. Hehd. Seances Acad. Sci., Ser. D* **271**, 2163–2166.

Clarac, F., Wales, W., and Laverack, M. S. (1971). Stress detection at the autonomy plane in the Decapod Crustacea. II. The function of receptors associated with the cuticle of the basi-ischiopodite. *Z. Vergl. Physiol.* **73**, 383–407.

Clarac, F., Vedel, J. P., and Bush, B. M. H. (1978). Intersegmental reflex coordination by a single joint receptor organ (CB) in rock lobster walking legs. *J. Exp. Biol.* **73**, 29–46.

Cohen, M. J. (1965). The dual role of sensory systems. Detection and setting central excitability. *Cold Spring Harbor Symp. Quant. Bio.* **30**, 587–599.

Davis, W. J. (1971). Functional significance of motoneuron size and soma position in swimmeret system of the lobster. *J Neurophysiol.* **34**, 274–288.

Evoy, W. H., and Beranek, R. (1972). Pharmacological localization of excitatory and inhibitory synaptic regions in crayfish slow abdominal flexor muscle fibers. *Comp. Gen. Pharmacol.* **3**, 178–186.

Evoy, W. H., and Cohen, M. J. (1969). Sensory and motor interaction in the locomotor reflexes of crabs. *J. Exp. Biol.* **51**, 151–169.

Evoy, W. H., and Cohen, M. J. (1971). Central and peripheral control of arthropod movements. *Adv. Comp. Physiol. Biochem.* **4**, 225–266.

Evoy, W. H., and Fourtner, C. R. (1973). Nervous control of walking in the crab *Cardisoma guanhumi*. III. Proprioceptive influences on intra- and inter-segmental coordination. *J. Comp. Physiol.* **83,** 303-318.

Evoy, W. H., and Kennedy, D. (1967). The central nervous organization underlying control of antagonistic ;muscles in the crayfish. I. Types of command fibers. *J. Exp. Zool.* **165,** 223-238.

Evoy, W. H., Kennedy, D., and Wilson, D. M. (1967). Discharge patterns of neurones supplying tonic abdominal flexor muscles in the crayfish. *J. Exp. Biol.* **46,** 393-411.

Field, L. H. (1974). Sensory and reflex physiology underlying cheliped flexion behavior in hermit crabs. *J. Comp. Physiol.* **92,** 397-414.

Fields, H. L. (1966). Proprioceptive control of posture in the crayfish abdomen. *J. Exp. Biol.* **44,** 455-468.

Fields, H. L. (1976). Crustacean abdominal and thoracic muscle receptor organs. In "Structure and Function of Proprioceptors in the Invertebrates" (P. J. Mill, ed.), pp. 65-114. Chapman & Hill, London.

Fields, H. L., and Kennedy, D. (1965). Functional role of muscle receptor organs in crayfish. *Nature (London)* **206,** 1235-1237.

Fields, H. L., Evoy, W. H., and Kennedy, D. (1967). Reflex role played by efferent control of an invertebrate stretch receptor. *J. Neurophysiol.* **30,** 859-875.

Fourtner, C. R., and Evoy, W. H. (1973). Nervous control of walking in the crab, *Cardisoma guanhumi*. IV. Effects of myochordotonal organ ablation. *J. Comp. Physiol.* **83,** 319-329.

Glantz, R. M. (1977). Visual input and motor output of command interneurons of the defense reflex pathway in the crayfish. In "Identified Neurons and Behavior of Arthropods" (G. Hoyle, ed.), pp. 259-274. Plenum, New York.

Grobstein, P. (1973a). Extension-sensitivity in the crayfish abdomen. I. Neurons monitoring nerve cord length. *J. Comp. Physiol.* **86,** 331-348.

Grobstein, P. (1973b). Extension-sensitivity in the crayfish abdomen. II. The tonic cord stretch reflex. *J. Comp. Physiol.* **86,** 349-358.

Hughes, G. M., and Wiersma, C. A. G. (1960). Neuronal pathways and synaptic connections in the abdominal cord of the crayfish. *J. Exp. Biol.* **37,** 291-307.

Jahromi, S. S., and Atwood, H. L. (1967). Correlation of structure, speed of contraction, and total tension in fast and slow abdominal muscle fibers of the lobster (*Homarus americanus*). *J. Exp. Zool.* **171,** 25-38.

Jansen, J. K. S., Nja, A., and Walloe, L. (1970a). Inhibitory control of the abdominal stretch receptors of crayfish. I. The existence of a double inhibitory feedback. *Acta Physiol. Scand.* **80,** 420-425.

Jansen, J. K. S., Nja, A., and Walloe, L. (1970b). Inhibitory control of the abdominal stretch receptors of the crayfish. II. Reflex input, segmental distribution and output relations. *Acta Physiol. Scand.* **80,** 443-449.

Jansen, J. K. S., Nja, A., Ormstad, K., and Walloe, L. (1971). On the innervation of the slowly adapting stretch receptor of the crayfish abdomen. An electrophysiological approach. *Acta Physiol. Scand.* **81,** 273-285.

Kahan, L. B. (1971). Neural control of postural muscles in *Callianassa californienis* and three other species of decapod crustaceans. *Comp. Biochem. Physiol. A* **40A,** 1-18.

Kennedy, D., and Takeda, K. (1965a). Reflex control of abdominal flexor muscles in the crayfish. I. The twitch system. *J. Exp. Biol.* **43,** 211-227.

Kennedy, D., and Takeda, K. (1965b). Reflex control of abdominal flexor muscles in the crayfish. II. The tonic system. *J. Exp. Biol.* **43,** 228-246.

Kennedy, D., Evoy, W. H., and Fields, H. L. (1966). The unit basis of some crustacean reflexes. *Symp. Soc. Exp. Biol.* **20,** 75-109.

Kovac, M. (1974). Abdominal movements during backward walking in crayfish I. Properties of the motor program. *J. Comp. Physiol.* **95**, 61-78.

Kuffler, S. W. (1954). Mechanisms of activation and motor control of stretch receptors in lobster and crayfish. *J. Neurophysiol.* **17**, 558-574.

Kuffler, S. W., and Eyzaguirre, C. (1955). Synaptic inhibition in an isolated nerve cell. *J. Gen. Physiol.* **39**, 155-184.

Larimer, J. L., and Eggleston, A. C. (1971). Motor programs for abdominal positioning in crayfish. *Z. Vergl. Physiol.* **74**, 388-402.

Laverack, M. S. (1976). External proprioceptors. In "Structure and Function of Proprioceptors in the Invertebrates" (P. J. Mill, ed.), pp. 1-63. Chapman & Hill, London.

MacMillan, D. L. (1975). A physiological analysis of walking in the american lobster (*Homarus americanus*) *Philos. Trans. R. Soc. London, Ser. B* **270**, 1-59.

MacMillan, D. L. (1976). Arthropod apodeme tension receptors. In "Structure and Function of Proprioceptors in the Invertebrates" (P. J. Mill, ed.), pp. 427-442 Chapman & Hill, London.

MacMillan, D. L., and Dando, M. R. (1972). Tension receptors on the apodemes of muscles in the walking legs of the crab, *Cancer magister*. *Mar. Behav. Physiol.* **1**, 185-208.

Mill, P. J. (1976). Chordotonal organs of crustacean appendages. In "Structure and Function of Proprioceptors in the Invertebrates" (P. J. Mill, ed.), pp. 243-297. Chapman & Hill, London.

Moody, C. (1970). A proximally directed intersegmental reflex in a walking leg of the crayfish. *Am. Zool.* **10**, 501.

Muramoto, A. (1965). Proprioceptive reflex of the PD organ of *Procambarus clarkii* by passive movement and vibration stimulus. *J. Fac. Sci., Hokkaido Univ., Ser. 6* **15**, 522-534.

Muramoto, A., and Shimozawa, T. (1970). Reflex response to passive movement of the two adjacent joints (PD and CP joints) in the cheliped of the crayfish. *J. Fac. Sci., Hokkaido Univ., Ser. 6* **17**, 411-421.

Nakajima, S., and Onodera, K. (1969). Adaptation of the generator potential in the crayfish stretch receptors under constant length and constant tension. *J Physiol. (London)* **200**, 187-204.

Pabst, H., and Kennedy, D. (1967). Cutaneous mechanoreceptors influencing motor output in the crayfish abdomen. *Z. Vergl. Physiol.* **57**, 190-208.

Page, C. H. (1975a). Command fiber control of crayfish abdominal movement. I. MRO and extensor motorneuron activities in *Orconectes* and *Procambarus*. *J. Comp. Physiol.* **102**, 65-76.

Page, C. H. (1975b). Command fiber control of crayfish abdominal movement. II. Generic differences in the extension reflex of *Orconectes* and *Procambarus*. *J. Comp. Physiol.* **102**, 77-84.

Page, C. H. (1978). Load compensation in the crayfish abdomen. *J. Comp. Physiol.* **123**, 349-356.

Page, C. H., and Sokolove, P. G. (1972). Crayfish muscle receptor organ: Role in regulation of postural flexion. *Science* **175**, 647-650.

Parnas, I., and Atwood, H. L. (1966). Phasic and tonic neuromuscular systems in the abdominal extensor muscles of the crayfish and rock lobster. *Comp. Biochem. Physiol.* **18**, 701-723.

Pilgrim, R. L. C. (1960). Muscle receptor organs in some decapod Crustacea. *Comp. Biochem. Physiol.* **1**, 248-257.

Pilgrim, R. L. C. (1974). Stretch receptor organs in the thorax of the hermit crab, *Pagurus bernhardus* (L.). *J. Mar. Biol. Assoc. U.K.* **54**, 13-24.

Pilgrim, R. L. C., and Wiersma, C. A. G. (1963). Observations on the skeleton and somatic musculature of the abdomen and thorax of *Procambarus clarkii* (Girard), with notes on the thorax of *Panulirus interruptus* (Randall) and *Astacus*. *J. Morphol.* **113,** 453–487.

Schöne, H. (1951). Die statische Gleichgewichts-orientierung dekapoder Crustaceen. *Verh. Dtsch. Zool. Ges.* **16,** 157–162.

Schöne, H. (1954). Statozystenfunktion und Statische Lageorientierung bei Dekapoden Krebsen. *Z. Vergl. Physiol.* **36,** 241–260.

Schöne, H., Niel, D. M., Stein, A., and Carlstead, M. K. (1976). Reactions of the spiny lobster, *Palinurus vulgaris*, to substrate tilt (I.). *J. Comp. Physiol.* **107,** 113–128.

Schöne, H., Niel, D. M., and Scapini, F. (1978). The influence of substrate contact on gravity orientation (Substrate orientation in spiny lobsters V). *J. Comp. Physiol.* **126,** 293–295.

Sokolove, P. G. (1973). Crayfish stretch receptor and motor unit behavior during abdominal extensions. *J. Comp. Physiol.* **84,** 251–266.

Sokolove, P. G., and Tatton, W. G. (1975). Analysis of postural motorneuron activity in crayfish abdomen. I. Coordination by premotoneuron connections. *J. Neurophysiol.* **38,** 313–331.

Tatton, W. G., and Sokolove, P. G. (1975). Analysis of postural motorneuron activity in the crayfish abdomen. II. Coordination by excitatory and inhibitory connections between motoneurons. *J. Neurophysiol.* **38,** 332–346.

Vedel, J. P., Angaut-Petit, D., and Clarac, F. (1975). Reflex modulation of motorneuron activity in the cheliped of the crayfish *Astacus leptodactylus*. *J. Exp. Biol.* **63,** 551–567.

Velez, S. J., and Wyman, R. J. (1978). Synaptic connectivity in a crayfish neuromuscular system. I. Gradient of innervation and synaptic strength. *J. Neurophysiol.* **41,** 75–84.

Wales, W., Clarac, F., Dando, M. R., and Laverack, M. S. (1970). Innervation of the receptors present at the various joints of the pereiopods and third maxilliped of *Homarus gammarus* (L.) and other macruran decapods (Crustacea). *Z. Vergl. Physiol.* **68,** 345–384.

Wales, W., Clarac, F., and Laverack, M. S. (1971). Stress detection at the autotomy plane in the decapod crustacea. I. Comparative anatomy of the receptors of the basi-ischiopodite region. *Z. Vergl. Physiol.* **73,** 357–383.

Whitear, M. (1962). The fine structure of crustacean proprioceptors. I. The chordotonal organs in the legs of the shore crab, *Carcinus maenas*, *Philos. Trans. R. Soc. London, Ser. B* **245,** 291–325.

Whitear, M. (1965). The fine structure of crustacean proprioceptors. II. The thoracico-coxal organs in *Carcinus*, *Pagurus* and *Astacus*. *Philos. Trans. R. Soc. London, Ser. B* **248,** 437–456.

Wiersma, C. A. G., and Pilgrim, R. L. C. (1961). Thoracic stretch receptors in crayfish and lobsters. *Comp. Biochem. Physiol.* **2,** 51–64.

Wiersma, C. A. G., and Ripley, S. H. (1952). Innervation patterns of crustacean limbs. *Physiol. Comp. Oecol.* **2,** 391–405.

Wiersma, C. A. G., Furshpan, E., and Florey, E. (1953). Physiological and phramacological observations on muscle receptor organs of the crayfish, *Cambarus clarkii* Girard. *J. Exp. Biol.* **30,** 136–150.

Wine, J. J., and Hagiwara, G. (1977). Crayfish escape behavior. I. The structure of efferent and afferent neurons involved in abdominal extension. *J. Comp. Physiol.* **121,** 145–172.

Wine, J. J., Mittenthal, J. E., and Kennedy, D. (1974). The structure of tonic flexor motoneurons in crayfish abdominal ganglia. *J. Comp. Physiol.* **93,** 315–335.

3

Locomotion and Control of Limb Movements

WILLIAM H. EVOY AND JOSEPH AYERS

I.	Introduction	62
II.	Limb Movement	63
III.	Intrasegmental Motor Programs	64
	A. Walking in Macrurans	64
	B. Walking in Brachyurans	68
	C. Beating of Swimmerets in Lobsters and Crayfish	70
	D. Swimming by Means of Exopodites	72
	E. Swimming by Means of Uropods	73
	F. Swimming Mediated by the Walking Legs	74
	G. Filtering Movements of Barnacles	74
	H. Movements of Specialized Appendages	74
IV.	Inter-Limb Coordination and Gaits	75
	A. Walking of Macrurans	76
	B. Walking of Brachyurans	76
V.	Segmental Oscillators for Rhythmic Limb Movements	77
	A. Macruran Swimmerets	77
	B. Anomuran Uropods	77
	C. Walking Legs	78
	D. Cirri of Barnacles	79
VI.	Intersegmental Control Systems	79
	A. Command Systems	79
	B. Coordinating Neurons	81
VII.	Sensory Modulation of Limb Motor Programs	83
	A. General Features of Reflexes	83
	B. Synaptic Pathways	83
	C. Negative Feedback Resistance Reflexes	84
	D. Positive Feedback and Variable Reflexes	88
	E. Responses to Changes in Load	90

	F.	Reflexes Involving Cuticular Stress Detectors	91
	G.	Reflexes Involving Apodeme Tension Receptors	91
	H.	Reflexes Involving Muscle Receptor Organs	91
	I.	Reflexes Involving Sensory Hairs	92
	J.	Distributed Reflexes	93
	K.	Intersegmental Reflexes	94
	L.	Exteroceptive Reflexes	96
VIII.	Summary and Conclusions		97
	References		98

I. INTRODUCTION

Limb movements play a major role in a variety of crustacean behaviors, ranging from simple rhythmic acts, such as walking and feeding, to elaborate agonistic and courtship displays. The means by which these movements are controlled and regulated to achieve remarkable examples of rapidity, precision, or strength have challenged observers for over a century. The speed and agility of the ghost crab, *Ocypode,* the ballistic strike of the stomatopod, *Squilla,* the defensive pinch of the blue crab, *Callinectes,* and the elaborate cheliped displays of the fiddler crab, *Uca,* fascinate and often dismay even the most casual observers and would-be predators of these animals. Crustacean motor systems provide excellent material for analysis of neuronal mechanisms underlying motor control.

Crustacean limbs possess features in common with those of other arthropods and of vertebrates: i.e., a system of joints and levers that is powered by antagonistically arranged groups of muscles. In addition, crustacean limbs are invested with an elaborate array of proprioceptors that are responsive to every aspect of movement and to the development of force, and these connect with appropriate integrative sites in the central nervous system to play a significant role in the control of motor activity. Neuronal control of appendages may involve regularly repeated cycles, maintained posture, precise and complex manipulatory movements, or all three. In no case is the entire neuronal connectivity or activity that underlies these movements fully understood, but pertinent aspects have been revealed and have provided important clues as to the principles by which motor control is achieved.

All movements of crustacean appendages involve coordinated sequences of contraction of several muscles. Within an appendage, individual muscles may participate in different types of movement: muscles that are synergistic for one type of movement (e.g., forward walking) may be antagonistic for another mode of movement, such as sideways walking, swimming, or display. Walking and swimming also necessitate phasing the movements of appendages in different segments of the body, requiring coupling between the motor outputs to the individual limbs. Evidence for oscillatory systems responsible for the movements of walking legs is indirect, since it is derived from observations of normal and perturbed limb movements in several different species. The most complete data on control of repetitive activity in appendages come from work on movements of respiratory appendages or swimmerets, which are more regular than most other limb movements and which appear to be less responsive to environmental contingencies. The current status of work on central control of rhythmic movements is covered in this volume by Wiens (Chapter 7), and by Wales (Chapter 6), and by McMahon and Wilkens in a later volume of this series. The functional morphology of appendages specialized for feeding and respiration are also treated in those chapters. Other specialized functions that can be compared to locomotory systems are dealt with briefly in the present chapter.

II. LIMB MOVEMENT

In order to understand the movements of appendages used in locomotion and other forms of behavior, it is necessary to present briefly the mechanical aspects of limb and joint movements. In arthropod limbs, skeletal muscle exerts its force on a mechanically constrained system of levers formed by apodemes and joints of the segments. Most arthropod limb joints operate in a single plane; rotational movements at a single joint have been described in a few instances, but they are uncommon (e.g., pivoting of the uropods in *Emerita;* Paul, 1971b). Usually, the movements of an individual joint are brought about by a pair of antagonistic muscles; but at some joints, a muscle may be made up of several heads that insert at different angles, and the parts of the muscle serve different functions in locomotion and posture. Successive joints along a particular appendage operate at an angle of approximately 90° to one another, and most movements result from various combinations of planar displacements at several joints (see Fig. 1).

Crustaceans also exhibit two major forms of swimming. Macrurans and other tailed forms can make use of rhythmic extension and flexion of the axial musculature of the abdomen to propel them through the water (see Wine and Krasne, Chapter 8, this volume). In some crustaceans, swimming is due to rhythmic sculling movements of the locomotory appendages. This latter form of swimming will be discussed in this chapter.

Swimming movements may be distinguished from walking movements by their occurrence in the absence of mechanical contact with the substrate; thus, they occur in a homogeneous medium, water. In later sections we will review the motor programs that underly four major types of limb-mediated

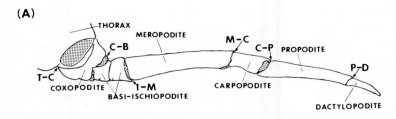

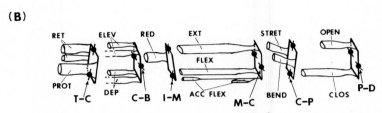

Fig. 1. The segments, joints, and major muscles of a walking leg of *Palinurus elephas*. (A) Anterior-lateral view of the third left leg. (B) Representation of the major groups of muscles showing the approximate orientation of the hinge points at the joints, as labeled, which are moved with the fulcra of the joints shown as black dots. Abbreviations of muscles: RET: thoracocoxopodite (T-C) retractors; PROT: T-C protractor; ELEV: coxobasipodite (C-B) elevators; DEP: C-B depressors; RED: ischiomeropodite (I-M) reductor; EXT: Merocarpopodite (M-C) extensor; FLEX: M-C flexor; STRET: carpopropodite (C-P) stretcher; BEND: C-P bender; OPEN: prodactylopodite (P-D) opener; CLOS: P-D closer; ACC FLEX: accessory flexor. (Adapted from Clarac et al., 1978.)

swimming: sculling by the thoracic exopodites in larvae of *Homarus* (Macmillan et al., 1976), beating of the uropods in the sand crab, *Emerita* (Paul, 1979), beating of the swimmerets in macrurans (Davis, 1973), and swimming mediated by the walking legs in *Callinectes* (Spirito, 1972).

III. INTRASEGMENTAL MOTOR PROGRAMS

A. Walking in Macrurans

Unlike the more specialized brachyurans, macrurans have retained the ability to walk in all directions with equal facility (e.g., Ayers and Davis, 1977a; Ayers and Clarac, 1978). Hence, in these forms, the central generators underlying walking must be able to generate and superimpose several motor programs, each of which underlies walking in a different direction. Several different investigators (e.g., MacMillan, 1975; Ayers and Davis, 1977a; Barnes, 1977; Ayers and Clarac, 1978; Evoy, 1977; Grote,

3. Locomotion and Control of Limb Movements

1981) have examined the motor output at different regions of the leg, so that it is now possible to assemble a fairly complete picture of the control of the legs, especially during forward walking.

The motor programs underlying walking have been examined under several different sets of experimental conditions. Some studies consist of an electromyographic analysis of unrestrained walking (Ayers and Clarac, 1978; Clarac and Ayers, 1977). Other studies have focused on walking in animals on a motor-driven treadmill, which was modulated in velocity to keep the animal in front of a video camera (MacMillan, 1975). Still other studies have examined walking in animals that were tethered over a treadmill. In one case, the animal was allowed to provide the propulsive force that moved the treadmill (Barnes, 1977). In another case, the treadmill was controlled by an electric motor of variable speed, while the animal was restrained from moving in the roll or yaw planes (Ayers and Davis, 1977a). Several of these studies (e.g., Macmillan, 1975; Barnes, 1977) have treated only forward walking, whereas others (e.g., Ayers and Davis, 1977a; Clarac and Ayers, 1977; Ayers and Clarac, 1978) have examined the patterns of coordination that underly walking in different directions. These latter patterns have been termed "metastable," for the organisms can spontaneously change their direction of walking and underlying mode of coordination on a cycle-by-cycle basis (Ayers and Davis, 1977a; Clarac and Ayers, 1977).

The cycle of stepping in *Homarus americanus* consists of a power stroke (during which the limb bears weight and provides propulsive force), followed by a return stroke (during which the limb is elevated and returned to its original position to initiate the subsequent power stroke). Temporal organization of the stepping cycle depends on the stepping frequency or experimental conditions, or both. In MacMillan's (1975) experiments with unrestrained lobsters, variation in the stepping frequency resulted from proportionate variations in duration of both power and return strokes. In contrast, in specimens tethered on a motor-driven treadmill, duration of the return stroke remained constant (Fig. 2C), and changes in the stepping frequency resulted from variation in the duration of the power stroke (Fig. 2D). The latter result was obtained for all directions of walking (Ayers and Davis, 1977a). Note that the walking movements analyzed in the two studies occurred over different ranges of walking speeds (MacMillan, 1975; Ayers and Davis, 1977a). Comparison of these results with those obtained in loaded and unrestrained crabs (Evoy and Fourtner, 1973a) suggests that different motor programs are utilized in loaded and in unrestrained walking.

The limb movemens underlying walking in the different directions occur predominately at three joints of the macruran leg (Fig. 3). Cyclic elevation and depression of the C-B joint underly the return and power strokes of walking in all directions. During forward and backward walking, propulsive

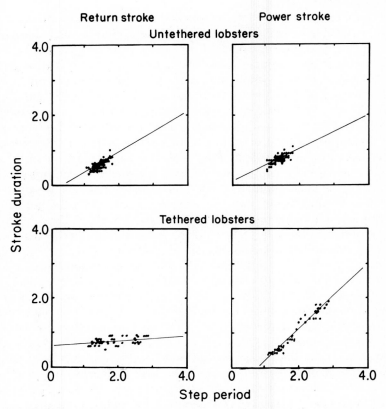

Fig. 2. Dynamic organization of the stepping cycle of the lobster *Homarus* spp. Each graph shows the duration of the relevant stroke of the step plotted against the total period of the cycle. The upper pair of graphs represents data replotted from MacMillan's (1975) study of walking in untethered lobsters. The lower pair represents data replotted from Ayers and Davis' (1977a) study of locomotion in lobsters tethered over a driven treadmill. In MacMillan's study, the regression slope for the return stroke was 0.57, while for the power stroke it was 0.48. In the study by Ayers and Davis, the regression slope for the return stroke was 0.08, while that for the power-stroke slope was 0.92.

forces are provided by protraction and retraction of the basalar T-C joint (Fig. 3). In contrast, during lateral walking, propulsive forces are produced predominately by extension and flexion of the more distal M-C joint (Fig. 3). Joint movements that are synergistic for walking in one direction (e.g., elevation and protraction during the return stroke of forward walking), become antagonistic for walking in the opposite direction (elevation and retraction during the return stroke of backward walking).

On the basis of timing criteria, motor neurons of the walking leg can be

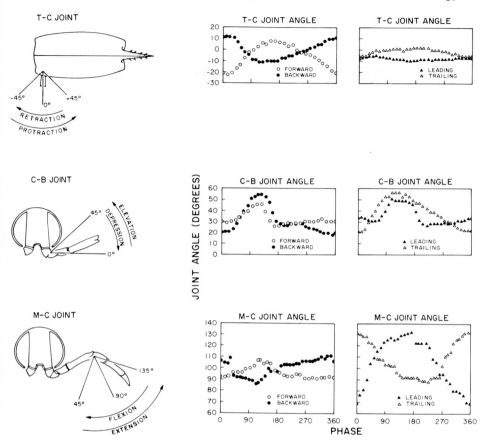

Fig. 3. Joint movements underlying walking in different directions in lobsters (*Homarus*). Each point represents the angle of the respective joint during a single frame of a motion picture (inter-frame interval 55.5 msec) plotted against the phase of the frame in the stepping cycle. Phase 0 and 360 correspond to the initiation of the return stroke in subsequent steps. Reference coordinates are shown at left. The middle column corresponds to forward and backward walking, while the right column corresponds to lateral walking in the two directions. The upper row represents thoracocoxal (T-C) joint movements, while the middle and bottom rows represent coxobasal (C-B) and merocarpopodite (M-C) joint movements, respectively. (From Ayers and Davis, 1977a.)

divided into three different functional classes (Ayers and Davis, 1977a). (1) Some motor elements discharge only during the return stroke. The duration of bursts in these elements remains constant during variations in stepping period, independent of the direction of walking. An example is the motor neurons of the anterior elevator muscle of the C-B joint. (2) Other motor elements, such as those of the anterior depressor muscle of the C-B joint, discharge only during the power stroke. The duration of bursts in these elements varies directly with the step period, and, therefore, one can attribute all variation in the stepping frequency to variation in the duration of power-stroke bursts except during the highest speeds of walking (e.g., MacMillan, 1975). During walking, the motor neurons responsible for the power and return strokes typically discharge in alternating reciprocal bursts (Ayers and Davis, 1977a). (3) Remaining motor neurons of the leg, termed "bifunctional" motor units, can discharge during either the power stroke or the return stroke (Ayers and Davis, 1977a; Ayers and Clarac, 1978). These motor elements may discharge tonically throughout the stepping cycle, or they may burst cyclically in synergism with the motor neurons of either the return stroke or the power stroke. When bifunctional motor elements burst, their burst duration varies with that of the motor neurons with which they are synergistic (Ayers and Clarac, 1978).

In freely moving animals, two mechanisms are utilized to differentiate the motor output during walking in the different directions (Ayers and Clarac, 1978; Clarac and Ayers, 1977). First, bifunctional motor neurons discharge at different frequencies during different types of walking behavior. For example, the discharge of the tonic extensor motor unit is highest when it serves a power-stroke function and lowest when it serves a return-stroke function (Ayers and Clarac, 1978). Second, the selective recruitment of phasic motor neurons is utilized, in part as a mechanism to produce different modes of coordination for the different directions of walking. Within each mode of coordination, the discharge frequency of a phasic motor neuron is invariant during changes in stepping frequency; gradation of muscle force is apparently achieved by changing the duration of the burst of motor impulses (Spirito et al., 1972).

B. Walking in Brachyurans

Motor programs underlying lateral walking in several crabs have engendered considerable interest as models for the proprioceptive control of locomotion (e.g., Evoy and Fourtner, 1973a; Clarac, 1977). Although crabs can walk in all directions, they usually walk laterally. Descriptions of walking in crabs must treat motor programs for walking in the two lateral directions: lateral leading (power stroke during flexion) and lateral trailing (power stroke during extension).

While the muscles of the T-C joint are the primary source of power in forward and backward movements of most crustacean limbs, lateral walking requires that most power be generated by the musculature of the M-C joint. The large mass of M-C flexor and extensor muscle in the meropodite reflects the

3. Locomotion and Control of Limb Movements

importance of this joint for brachyuran locomotion. *Carcinus maenas* derives most of its propulsive force from contraction of M-C flexors of legs on the leading side and M-C extensors on the trailing side (Clarac and Coulmance, 1971), as during lateral walking in macrurans. These movements are synchronized with depression of the C-B joint during the power stroke and elevation of the same joint during the return stroke.

The P-D joint participates to some extent in lateral walking; its movements are appropriately phased with those of the M-C and C-B joints (Evoy and Fourtner, 1973a; Barnes et al., 1972; Clarac and Coulmance, 1971). However, its distal location and relatively small muscles limit it to a primarily supportive function. Relative angular changes of the several joints during lateral walking in *Carcinus* sp. are shown in Fig. 4 (adapted from Clarac and Coulmance, 1971). In this study, there were definite differences in the movements of the individual joints and in the activation of muscles on both sides of the animal during unrestrained walking on a level surface. On the trailing side, the range of movement of the M-C joint was 130–160°, whereas on the leading side it was 100–130°; thus, pushing movements of extension on the trailing side provided a greater contribution to overall propulsion.

In *Ocypode ceratophthalma*, even greater disparity exists between the

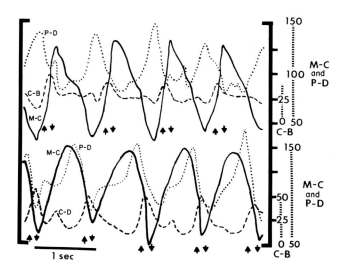

Fig. 4. Angular changes at joints of two walking legs on opposite sides of the same segment of a crab during unperturbed lateral walking. Top: Leg on the leading side. Bottom: Leg on the trailing side. Movements of three joints (C-B, M-C, and P-D) are shown as measured from motion pictures. Scales for the angular changes of the joints are shown at the right as labeled. Upward arrows indicate lifting of the dactylopodite tip from the surface; downward arrows show the time of contact. (Adapted from Clarac and Coulmance, 1971.)

activities of the legs on the two sides (Hafemann and Hubbard, 1969; Burrows and Hoyle, 1973). During the most rapid running (slightly greater than 2 m/sec), only the third and fourth pereiopods on the trailing side cycle. The second and fifth trailing legs are raised and the animal leaps forward bipedally, skidding periodically on the dactyl of legs on the leading side. The leaping mode of locomotion allows a considerable increase in speed without necessitating high cycling rates of muscle contraction. In *Ocypode,* the dactylopodite makes up 20-65% of the total length of leg. Relatively long dactylopodites are found in many crabs that have become specialized for rapid walking and running, in contrast to the shorter dactylopodites of slower species (e.g., 13-20% of total leg length in *Pachygrapsus transversus;* Hafemann and Hubbard, 1969). Thus, P-D arcs of approximately 30° measured during walking in *Carcinus* and *Cardisoma* may provide a sizeable contribution to the movement, even though the muscles operating this joint are unlikely to develop much force.

Temporal organization of the stepping cycle in crabs appears to depend on experimental conditions. Some studies have focused on unrestrained walking, either in air or under water (Clarac and Coulmance, 1971; Evoy and Fourtner, 1973a), whereas other studies have examined locomotion in animals that were tethered over a treadmill that the animal was able to move (Barnes et al., 1972). Differences in timing of motor output in the two experimental conditions appear to result from differences in loading of the animal and in walking speed. For example, Evoy and Fourtner (1973a) found from cinematographic analysis that variation in stepping frequency results from variation in the duration of both power and return strokes (Fig. 5a,b). However, when these same crabs are forced to bear an added load, these temporal relations are changed: the return stroke is constant in duration and the power stroke is variable (Fig. 5c,d). Furthermore, these two modes of dynamic organization occur over different ranges of stepping period; loaded crabs walk much more slowly than do unrestrained specimens. Thus, in both macrurans (Fig. 2) and brachyurans, the dynamic organization of the stepping cycle depends on the frequency of stepping as well as on the degree of loading.

C. Beating of Swimmerets in Lobsters and Crayfish

In swimmerets of the lobster (*Homarus americanus*), Davis (1968a) identified 26 separate muscle bundles on morphological grounds, but on the basis of their sharing of innervation and their roles in producing discrete movements, he was able to categorize the muscles into 12 functional groups. Electromyographic recordings from the muscles during the beating

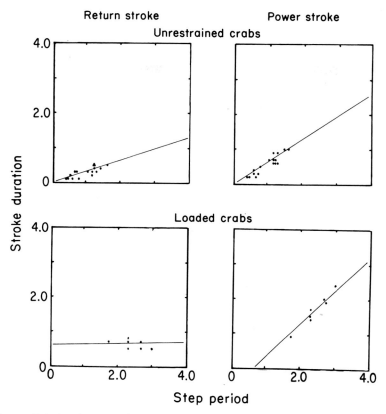

Fig. 5. Relative duration of power and return strokes of a walking leg of *Cardisoma guanhumi* plotted against duration of the total step cycle, as analyzed from motion pictures of a freely walking crab. In the upper row of graphs ("unloaded"), the animal walked sideways on a level surface; in the lower row, it walked up a 45° incline. Under unloaded conditions, the duration of both power and return strokes increased as duration of stepping increased. Under loaded conditions, stepping frequency was lower, and only duration of the power stroke showed a significant correlation with duration of stepping, whereas duration of the return stroke was relatively constant. Compare this figure with the analogous findings in the lobster (Fig. 2). In unloaded crabs the regression slope for the return stroke was 0.32, while for the power stroke it was 0.66. In loaded crabs, regression slope for the return stroke was 0.03, while for the power stroke slope it was 0.95. (Replotted from Evoy and Fourtner, 1973a.)

of the swimmerets provided a further breakdown of the muscles into power- and return-stroke groups.

The swimming movements of the abdominal swimmerets consist of alternating power and return strokes. Accessory movements consist of the opening and curling of the rami during the power stroke and the closing and uncurling of the rami during the return stroke. The muscles that function during the beating of the swimmerets are distinguishable as power-stroke and return-stroke groups that approximate reciprocity in their discharge, although there is overlap in their motor bursts. The waveform of the beat is nearly symmetrical (Davis, 1968b,c). Variations in the period of the beat cycle are produced by proportionate increases in the duration of both power and return strokes. At the level of the motor neuron however, the oscillation is somewhat asymetrical, with the power-stroke burst being relatively constant in duration, and the return-stroke burst varying proportionately with duration of the total cycle.

The amplitude of swimmeret movement varies inversely with period; i.e., movement is of greater amplitude during more rapid beating (Davis, 1971). This variation in amplitude is produced in two ways. First, as the frequency of beating increases, the number of active motor neurons increases in the order of increasing size and neuromuscular effect (Davis, 1971). Second, as the frequency of beating increases, individual motor neurons increase their frequency of discharge (Davis, 1971).

Different species vary in the number of swimmeret muscles: seven in *Penaeus setiferus* (Young, 1959), nine in *Astacus astacus* (Schmidt, 1915), and six in *Pagurus pollicaris* (Bent and Chapple, 1977). Direct homologies are therefore unclear, especially since the Paguroidea lack intrinsic muscles of the exopodite and endopodite rami. Some species (*Homarus americanus* and *Procambarus clarkii*) are capable of asymmetrically beating the swimmerets, in which the power stroke is directed laterally during turning, righting, or sideways walking (Davis, 1968b; Bowerman and Larimer, 1974a). During the righting response of a lobster tilted to one side, the basipodites on the raised side are rotated outward before and during the power stroke (Davis, 1968b). Specimens of *Penaeus* are capable of a 90° rotation at the basipodite–exopodite joint to permit feathering during the return stroke (Young, 1959). Numbers of swimmeret motor neurons, determined by cobalt backfill of the first root, also show variation between species in general correspondence with the numbers of muscles (Bent and Chapple, 1977).

D. Swimming by Means of Exopodites

One of the more primitive forms of swimming consists of rhythmic power- and return-stroke movements of thoracic setaceous appendages (Storch, 1929; Lent, 1971, 1977; Lochhead, 1961). Perhaps the most detailed study of this type of locomotion is a cinematographic analysis of swimming of larval *Homarus* (Neil et al., 1976; MacMillan et al., 1976; Laverack et al., 1976), which move about by rhythmic rowing movements of the exopodites of the thoracic appendages. The movements of the exopodites consist of a return stroke, during which the setae are flattened relative to the body of the exopodite; followed by a power stroke, during which the setae are passively expanded to increase the surface area (Neil et al., 1976). In addition to the major power- and return-stroke muscles that originate in the basipodite and insert on the inner segment of the exopodite, three muscles that originate in the inner segment adjust pitch by rotating the flagellum. Thus, the animal can swim up and forward, horizontally forward or backward, or vertically upward; or it can hover. During the tailflick of the escape response, the limbs are held protracted in the same manner as are crayfish legs (Wine and Krasne, 1972) to reduce drag (Pond, 1975).

E. Swimming by Means of Uropods

The swimming movements of the mole crab, *Emerita analoga*, consist of alternating power and return strokes of the abdomen and uropods (Paul, 1971a,b,c). The hinging of the uropods allows them to be moved in three axes to produce rotation, extension/flexion, and protraction/retraction. The protopodite (probably a fusion of coxa- and basipodites) is hinged so that it can be moved in both horizontal and vertical planes, as well as rotated about its axis. In *Emerita*, the protopodite can move 70° vertically with respect to the sixth abdominal segment and can rotate 40° around its axis and extend slightly laterally. The entire uropod is held forward along the abdomen rather than along the telson as in forms that use it mainly for steering (Paul, 1971b). Cyclic movements result from four separate but interacting components: (1) cyclic power and return strokes of the protopodite; (2) spreading and folding of the uropod rami; (3) supination and pronation of the whole uropod; and (4) extension and flexion of the uropod (Paul, 1971a). The cyclic movements have been analyzed cinematographically and by electromyography; they involve the coordinated activity of 11 distinct muscles (Paul, 1971a,b,c). Two forms of uropod-mediated behavior have been identified (Paul, 1976). Swimming, the more rapid of the two, is characterized by the maintenance of a constant phase of the power stroke's motor burst in the interval between return strokes. Power-stroke relations similar to those of intact animals are seen in an isolated preparation, although the timing of the return stroke is less predictable (Paul, 1979). During "treading water," however, the latency from return stroke to power stroke remains constant, and phase of the power stroke becomes variable. The similarity of these movements to the rhythmic movements of the telson and uropods of *Procambarus clarkii* (Larimer and Kennedy, 1969b), and the apparent homology of flexors and extensors with axial abdominal muscles (Larimer and Kennedy, 1969a), suggest that the control mechanisms for movements of the uropods in *Emerita* are similar to those that mediate swimming in crayfish (Krasne and Wine, 1977).

F. Swimming Mediated by the Walking Legs

In some specialized brachyurans, pereiopods are flattened so that they can act as effective paddles; in other crustaceans, such as *Homola barbata*, there is little morphological specialization beyond the ability to rotate the leg at C-B, although the motor program may be modified so that walking legs are effective in propelling the animal through the water (Hartnoll, 1970, 1971). Extreme morphological adaptations, such as the flattening of the pro- and dactylopodites to form paddles and the increased length of fringes of setae, result in increased surface area, as seen in the Portunidae. Streamlining of the carapace is also seen.

The motor program underlying swimming in the blue crab, *Callinectes sapidus*, has been examined by Spirito (1972). During swimming, *Callinectes* uses only leading legs two through four and the fifth legs of both sides (Spirito, 1972). The flattened fifth legs are adapted for sculling: rotation occurs toward the end of both up- and downstrokes. The musculature of the fifth leg is specialized. Two components of the remotor group and the promotor serve to adjust the angle of the limb, while the elevator and depressor systems are responsible for the beat during sculling (White and Spirito, 1973). The basic power stroke is in a dorso-ventro-caudal plane because of rotation of the limb. Similar specializations are present in power-stroke muscles of *Portunus sanguinolentus* (Hoyle and Burrows, 1973).

G. Filtering Movements of Barnacles

Sessile adults of Cirripedia have modified thoracic appendages, the cirri, which are used primarily in feeding movements, although the innervation and position of these appendages appear to be homologous with the pereiopods and other thoracic appendages of motile forms. In species such as *Balanus*, that feed by rhythmic beating of the cirri, flexor muscles curl the rami during withdrawal, but extension is due to pumping of the hemolymph by the somatic musculature (Gwilliam and Bradbury, 1971; Gwilliam, 1976; Clark and Dorsett, 1978).

H. Movements of Specialized Appendages

Movements of appendages during a variety of behavioral acts often involve motor programs quite different from those of locomotion, although in many cases the neuromuscular systems are essentially homologous. For instance, muscles that appear to be functional antagonists in other limbs are co-activated during feeding and cleaning movements of chelipeds in *Pagurus ochotensis* (Field, 1974a, 1977; see Section VII,C). In two species of alpheoid shrimp, different specializations are found in the mechanism that produces rapid closing of the P-D joint to produce a water jet. In *Alpheus californiensis,* tension is built up in the P-D closer muscle until it overcomes the cohesive forces of water between the surfaces of two cuticular disks (Ritzmann, 1973). In *A. heterochelis,* rapid closing depends on the mechanical orientation of the apodemes of two heads of the closer; tension in one head holds the joint open until tension in the other head increases sufficiently to change the position of the first and release the closing movement (Ritzmann, 1974).

Striking movements of the enlarged maxillipeds of the stomatopod *Hemisquilla ensigera* are due to co-contraction of flexor and extensor muscles of the M-C joint; tension built up in the extensors is released when the flexors relax to unlock a click-joint (Burrows and Hoyle, 1972).

Complex movements of appendages during visual and auditory displays in a variety of Crustacea involve yet other motor programs, probably more complex than those that have been examined to date (see chapter by Salmon and Hyatt in a later volume of this series).

IV. INTER-LIMB COORDINATION AND GAITS

The sequence in which the movements of different appendages are activated defines the locomotory gait of the animal. Factors governing inter-appendage coordination depend in large part on the nature of locomotion; e.g., swimming or walking. In walking, a sufficient number of legs must contact the ground and bear weight so that the animal is able to maintain its primary orientation relative to gravity. If a gait were to allow all of the right walking legs to be elevated at once, the animal would roll to the right. At least one leg on each side must bear weight at any given time. Swimming forms, while not always using legs to bear weight, nevertheless maintain observable gaits.

By far the most common gait seen in crustaceans is the metachronal rhythm, in which the limbs are activated sequentially. Activation progresses either from front to rear or from rear to front. The actual gait depends largely on the overall frequency of the rhythm. If the period of repetitive movements is less than the inter-limb latency, the metachronal waves may overlap (Wilson, 1966). A metachronal rhythm requires that all of the ipsilateral limbs move at the same frequency and that intersegmental latencies between all adjacent limbs are equivalent.

In walking forms, requirements of support may dictate a more rigid form of inter-leg coordination, e.g., the alternating tripod gait, in which two non-adjacent legs are bearing weight with the interposed contralateral leg. In forms that utilize eight legs at once, this may be extended to the alternating tetrapod gait.

These two types of gait require fundamentally different forms of inter-leg coordination. In metachronal systems, the controlled variable is the intersegmental latency, and different gaits result from variation in period of the cycle. In the more rigid alternating tripod or tetrapod systems, the controlled variable is the phase angle between movements of different appendages. Thus, adjacent ipsilateral or contralateral limbs must be activated in antiphase, so that when one is up, the other is down. Many systems appear to utilize combinations of the two forms of coordination, athough in most studies there is insufficient variation in repetition frequency for the two to be distinguishable.

A. Walking of Macrurans

Crayfish and lobsters typically walk either forward or backward, although sideways walking is also commonly observed in some species (Clarac and Ayers, 1977). Thus, these animals display motor programs in which both intra- and intersegmental coordination varies for different modes of walking. In adults of *Homarus americanus,* phase relations between legs approximate those of the alternating tetrapod model; phase relations remain constant as stepping frequency varies, reflecting the dominant sequence of leg movements (MacMillan, 1975). Nevertheless, variation from the strict alternating tetrapod situation also reflects transient deviation from strictly metachronal latency-locked sequencing. Although not analyzed in terms of discrete intersegmental phase relations, instability in the sequences of stepping was noted in the changes in order of usage of individual legs in *Procambarus blandingii* and *Orconectes virilis* during terresterial walking (Parrack, 1964). While in *Procambarus clarkii* the first three pairs of walking legs (pereiopods 2-4) show fairly constant phase relations, the most posterior leg is less tightly coupled to the movements of the others (Evoy and Fourtner, 1973b; Grote, 1981). Aspects of metachronal coordination are retained in adult macrurans; yet, there are not sufficient data on timing relations between legs to determine whether phase coupling is sufficiently strong to suggest an efference copy-coordinating system, as in the swimmeret system.

B. Walking of Brachyurans

Aspects of metachrony in the stepping of ipsilateral legs occur during normal sideways walking in *Carcinus maenas, C. meditterraneus,* and *Uca pugnax* (Clarac and Coulmance, 1971; Barnes, 1975a,b). Phase changes with the stepping rate in *Uca,* although such changes are neither large nor consistent for all inter-leg relations. Changes in gait occur, and conditions strictly satisfying the alternating tetrapod gait are rarely seen. An extreme variation is seen in the high-speed running of *Ocypode ceratophthalma*, where only two adjacent legs on the trailing side alternate with considerable variation in phase (Burrows and Hoyle, 1973). Nevertheless, at lower speeds, sequences of leg movements are similar to those of other brachyurans.

Variation appears to be the rule in brachyuran locomotion, whether between different speeds of walking or in walking–swimming transitions. Proprioceptive input is a major factor in this variation (see below); however, in the absence of experimental evidence from deafferented preparations, it must be anticipated that changes in central coupling between the walking legs of higher crustaceans may contribute to the modifiable intersegmental timing relations.

V. SEGMENTAL OSCILLATORS FOR RHYTHMIC LIMB MOVEMENTS

In several cases, central motor programs have been demonstrated in crustacean limb systems (Mendelson, 1971; Hughes and Wiersma, 1960; Ikeda and Wiersma, 1964; Simmers and Bush, 1980), but the list is far from complete. The fundamental criterion for the existence of a central oscillator is that the central nervous system must generate a replica of the behaviorally significant motor output in the absence of sensory feedback.

A. Macruran Swimmerets

The earliest example of a crustacean locomotory pattern generator was provided by Hughes and Wiersma (1960). The swimmeret rhythm can be produced by an isolated abdominal hemiganglion. The central program consists of alternating discharge in reciprocal power-and-return-stroke motor neurons. Although in this system, the mechanism whereby the rhythm is generated is far from understood, Heitler (1978) has recently provided evidence that at least some motor neurons of the swimmerets are integral to the pattern generator. More recent findings (Heitler and Pearson, 1980), have indicated that there are at least three classes of non-spiking interneurons that have intimate access to the rhythm generating system.

B. Anomuran Uropods

Paul (1979) has recently provided evidence that the motor pattern underlying uropod swimming in *Emerita* can be generated by the isolated CNS. The central program consists of a power-stroke burst of constant duration, which occurs at a lower repetitive frequency than in intact animals. Alternating return-stroke bursts are observed, but they are brief and sporadic. She concludes that the central pattern generator drives the power-stroke motor neurons and may concurrently inhibit the return-stroke elements.

C. Walking Legs

The crustacean walking system is one of the few invertebrate motor pattern generators in which there has yet to be a demonstration of a segmental CNS oscillator that can operate in the absence of sensory feedback. One of the major problems in this regard is methodological: the walking legs are inherently difficult to deafferent. Many of the nerves to the walking legs are mixed, containing both sensory and motor fibers. This fact, combined with the presence of a multiplicity of peripheral sense organs, has made it extremely difficult to eliminate peripheral feedback, as is possible in vertebrates by dorsal root transaction (e.g., Grillner, 1975). Furthermore, there is no known analog to curare for crustacean meuromuscular systems that would allow muscular contractions to be disabled while allowing the CNS oscillators to function, as is possible in vertebrates (e.g., Grillner, 1975).

Another problem with the walking leg system is that the thoracic nerve cord is rather intolerant to surgical intervention, due perhaps in large part to the extensive vascularization of this region and the dependence of the central integrative units on maintained capillary circulation (Brown et al., 1979). In any event, there has been no demonstration that the isolated thoracic chain can generate the walking motor program in the absence of sensory feedback. Several investigators have taken this fact to indicate that cyclic sensory feedback is necessary for the production of the walking program. The most obvious requirement has been the need for the limb to actually contact the ground and bear weight (Atwood and Wiersma, 1967; Bowerman and Larimer, 1974b; Evoy, 1977). Clarac (1978) has demonstrated that the stump of an autotomized leg will generate patterns of discharge to the basalar leg muscles appropriate for forward and backward walking if the other legs are allowed to step normally. Thus, if proprioceptive feedback is necessary for the generation of the motor program, it can clearly be distributed intersegmentally.

Another potentially important factor is input from other centers in the CNS. For example, coordinated walking movements are not observed in the absence of input from the subesophageal ganglion (Ward, 1879). Inputs from other centers might operate by two mechanisms. First, they might provide extrinsic network elements that participate in the generation of the rhythm (e.g., Russell, 1976; Stent et al., 1979). Second, they might provide tonic drive, which serves to activate the rhythm generators in a non-phasic fashion (e.g., Wilson and Wyman, 1965; Russell and Hartline, 1978).

D. Cirri of Barnacles

The intact ventral ganglion of the barnacle, *Balanus,* can produce repetitive patterns of discharge in motor neurons that normally would underlie rhythmic filtering movements of the thoracic appendages. The participating motor neurons are quite amenable to intracellular recording (e.g., Gwilliam, 1976; Clark, 1979); they consist of alternating inhibited bursters in the midline and lateral excited bursters. The rhythm apparently is generated by interneurons that are antecedent to the motor neurons; the interneurons excite one of the antagonistic pools and inhibit the other (Clark, 1979).

VI. INTERSEGMENTAL CONTROL SYSTEMS

A. Command Systems

Crustacean nervous systems are composed of segmental ganglia linked by connectives of neuronal axons. Due to this favorable anatomy, it has been possible to subdivide the connectives into fine bundles and ultimately into single axons, which can be stimulated to test for behavioral effects. As early as 1938, Wiersma demonstrated that stimulation of single giant fibers could elicit tail-flip escape behavior. More recent investigations have demonstrated the ability of central interneurons to control both posture and rhythmic movements of the appendages (Atwood and Wiersma, 1967; Davis and Kennedy, 1972a,b,c; Bowerman and Larimer, 1974a,b).

1. SWIMMERET SYSTEM

The swimmeret system of crayfish and lobsters is controlled in part by approximately five pairs of command elements, which can be selectively stimulated in the connectives between the last thoracic and first abdominal ganglion (Wiersma and Ikeda, 1964; Davis and Kennedy, 1972a,b,c). An example of swimmeret activity induced by command fibers is shown in Fig. 6. Several differences are observed between command elements of various swimmerets. First, there are both excitatory as well as inhibitory command elements, which can be demonstrated by the stimulation of pairs of neurons. Second, the different excitatory elements differ in the range of frequency of swimmeret beating that they evoke (i.e., they exhibit "range fractionation"). Lastly, the different command elements differ in segmental distribution of their effects, some selectively exciting anterior segments and others exciting

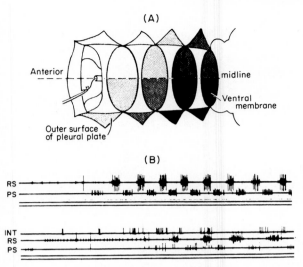

Fig. 6. Command neurons of swimmerets. The upper diagram (A) represents the receptive field of a command neuron of a swimmeret; its activity is shown below in (B). Regions that evoke stronger effects upon stimulation are indicated by darker stipple. Note the bilateral effects (strongest ipsilaterally), which decline from rear to front. In (B) the efferent effects upon natural stimulation (upper record) and natural tactile stimulation (lower record) on the discharge of return-stroke (RS) and power-stroke (PS) motor neurons of a single deafferented abdominal segment are demonstrated. INT represents the response of the command neuron to tactile stimulation (lower record). The lower two traces of each record are a stimulus marker and a time mark (0.1 msec). (From Davis, 1973.)

more posterior segments. Due to these differences, there is no single command element that can evoke the rull range of swimmeret outputs seen in normally moving animals (Davis, 1973). This latter finding suggests that the command fibers are normally used in concert rather than singly (see Chapter 9, this volume).

2. WALKING LEGS

Two major classes of command cells have been identified for the walking legs (Bowerman and Larimer, 1974b). The first class includes those which induce forward walking upon stimulation, while a second class includes those which induce backward walking. Command neurons that produce lateral walking in macrurans have not been demonstrated; this is probably due to the fact that the experimental chamber used in these studies did not allow movement in lateral directions (Bowerman and Larimer, 1974b). Another type of intersegmental command has been demonstrated by more indirect means. Evoy and Fourtner (1973a) found that a leg that was tied up proximally so that the dactyl could not contact the ground, although the peripheral segments were free to move, would make locomotory movements only if the animal was forced to bear an additional load. This finding was confirmed for forward and backward walking in lobsters (J. Ayers and F. Clarac, unpublished) and may represent the activation of intersegmental load-compensating elements.

Another difference among command cells of walking legs is the posture which they evoke. Some elements cause the claws to be elevated in a posture reminiscent of the "defense posture," whereas others cause the claws to be extended in front of the body (Wiersma, 1952; Bowerman and Larimer, 1974b). Similarly, some command cells of walking legs cause the abdomen to be tightly folded under the thorax, whereas others cause it to be rigidly extended. Other differences are found in the tendency for the stimulation of command fibers of walking legs to cause the beating of swimmerets.

B. Coordinating Neurons

The rear-to-front progression of the beating of swimmerets in crayfish (*Procambarus clarkii*) is abolished when the medial half of each intersegmental connective is transected (Stein, 1971). This region contains interneuron axons that carry an "efferent copy" of the discharge of the swimmeret's power-stroke motor neuron to the next anterior segmental ganglion. These interneurons preserve normal intersegmental phasing of discharge from pattern generators and have been shown to cause phase shifts when stimulated at different points in the ongoing cycle of activity (Fig. 7). Stein (1974, 1977) suggests that such coordinating neurons are a necessary feature of all systems that show stable interlimb phase relations.

Coordination of neuronal oscillators has been examined at the synaptic level in other systems. Entrainment of oscillators may occur by either excitatory or inhibitory inputs, as in the pyloric system of the stomatogastric ganglion of lobsters (Ayers and Selverston, 1977, 1979). The relative phase between a governing and a governed oscillator depends both on the nature of the phase-response curve (a property of the governed oscillator) and on the ratio of the periods of the two oscillations. For example, EPSP inputs to the pyloric system of lobsters can produce two different phase relations of coupling, depending on whether the cyclic stimulus is faster or slower than the governed oscillation. The governing inputs do not necessarily need to be central "efference copy," as in the swimmeret system. For example, intersegmental sensory feedback from other limbs could, in principle, provide the cyclic stimulus.

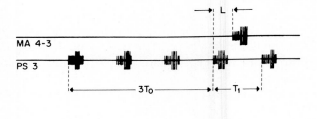

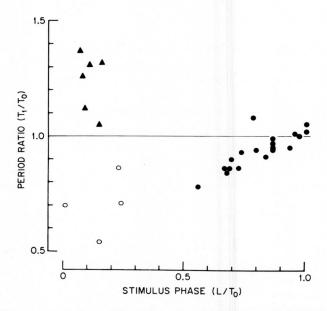

Fig. 7. Phase response curve: swimmeret power-stroke discharge in the third abdominal ganglion responding to coordinating neuron discharge originating in the fourth abdominal ganglion (Stein, 1974; cut command fiber preparation). Power-stroke bursting was evoked in the third ganglion by stimulation of the constant command neuron to the anterior segment. Coordinating neuron bursts were evoked by brief command stimulation to the posterior segment. Triangles represent points where the coordinating neuron burst began and ended during a power-stroke burst. Filled circles represent points where coordinating neuron bursts began between power-stroke bursts and ended during a power-stroke burst. Open circles are points at which the coordinating neuron burst began during a power-stroke burst and ended during a silent period. The phase shift in the power-stroke pattern is measured as the ratio of the perturbed inter-burst interval (T_1) to the mean of the three prior unperturbed intervals (T_0). Stimulus phase is the ratio of latency (L) to T_0, where L is the interval between the coordinating neuron burst and the immediately preceding power-stroke burst. Note that coordinating neuron bursts that arrive during the silent period between power-stroke bursts cause an advance in the onset of the next power stroke, whereas those that occur during a power stroke delay the onset of the subsequent burst. (From Stein, 1974.)

VII. SENSORY MODULATION OF LIMB MOTOR PROGRAMS

A. General Features of Reflexes

Sensory information plays a major role in the initiation and modulation of the motor activities of all crustacean appendages. Some sensory inputs trigger discrete behavioral responses, such as escape or defense reactions. Others serve to modulate ongoing locomotory or postural activities. Compensatory adjustments to gravitational load and negative feedback resistance to movement are known to be involved. In many cases, it is not possible to state whether a particular sensory input operates through a discrete command system, by feedback to modulate an ongoing oscillatory motor pattern, or by overriding or gating one response in favor of another.

Movements of the joints of crustacean appendages are monitored by a variety of proprioceptors responsive to angular change, joint position, muscular tension, or cuticular deformation produced by muscle contraction. Experimental manipulation by selective stimulation, ablation, or blockage of joint movements sometimes results in measurable changes in motor activity, but this may depend on the direction of locomotion. Proprioceptive reflexes involved in postural control of pereiopods and the nature of the receptors involved are discussed in Chapter 2 of this volume. Mechanoreceptor inputs are summarized in Chapter 9 of Volume 3.

Whatever the function of the various reflexes associated with the limbs may be, they are rarely involved in the initiation of movements. However, proprioceptive inputs that serve a postural function may interact with repetitive or directed movements which are part of locomotory or other behaviors. Although there have been many speculative models, no system has been fully analyzed as to the role of proprioceptive inputs in the generation of motor programs for limb and other movements. The key to this question appears to involve an understanding of the ways in which individual proprioceptive feedback loops are integrated with central nervous and other sensory components involved in the same movements.

B. Synaptic Pathways

In order to understand the integration of proprioceptive reflexes with motor programs, it is necessary to have a picture of the synaptic connections of afferents with motor and premotor components of the central nervous system. A start on this problem has been made, and in several cases evidence for monosynaptic connections between proprioceptors and motor neurons has been found. Wiens and Gerstein (1976) calculated central delays on the order of 1.5 msec for the negative feedback resistance reflex between P-D joint-opening movements and the slow closer excitatory motor neuron in the cheliped of *Procambarus clarkii.* The existence of probable monosynaptic connections of single P-D afferents onto slow closer motor neurons was further substantiated by fixed latency responses in the same preparation (Lindsey and Gerstein, 1977) and by fixed latencies of M-C afferent connections with extensor and depressor motor neurons in walking legs of *P. clarkii* (Evoy and Crabtree, 1978). Field (1974a) measured integration periods of 14 msec of this reflex, which were partially due to the rise time of EPSP's in the motor neuron. Monosynaptic connections appear to be responsible for the reflex set up by graded depolarization of the T fiber of the non-spiking coxal muscle receptor of *Carcinus maenas;* cobalt staining shows a close overlap between the neuropil processes of the receptor terminals and branches of promotor neurons (Bush, 1977). Pre- and postsynaptic intracellular recording from the T fiber and promotor neurons in *Callinectes sapidus* revealed the existence of a chemical monosynaptic junction (Blight and Llinás, 1980).

C. Negative Feedback Resistance Reflexes

One commonly observed response to passive joint movements is the negative feedback resistance reflex, in which passive joint motion causes activation of the motor neurons to muscles that oppose the movement. Resistance reflexes rarely occur during natural spontaneous movements. Otherwise, normal cyclic activation of antagonistic muscle groups would not be possible. One possible explanation for the integration of negative feedback resistance reflex connections with central pattern generators is that the central network provides an inhibitory "efference copy" to gate out the inappropriate sensory input when it is temporally and quantitatively matched to the intended activity (Barnes, 1975a, 1977). Inhibitory gating of sensory input has been demonstrated in the abdominal tail-flip system of the crayfish (Krasne and Bryan, 1973; see Chapter 8 of this volume) and has been suggested in some command-stimulated movements of swimmerets (West et al., 1979).

A reflex in the swimmerets of *Homarus americanus,* initiated by activity in the chordotonal organs of the coxopodite during passive retraction of the swimmerets, excites motor output to muscles of the return stroke as well as reinforcing motor neuron activity of the power stroke (Davis, 1969a,b, 1973; Fig. 8A). However, inputs from setae that are deflected during the power-stroke activity provide an inhibitory override of the reflex to motor neurons of the return stroke and enhance power-stroke activity (Fig. 8B; see also Section VII,I). Motor neurons of the return stroke probably receive re-

3. Locomotion and Control of Limb Movements

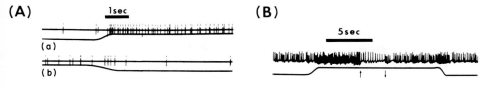

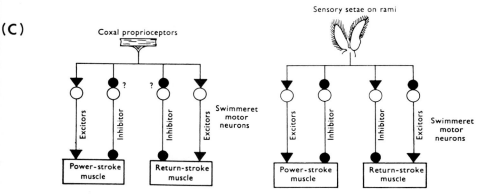

Fig. 8. Reflexes and their suggested pathways in the swimmeret system of *Homarus americanus*. (A) Recordings from power-stroke motor neurons of a swimmeret during stepwise retraction (indicated by upward deflections of the lower trace). Upper and lower records are the beginning and end of the same record; several seconds of activity are omitted between them. With this input alone, motor activity to both power-and-return-stroke muscles is increased. (B) Motor activity in a return-stroke muscle of a swimmeret. Upon upward deflection of the lower trace, the swimmeret is retracted as in A. Between the two arrows, both rami are gently squeezed, causing discharge in the sensory axons from the setae and evoking transient inhibition of the reflex to the return stroke muscles. See text for further description. (C) (Left) Diagram symbolizing the reflex pathways activated by retraction of the swimmeret and probably controlled by coxal proprioceptors. Black arrowheads denote excitatory influences while black circles indicate inhibitory ones. The major features illustrated are: (1) similarity of the effect of proprioceptive input on motor neurons of the power stroke and return stroke; and (2) reciprocal effect of this input on excitor and peripheral inhibitor axons to the same muscle. The pathways designated by question marks could be replaced by different ones that are equally capable of accounting for the available data. (Right) Diagram symbolizing reflex pathways activated by stimulation of the sensory setae that border the rami of the swimmeret. Black arrowheads represent excitatory influences, while black circles represent inhibitory ones. The major features illustrated are: (1) opposite effects of setal stimulation on neurons of the power stroke and return stroke; and (2) reciprocal effects of setal stimulation on excitor and peripheral inhibitor axons to the same muscle. (From Davis, 1969a.)

duced inhibitory input toward the end of the power stroke as resistive drag reduces the momentum of the movement, so that the resistance reflex can enhance the beginning of the next return stroke by interaction with the local control center for cyclic limb movements (Davis, 1969a). In this case, interaction of appropriately tuned sensory inputs with the central generator appears to be sufficient to provide the necessary override of the negative feedback resistance reflex without invoking a central inhibitory control mechanism (Davis, 1969b). Suggested connections for interaction of these reflexes are represented in Fig. 8C.

Movements applied to the P-D joint of *Cardisoma guanhumi* during walking evoke reflex responses comparable to those seen during quiescence (Barnes et al., 1972). Such responses are stronger when they oppose a movement than when they assist it. However, negative feedback resistance reflexes of the sort recorded in restrained animals (Spirito et al., 1972) are absent when the joint movements in walking animals are not perturbed. Similarly, some negative feedback resistance reflexes of the more proximal segments, such as the depressor response to passive elevation in *Homarus*, are absent during active elevation movements when the animal walks on a motor-driven treadmill (Ayers and Davis, 1977a,b, 1978). The reflexes are rate sensitive, but the discharge stops when the imposed movement stops, so they are not appropriate to cause reflex cycling of motor activity. However, the possibility still exists that some inappropriate reflexes may be suppressed by inhibition derived from command, oscillatory, or other reflex systems.

One possible indication of a compensatory mechanism involving negative feedback resistance reflexes comes from experiments with *Astacus leptodactylus* walking on a passive treadmill (i.e., one moved by the animal rather than by a motor). A negative-feedback resistance reflex to the M-C flexor occurs when the dactyl slips on a smooth surface during an ongoing extension movement (Barnes, 1977; see also Fig. 9A).

Although the C-P stretcher and bender muscles in the cheliped of *Pagurus ochotensis* possess resistance reflexes similar to those of many crustacean limb muscles (Fig. 9B,C), the antagonistic muscles do not operate in a reciprocal manner during the rhythmic movements of cleaning behavior (Field, 1974b, 1976; see Section III). The shared stretcher P-D opener motor neuron is co-activated with the bender motor neurons as the tip of the cheliped is moved medially, although contraction of the stretcher muscle is reduced by action of a peripheral inhibitor (Fig. 9D). Return movements do not involve stretcher bursts, and passive stretching of the C-P joint does not evoke a bender burst. Central inhibition of the proprioceptive input during the activity of the central program for cleaning behavior is implicated.

One way in which resistance reflexes can be separated from the systems generating motor programs is by addressing different elements of the motor neuron population. In the antennae of *Homarus*, inputs from a chordotonal organ spanning the two distal joints evoke negative feedback resistance reflexes to the segment-specific flexor and extensor motor neurons, but to only one of the two motor neurons that innervate the extensors of both

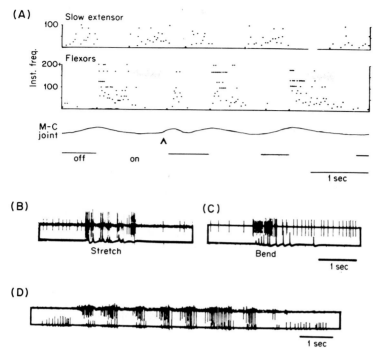

Fig. 9. Negative feedback resistance reflexes that do not normally participate in the program for limb movements. (A) Ongoing activity of the M-C extensor and flexor motor neurons while *Astacus leptodactylus* walks on a freely moving treadmill. Instantaneous frequency is the reciprocal of each interval between potentials. Upward movement of the trace labeled M-C joint is extension. Only when the leg encountered a "slippery patch" so that it extended rapidly was a negative feedback resistance reflex seen in the flexor to oppose the unexpected movement (from Barnes, 1977). (B and C) Negative feedback resistance reflexes recorded at the C-P joint of the cheliped of *Pagurus ochotensis* in response to movements imposed upon the joint. Upper traces are recordings from the chordotonal organs responding to joint movements. Lower traces are from the bender (B) and stretcher (C) muscles. Note that in each case, the motor response is appropriate to oppose the applied movement (from Field, 1974b). (D) Activity recorded from bender (upper) and stretcher (lower) muscles of the cheliped of *Pagurus ochotensis* during cleaning flexion movements induced by pipetting glycine at the cheliped. Activity to the two antagonistic muscles occurs simultaneously, although the potentials to the stretcher muscle are attenuated by the action of a peripheral inhibitor. Active participation of the negative feedback resistance reflexes observed in B and C could be expected to produce reciprocal bursting of motor neurons to these muscles (from Field, 1974a).

joints. Discharge of the single flexor motor neuron shared by the two segments is independent of the proprioceptive input (Sigvardt, 1977). However, activity in the motor neurons that are insensitive to reflex feedback was observed during phasic movements of the freely behaving animal. Much the same pattern was found in *Palinurus elephas*, although the shared fibers were thought to be inhibitory (Vedel and Clarac, 1974).

Although feedback from negative feedback resistance and other reflexes generally appears to interact with the central program to modulate the amplitude or, in some cases, the phase of the output, there are cases in which reflexes have been demonstrated to affect subsequent phases of movement. In one of the two modes of uropod beating in *Emerita*, "treading water" is lost following the ablation of non-spiking receptors, leaving only a pattern characteristic of the "swimming" mode of locomotion (Paul, 1976). The same receptors interact with the "swimming" mode, but only to modulate its excitability. Reafference from the receptors is therefore necessary to generate or maintain the power stroke of the "treading water" mode.

The resistance reflex from the non-spiking T fiber of the coxal muscle receptor of *Carcinus maenas* to various members of the promotor neuron population has both phasic and tonic components, but its possible involvement in locomotion has not been worked out (Cannone and Bush, 1980a,b). Ablation of a similar receptor in *Procambarus clarkii* results in alterations of leg positioning during both forward and backward walking, but does not greatly interfere with expression of the motor program (Crabtree and Evoy, 1980).

D. Positive Feedback and Variable Reflexes

In several instances, responses that provide a positive feedback to aid the ongoing movement have been observed. By adjusting the rates of joint movement to correspond to the angular velocities observed during walking, Ayers and Davis (1977b, 1978) found several instances in which applied joint movements evoked excitatory output to elevator, depressor, protractor, and retractor muscles of the C-B and T-C joints of walking legs of *Homarus*. These reflexes were tuned to the ranges of angular velocity observed in the joints during walking behavior (Table I).

Shifts can occur in the nature of motor responses to applied movements, either spontaneously or as a result of other sensory or central stimuli, so that the sign of a reflex varies between negative and positive. Changes in gain, measured as the slope of the relation between imposed angular velocity of M-C joint movements and the instantaneous frequency of motor neuron responses, were observed in response to tactile stimuli in the cheliped of *Procambarus clarkii* or electrical stimulation of fibers in the circumesophageal connectives (Marrelli and Evoy, 1976; Evoy, 1977). Stimuli that caused the animal to attempt a flexion movement enhanced the gain of responses via flexor motor neurons to cyclically applied joint movements and reduced the gain of responses in extensor motor neurons. The reverse of

3. Locomotion and Control of Limb Movements

TABLE I

Mean Angular Velocities of Locomotory and Passive Movements That Evoke Reflex Discharge in Lobster Walking Legs[a,b]

	Thoracocoxal joint		Coxobasal joint	
Movements	Protraction	Retraction	Elevation	Depression
Locomotory				
Forward walking	48.6	38.1	36.9	57.8
Backward walking	40.3	64.6	105.5	84.1
Passive				
Elevator reflexes	38.4	43.7	32.9	42.0
	distributed	distributed	(+) feedback	(−) feedback
Depressor reflexes	33.5	45.9	58.4	40.3
	distributed	distributed	(−) feedback	(+) feedback
Protractor reflexes	—	—	43.8	48.0
			distributed	distributed
Retractor reflexes	39.7	40.3	84.5	63.8
	(+) feedback	(−) feedback	distributed	distributed
Flexor reflexes	36.1	31.4	60.3	45.3
	distributed	distributed	distributed	distributed

[a] Adapted from Ayers and Davis, 1977b, 1978.
[b] Each column represents a particular joint movement and either the velocity (in degrees/sec) at which it occurs in treadmill-induced locomotion or the passive movement velocity at which it induces maximal reflex discharge in the corresponding muscle. The terms below the values indicated under passive movements indicate the reflex type. (+) Feedback indicates a positive feedback reflex. (−) Feedback indicates a resistance reflex. Distributed reflexes are indicated by the term distributed.

this effect was found when the central program was biased to produce extension movements. "Spontaneous" reversal of reflexes also occurred. After a period without applied stimulation, the first few cycles of M-C joint movement produced purely positive feedback "aiding" responses. Subsequently a gradual change in gain occurred following which the negative feedback resistance reflex became reliable. In the coxal muscle receptor–promotor reflex of Carcinus maenas, changes from the resistance reflex to a positive feedback reflex occurred spontaneously or following periods of spontaneous motor activity (DiCaprio and Clarac, 1981). Shifts from a "positive feedback" or synergistic reflex to a negative feedback resistance reflex were seen when the chordotonal organs of the flagellum joint in the carpopodite of antennae in Palinurus elephas were stimulated repetitively (Clarac and Vedel, 1975). Stimulation of other parts of the animal produced changes in the feedback responses to applied joint movements (Vedel and Clarac, 1975). The positive feedback reflex is seen when the animal becomes more active and serves to reinforce antennal extension as would occur during centrally-commanded defense displays (Vedel, 1980).

The integrative mechanisms responsible for reflex variability are not well understood. The summation of local and distributed inputs at the motor neurons and perhaps elsewhere definitely occurs (Bush et al., 1978; Evoy and Crabtree, 1978; in preparation). Tension and deformation-sensitive inputs may be integrated with the feedback from joint receptors (Clarac and Dando, 1973; Clarac, 1976; Vedel and Clarac, 1979). However, reversal or switching of reflex responses, also seen in insects and mammals, (Camhi, 1977; Forssberg et al., 1975), appears to involve complex synaptic integration.

E. Responses to Changes in Load

Compensatory responses to changing or resistive gravitational load are often seen in the intensity, duration, recruitment, and patterning of the motor output (see also Chapter 2, this volume). When the crab, *Cardisoma guanhumi,* walked up a steep incline or was subjected to the braking forces of a passive treadmill, the duration of the power stroke increased with little change in duration or frequency of the return stroke; slower walking resulted (Evoy and Fourtner, 1973a; see also Section III,B). Similarly, in lobsters walking on a motor-driven treadmill, the frequency of stepping can be controlled by the passive traction provided by motion of the belt. Here again, the controlled variable is the duration of power-stroke bursts, independent of the direction of walking (Ayers and Davis, 1977a). Loading the crayfish, *Astacus leptodactylus,* with weights while it walks on a passive treadmill increases the frequency within power-stroke bursts, concomitant with increases in the period of the cycle and in relative duration of the power-stroke bursts (Barnes, 1977). Adding weight to *Procambarus clarkii* while it walks in water or when it walks in air, results in recruitment of activity in the C-B depressor—an antigravity response (Grote, 1981). However, further addition of weight during walking on land only decreases the stepping frequency and increases the relative duration of the power stroke, similar to the situation seen in *C. guanhumi.*

Both negative feedback (resistance) and positive feedback (aiding) reflexes may participate in the responses to loading when there is a mismatch between the proprioceptive input and the ongoing central program (Barnes, 1975a; Ayers and Davis, 1978). Several other proprioceptive reflexes have been implicated in load compensation during centrally initiated postural changes or repetitive limb movements (Clarac, 1977). These include responses to loads, as mediated by cuticular stress detectors (CSD's), and responses to resistive loads encountered during active muscle contraction, mediated by apodeme tension receptors.

F. Reflexes Involving Cuticular Stress Detectors

The CSD's of the basi-ischiopodite region of decapod pereiopods (see Chapter 9 of Volume 3) are involved in the control of autotomy (McVean, Chapter 4). They also register the shearing force on the cuticle when a leg touches the ground; both local and distributed reflexes are altered by these inputs (Clarac, 1977).

In *Homarus americanus* and *Procambarus clarkii,* restoration of normal activity can occur in a leg in which the distal segment is removed and replaced by a rigid strut. This finding implicates the CSD's as likely candidates to register the force encountered by the proximal leg segments (MacMillan, 1975; Grote, 1981). In walking legs of *Carcinus maenas,* CSD1 does not discharge during an unimpeded movement, but bursts when the movement is resisted, an indication that the receptor does respond to added force on the proximal segments (Findlay, 1978). However, definitive experiments involving selective ablation of the CSD's or their phasic stimulation during walking are lacking.

G. Reflexes Involving Apodeme Tension Receptors

Apodeme tension receptors, described in Brachyura by MacMillan and Dando (1972) and MacMillan (1976), probably register forces due to muscle contraction against a load and modulate ongoing motor activity. However, their reflex connections and interactions with other responses are variable. In *Cancer pagurus,* inputs from the tension receptors associated with the M-C joint suppress the negative feedback resistance reflexes set up by applied joint movement (Clarac and Dando, 1973). However, high intensity stimulation of the apodeme sensory nerve excites both antagonists. The binding of joints in *Pagurus ochotensis* causes increased duration of bursts in muscles in which contraction is opposed during cleaning movements evoked by chemical stimulation (Field, 1976). Positive feedback from tension receptors is implicated in this response.

H. Reflexes Involving Muscle Receptor Organs

Muscle receptor organs (proprioceptors coupled to joint movements by specialized muscles that can often be activated independently of the muscles responsible for most of the movement) have been implicated in load compensation. The myochordotonal organ of the meropodite and the non-spiking thoracocoxal muscle receptor organ of decapod pereiopods have been the most intensely investigated. In both of these muscle receptors, efferent control is linked to that of the muscles with which they are in parallel. In the case of the myochordotonal organ, the receptor muscle is co-activated with the M-C flexor during lateral walking and distributed reflexes in *Palinurus* (Clarac, 1977; Ayers and Clarac, 1978) and in the cheliped of *Astacus leptodactylus* (Bush et al., 1978). In motor neurons, close coupling of individual impulses to receptor and M-C flexor muscles, suggestive of common premotor excitation, has been observed in *Astacus leptodactylus* (Angaut-Petit and Clarac, 1976) and in *Palinurus elephas* (Bush et al., 1978). In *Carcinus maenas,* spontaneous bursts of activity to coxopodite protractor and receptor muscles are nearly coincident (Bush, 1976). Positive feedback occurs in each case onto the major muscle with which the receptors lie in parallel (Bush, 1965; Evoy and Cohen, 1969; Bush and Cannone, 1973; Cannone and Bush, 1981a,b; See Fig. 3, Chapter 2. Thus, co-activation of receptor muscles and major locomotory muscles would seem to be a mechanism to maintain minimal differences in length between the two muscles and to make them continually sensitive to imposed opposition to an ongoing movement, as has been proposed for abdominal MRO's (Kennedy, 1969; Page, Chapter 2). However, severance of the loop by partial or complete ablation of the myochordotonal organs has failed to disturb compensatory responses to load during walking, although changes in posture or coordination were observed (Fourtner and Evoy, 1973; Clarac and Ayers, 1977). As opposed to a rapid damping in the intact state, ablation of the myochordotonal organ in chelipeds of *Procambarus clarkii* resulted in oscillations of the motor output when an ongoing movement was terminated abruptly (Marrelli and Larimer, 1978). In the absence of an obvious role for the muscle receptors in load compensation, CSD's and apodeme tension receptors remain as the major candidates, although the mechanism by which either might operate remains to be demonstrated.

I. Reflexes Involving Sensory Hairs

Setae of several types are found on locomotory and other appendages and serve to detect movements of the limb in water, as well as movement of one limb segment upon another. Sensory hairs and cuticular articulated pegs at limb joints are so situated that they serve as proprioceptors to indicate joint movements and position (Wales et al., 1970; Alexandrowicz, 1972; Laverack, 1976). Stroking of hairs on the inner surface of the dactyl in chelipeds of crayfish evokes a closing response; the opener excitor and inhibitor also discharge, but the inhibitory effect predominates (Wilson and

Davis, 1965; Smith, 1974). Simultaneous stimulation of setae that evoke closing and other hair fields that evoke opening results in a marked suppression of opener excitor activity while the closer excitor and, to a lesser extent, the opener inhibitor are driven (Wiens and Gerstein, 1976). Tactile stimulation of hairs on chelipeds of *Pagurus ochotensis* reinforces chemical inputs in evoking feeding flexions and rhythmic cleaning activity (Field, 1974a).

Locomotory appendages are also well-endowed with setae of various types. Positive feedback from setae on the swimmerets of *Homarus americanus* is an important factor in controlling the strength of the power-stroke, although it does not appear to affect the timing of the beat cycle (Davis, 1969b). Activation of setal afferents excites retractor motor neurons and suppresses activity in a single peripheral inhibitor; the effect on the protractor motor pool is the opposite (Fig. 8). A potentially important aspect of this reflex is that it may serve to override the excitation of return-stroke motor neurons by coxal proprioceptive inputs, which might otherwise interfere with the central program for reciprocal outputs (see Section VII,C).

It also seems possible that the setae of swimmerets can detect water currents from the rear when the swimmerets are held below or to the side of the animal (Davis and Kennedy, 1972a,b). Projections of sensory fibers and interneurons integrating these inputs are well represented in the ventral nerve cord (Wiersma and Bush, 1963).

J. Distributed Reflexes

Distributed proprioceptive reflexes, in which central connections of receptors associated with one segment of an appendage affect or evoke motor activity in another limb joint (usually one that produces movement in the same plane), have been implicated in the coordination of the entire limb (Ayers and Davis, 1977b). The reflex connections that have been demonstrated are summarized by Page (Chapter 2, this volume). Distributed reflexes in the walking legs of *Homarus americanus* are well matched to their function in walking by a selective tuning of joint velocities at which they are effective to normally occurring velocities of movement observed in the walking animal (Table I). The coupling of other leg movements to the cycle of elevation and depression at the C-B joint seems to be particularly strong and is thought to ensure that other muscles are coordinated with the regular cycles of activity in the C-B joint that are characteristic of forward, backward, and lateral walking. During the power-stroke of forward walking, when T-C retraction and C-B depression are simultaneous, the reflexes reinforce the synchrony of the synergistic depressor, flexor, and retractor muscles (Ayers and Davis, 1978). Coupling also occurs between the power-stroke muscles for backward walking: depressor, retractor, and flexor (Table

I). Similar correlations appropriate to the activation patterns of leg muscles were seen in *Palinurus elephas* in response to controlled stimulation of the CB chordotonal organ (Vedel et al., 1975b). A spread of the effects from the C-B joint movements to the motor output of the M-C joint also occurs; relaxation of the CB chordotonal organ excites the flexor and accessory flexor muscles and suppresses the extensor. Stretching the receptor has the opposite effects (Clarac et al., 1978; Bush et al., 1978; Vedel and Clarac, 1979). Retraction of the pleopods of *Homarus americanus* causes the rami to open and curl to the rear, as during the power-stroke of the swimmerets (Davis, 1969a,b). This is probably due to a distributed reflex similar to those of the pereiopods (Ayers and Davis, 1977b).

Similarly, in *Carcinus mediterraneus* and *C. maenas*, the fixing of the M-C joint in various positions during lateral walking causes perturbations in C-B and P-D movements that are at least partially due to changes in distributed reflexes (Clarac and Coulmance, 1971). Distributed reflexes from CSD's and coxal MRO's, although investigated primarily with a view to their possible roles in posture and load compensation, are also appropriately organized to contribute to coordination of joint movements during locomotion and other movements.

Angaut-Petit et al. (1974) and Vedel et al. (1975a) found that in *Astacus leptodactylus*, local stimulation of CSD2 excited tonic motor neurons of the M-C flexor and accessory flexor and enhanced or summed with their responses to the cyclic movement of the M-C joint. CSD1 excites the accessory flexor as well as the C-B elevator of *Carcinus maenas* (Clarac, 1976). Both C-B depressor and elevator activity are excited by CSD1 stimulation in *Palinurus elephas*, and local reflex responses of these muscles are enhanced by or sum with this input (Vedel and Clarac, 1979).

The interaction of reflexes may be additive or subtractive. The stimulation of CSD's in the cheliped of *Astacus leptodactylus* in conjunction with applied extension of the M-C joint results in a response frequency that is approximately the sum of that obtained by stimulating the two inputs independently (Vedel et al., 1975a). In the absence of discrete information on synaptic integration and other pathways involved in all of these reflexes, it is not possible to determine whether or not true facilitation of one input by another occurs, or whether these combined effects are the result of a simple summation of inhibitory and excitatory inputs to the motor neurons or premotor components.

K. Intersegmental Reflexes

Reflex-like interactions between similar and dissimilar limbs, referred to here as intersegmental reflexes, have been examined both behaviorally and neurophysiologically. In the swimmerets of *Homarus americanus,* intersegmental coupling of power-strokes biased rostrally has been mentioned briefly (Davis, 1969a, 1973). In *Cancer magister,* intersegmental reflexes between the M-C joints of adjacent walking legs generally follow the pattern of the intrasegmental reflexes, although they are weaker and more transitory (Evoy and Cohen, 1969). Ipsilateral intersegmental reflexes evoked by coxal receptors of *Jasus lalandii* have opposite effects in rostrocaudal and caudorostral directions. Remotion of an anterior leg evokes responses in posterior legs that are in phase with the resistance reflex; out-of-phase responses occur in an anterior leg when a posterior leg is remoted. There are both static and dynamic components of the motor bursts suggestive of a role in intersegmental control of posture and possibly movement (Clarac, 1981). Complex, but fairly stereotyped responses are evoked in other appendages and body segments by movements of individual leg joints in *Procambarus clarkii* (Page, 1981). In the chelipeds of *Homarus americanus,* postsynaptic responses of motor neurons to contralateral sensory stimulation are weaker in amplitude and longer in latency than are those to ipsilateral stimulation (Govind et al., 1979). In *Homarus americanus,* tying a leg so that it could no longer contact the ground resulted in changes in the power strokes of the unrestrained legs, apparently in compensation for the inactivated leg (MacMillan, 1975). Removal of a leg of *Uca pugnax* at the autotomy plane resulted in phase shifts of the stepping pattern of remaining legs during lateral walking: alternate ipsilateral legs, which normally step in phase, shifted to an alternating pattern similar to that of neighboring legs in normal animals (Barnes, 1975b). Removal of a leg in these experiments also tended to recruit coordinated stepping patterns of the chelipeds, which are rarely employed in walking in normal animals. Similar, but less striking changes in the interleg stepping phase were observed in *Cardisoma guanhumi* when one leg was tied (Evoy and Fourtner, 1973a); the tied leg showed only weakly rhythmic movements. If, however, the animal walked up an incline (loaded condition), nearly normal movement and electrical activity in distal muscles was seen. The addition of a strut to an amputated leg in *Procambarus clarkii* to allow normal movement and contact with the ground restored to the normal pattern the altered phase relationship between that leg and its unperturbed neighbor (Grote, 1981).

Connections also exist between dissimilar appendages, some of which appear to be responsible for quite discrete responses and interactions. Inputs set up by the CB chordotonal organs of walking legs in *Palinurus elephas* evoke predictable movements of the antennae (Clarac et al., 1976). Substrate movements registered by receptors of walking legs are integrated with statocyst inputs to cause movements of both antennae and eyestalks (Schöne et al., 1976; Scapini et al., 1978; Neil et al., 1979). Neil (1977) found that movements of the C-B joint in pereiopods enhanced the negative feedback resistance reflex of the antennal rotator muscle when the two stimuli were in phase and reduced it when they were out of phase. General stimulation of the pereiopods excited both antagonists of the two distal antennal flagellar segments, whereas stimulation of the uropods increased flexor but suppressed extensor activity (Clarac and Vedel, 1975; Vedel and Clarac, 1975). The chelipeds of *Astacus leptodactylus* are also influenced by inputs from many parts of the animal, including pleopods and antennae, that tend to evoke flexion and sum with the local reflexes (Vedel et al., 1975a).

There is really very little definitive information as to the pathways or significance of the role of proprioceptive inputs in shaping intersegmental gaits. An intersegmental coordinating system has been demonstrated in swimmerets but not in walking legs. In *Procambarus clarkii,* Wiersma (1958) and Wiersma and Bush (1963) found interneurons in the central nervous system that responded to movements of the T-C, C-B, and M-C joints, although more distal joints did not seem to be represented. However, the possible outputs of these interneurons to different segmental motor systems is not known.

L. Exteroceptive Reflexes

Coordinated behavioral responses to discrete sensory stimuli share with command control many properties of systems like walking, swimmeret beating, escape, and so on, although the central connectivity involved in initiating most behaviors has not yet been worked out.

1. OPTOMOTOR RESPONSES

Crustaceans exhibit many reactions to moving visual stimuli, including escape reactions (Wine and Krasne, 1972), antennal pointing, and optomotor reactions. Crustacean optomotor responses fall into two general classes: (1) optokinetic nystagmus of the eye cups in response to moving visual fields (Neil, Chapter 5); and (2) whole body compensatory reactions (Davis and Ayers, 1972). The nature of the latter responses depends on the visual stimulus. If both eyes detect relative angular motion of the surroundings, as in a striped drum, the animal attempts to turn the whole body to minimize the imposed displacement (Kennedy and Davis, 1977). These responses constitute the familiar negative feedback responses. If the stimulus appears to go in opposite directions in the two eyes, the organism tends to walk in the direction opposite the imposed motion; a positive feedback reaction (Davis and Ayers, 1972). Both of these response types can be explained by a relatively simple causal nexus (Kennedy and Davis, 1977).

3. Locomotion and Control of Limb Movements

2. RESPONSES TO CHEMOSENSORY INPUTS

Specific motor responses to dissolved substances form an important component of the behavior of most crustaceans, many of which are scavengers. Chemical cues, primarily amino acids, are significant, perhaps predominat, in such aspects of behavior as feeding, cleaning, mating, and orientation to food sources. Chemoreceptor hairs are concentrated on antennules, mouthparts, and dactylopodites of pereiopods (Chapter 8 of Volume 3). In free-living forms, the walking legs are involved in probing for food, and chemoreceptor inputs from the chelipeds trigger discrete movements involved in feeding and cleaning (Field, 1976). Cleaning and feeding movements of *Pagurus ochotensis* are often initiated by chemical stimulation, although other modalities are also involved (Field, 1976). Individual movements of chelipeds in the feeding response are initiated by concentrations of glycine between 10^{-3} and 10^{-5} M, while repetitive cleaning flexions are evoked by higher concentrations. The cleaning response is correlated with longer bursts of chemoreceptor activity at the higher concentrations (Field, 1974b).

3. RESPONSES TO GRAVITY

Changes in primary orientation to gravity evoke compensatory responses of virtually all crustacean appendages (see Chapter 5). In *Homarus americanus,* Davis (1968b) has analyzed the compensatory responses of the abdominal swimmerets to roll movements. The swimmeret on the up side beats out to the side, while the swimmerets on the down side beat with a rearward-directed power stroke. These responses can be abolished by ablation of the statocysts.

VIII. SUMMARY AND CONCLUSIONS

Several important concepts about neuronal organization of motor systems have emerged from analysis of the movements of crustacean appendages. Control systems found in the Crustacea have suggested heretofore unsuspected possibilities which have subsequently been tested and found relevant for other animal movements. Indeed, crustacean limb systems have proved to be important models for the study of locomotion (e.g., Stein, 1977, 1978).

The list of general principles that have originated in crustacean systems is extensive. For example, the underlying basis of endogenous rhythmicity can be due to at least two different sources: cellular oscillators and network connections of cells (see Wiens, Chapter 7). Non-spiking neuronal oscillators were first demonstrated for the drive of paguran scaphognathites by fluctuations in the membrane potential of neurons in the same ganglion (Mendelson, 1971). Network oscillators that include motor neurons appear to be the basis of the segmental cycle of swimmeret movements (Heitler, 1978; Heitler and Mulloney, 1978). The possibility that individual neurons or limited combinations of neurons could act as command systems to initiate and maintain the activity of discrete motor outputs was first suggested by the role of interneurons in the defense response (Wiersma, 1938; 1952) and in the beating of swimmerets (Wiersma and Ikeda, 1964). Intersegmental phasing was subsequently found in the same system to be due to interneurons linking the segmental oscillatory networks (Stein, 1971). Proprioceptive inputs acting by graded membrane responses have been found in several instances (Ripley et al., 1968; Paul, 1972). Other proprioceptive systems found in control of crustacean limb movements are involved in the integration of responses to encountered loads and to muscle tension with the ongoing central program. Crustacean motor systems provide a wide-open field for the exploration of the role of feedback control in the modulation of central programs. Plasticity, gain changes, and switching of responsiveness are found in many of these systems. In some cases, proprioceptive feedback controls the timing; in others the feedback controls the amplitude of output. Unraveling the synaptic mechanisms by which many of these systems operate is possible with current intracellular techniques, and appropriate morphological correlations are emerging through the application of appropriate cellular markers. The system of joints and levers of the cuticular exoskeleton of the appendages and attachments of the internal proprioceptors lend themselves to controlled mechanical stimulation.

The evolution of usage of appendages in various crustaceans appears to have been convergent, in that different appendages have become adapted to serve similar needs in display and manipulation, and also divergent, in that the same appendage may be specialized to serve quite different functions in different animals and may be controlled by more than one motor program utilizing the same final pathways.

In the behavior of crustaceans, the range of control regarding movements of appendages appears to be nearly as advanced as in vertebrates and to employ similar if convergent principles of operation. Analysis of the neuronal, sensory, and neuromuscular bases of control of movement in this group continues to be an important part of the overall understanding of the ways in which coordinated movements are organized.

REFERENCES

Alexandrowicz, J. S. (1972). The comparative anatomy of leg proprioceptors in some decapod Crustacea. *J. Mar. Biol. Assoc. U. K.* **52**, 605-634.

3. Locomotion and Control of Limb Movements

Angaut-Petit, D., and Clarac, F. (1976). A study of a temporal relationship between two excitatory motor discharges in the crayfish. *Brain Res.* **104,** 166-170.

Angaut-Petit, D., Clarac, F., and Vedel, J. P. (1974). Excitatory and inhibitory innervation of a crustacean muscle associated with a sensory organ. *Brain Res.* **70,** 148-152.

Atwood, H. L., and Walcott, B. (1965). Recording of electrical activity and movement from legs of walking crabs. *Can. J. Zool.* **43,** 657-665.

Atwood, H. L., and Wiersma, C. A. G. (1967). Command neurons in the crayfish central nervous system. *J. Exp. Biol.* **46,** 249-261.

Ayers, J. L., and Clarac, F. (1978). Neuromuscular strategies underlying different behavioral acts in a multifunctional crustacean leg joint. *J. Comp. Physiol. A* **128A,** 81-129.

Ayers, J. L., and Davis, W. J. (1977a). Neuronal control of locomotion in the lobster, *Homarus americanus*. I. Motor programs for forward and backward walking. *J. Comp. Physiol. A* **115A,** 1-27.

Ayers, J. L., and Davis, W. J. (1977b). Neuronal control of locomotion in the lobster *Homarus americanus*. II. Types of walking leg reflexes. *J. Comp. Physiol. A* **115A,** 29-46.

Ayers, J. L., and Davis, W. J. (1978). Neuronal control of locomotion in the lobster, *Homarus americanus*. III. Dynamic organization of walking leg reflexes. *J. Comp. Physiol. A* **123A,** 289-298.

Ayers, J. L., and Selverston, A. I. (1977). Synaptic control of an endogenous pacemaker network. *J. Physiol (Paris)* **73,** 453-461.

Ayers, J. L., and Selverston, A. I. (1979). Monosynaptic entrainment of an endogenous pacemaker network: A cellular mechanism for von Holst's magnet effect. *J. Comp. Physiol. A* **129A,** 5-17.

Barnes, W. J. P. (1975a). Nervous control of locomotion in crustacea. *In* "Simple Nervous Systems" (P. N. R. Usherwood and D. R. Newth, eds.), pp. 415-441. Arnold, London.

Barnes, W. J. P. (1975b). Leg coordination during walking in a crab, *Uca pugnax*. *J. Comp. Physiol. A* **96A,** 237-257.

Barnes, W. J. P. (1977). Proprioceptive influences on motor output during walking in the crayfish. *J. Physiol. (Paris)* **73,** 543-563.

Barnes, W. J. P., Spirito, C. P., and Evoy, W. H. (1972). Nervous control of walking in the crab, *Cardisoma granhumi*. II. Role of resistance reflexes in walking. *Z. Vergl. Physiol.* **76,** 16-31.

Bent, S. A., and Chapple, W. D. (1977). Simplification of swimmeret musculature and innervation in the hermit crab, *Pagurus pollicarus* in comparison to macrurans. *J. Comp. Physiol. A* **118A,** 61-74.

Blight, A. R., and Llinás, R. (1980). The non-impulsive stretch-receptor complex of the crab-a study of depolarization-release coupling at a tonic sensorimotor synapse. *Philos. Trans. R. Soc. London, Ser. B* **290,** 220-276.

Bowerman, R. F., and Larimer, J. L. (1974a). Command fibers in the circumoesophageal connectives of the crayfish. I. Tonic fibers. *J. Exp. Biol.* **60,** 95-117.

Bowerman, R. F., and Larimer, J. L. (1974b). Command fibers in the circumoesophageal connectives of the crayfish. II. Phasic fibers. *J. Exp. Biol.* **60,** 119-134.

Brown, S. K., Sherwood, D. N., and Wine, J. J. (1979). Vascularization of the abdominal central nervous system of the crayfish: Consequences for physiology. *Neurosci. Abstr.* **5,** 242.

Burrows, M., and Hoyle, G. (1972). Neuromuscular physiology of the strike mechanism of the mantis shrimp, *Hemisquilla*. *J. Exp. Zool.* **179,** 379-394.

Burrows, M., and Hoyle, G. (1973). The mechanism of rapid running in the ghost crab, *Ocypode ceratophthalma*. *J. Exp. Biol.* **58,** 327-349.

Burrows, M., and Willows, A. O. D. (1969). Neuronal coordination of rhythmic maxilliped beating in brachyuran and anomuran Crustacea. *Comp. Biochem. Physiol.* **31,** 121-135.

Bush, B. M. H. (1965). Leg reflexes from chordotonal organs in the crab, Carcinus maenas. Comp. Biochem. Physiol. **15**, 567-587.

Bush, B. M. H. (1976). Non-impulsive thoracic-coxal receptors in crustaceans. In "Structure and Function of Proprioceptors in the Invertebrates" (P. J. Mill, ed.), pp. 115-151. Chapman & Hall, London.

Bush, B. M. H. (1977). Non-impulsive afferent coding and stretch reflexes in crabs. In "Identified Neurons and Behavior of Arthropods" (G. Hoyle, ed.), pp. 439-460. Plenum, New York.

Bush, B. M. H., and Cannone, A. J. (1973). A stretch reflex in crabs evoked by muscle receptor potentials in non-impulsive afferents. J. Physiol (London) **323**, 95P-97P.

Bush, B. M. H., Vedel, J. P., and Clarac, F. (1978). Intersegmental reflex actions from a joint sensory organ (CB) to a muscle receptor (MCO) in decapod crustacean limbs. J. Exp. Biol. **73**, 47-63.

Camhi, J. M. (1977). Behavioral switching in cockroaches. Transformations of tactile reflexes during righting behavior. J. Comp. Physiol. A **113A**, 283-301.

Cannone, A. J., and Bush, B. M. H. (1980a). Reflexes mediated by non-impulsive afferent neurones of thoracic-coxal muscle receptor organs in the crab, Carcinus maenas. I. Receptor potentials and promotor motoneurone responses. J. Exp. Biol. **86**, 275-303.

Cannone, A. J., and Bush, B. M. H. (1980b). Reflexes mediated by non-impulsive afferent neurones of thoracic-coxal muscle receptor organs in the crab, Carcinus maenas. II. Reflex discharge evoked by current injection. J. Exp. Biol. **86**, 305-331.

Cannone, A. J., and Bush, B. M. H. (1981a). Reflexes mediated by non-impulsive afferent neurones of the thoracic-coxal muscle receptors in the crab, Carcinus maenas. III. Positive feedback to the receptor muscle. J. Comp. Physiol. **142A**, 103-112.

Cannone, A. J., and Bush, B. M. H. (1981b). Reflexes mediated by non-impulsive afferent neurones of thoracic-coxal muscle receptors in the crab, Carcinus maenas. IV. Motor activation of the receptor muscle. J. Comp. Physiol. **142A**, 113-125.

Clarac, F. (1976). Crustacean cuticular stress detectors. In "Structure and Function of Proprioceptors in the Invertebrates" (P. J. Mill, ed.), pp. 229-321. Chapman & Hall, London.

Clarac, F. (1977). Motor coordination in crustacean limbs. In "Identified Neurons and Behavior of Arthropods" (G. Hoyle, ed.), pp. 167-186. Plenum, New York.

Clarac, F. (1978). Locomotory programs in basal leg muscles after limb autotomy in the Crustacea. Brain Res. **145**, 401-405.

Clarac, F. (1981). Postural reflexes co-ordinating walking legs in a rock lobster. J. Exp. Biol. **90**, 333-337.

Clarac, F., and Ayers, J. (1977). La marche chez les crustaces: Activité motrice programmée et régulation périphérique. J. Physiol (Paris) **73**, 523-544.

Clarac, F., and Coulmance, M. (1971). La marche latérale du crabe (Carcinus): Coordination des movements articulaires et régulation proprioceptive. Z. Vergl. Physiol. **73**, 408-438.

Clarac, F., and Dando, M. R. (1973). Tension receptor reflexes in the walking legs of the crab Cancer pagurus. Nature (London) **243**, 94-95.

Clarac, F., and Vedel, J. P. (1975). Neurophysiological study of antennal motor patterns in the rock lobster, Palinurus vulgaris. I. Reflex modulation of extensor and flexor motor neuron activities. J. Comp. Physiol. A **102A**, 201-221.

Clarac, F., Neil, D. M., and Vedel, J. P. (1976). The control of antennal movements by leg proprioceptors in the rock lobster, Palinurus vulgaris. J. Comp. Physiol. A **107A**, 275-292.

Clarac, F., Vedel, J. P., and Bush, B. M. H. (1978). Intersegmental reflex coordination by a single joint receptor organ (CB) in rock lobster walking legs. J. Exp. Biol. **73**, 29-46.

Clark, J. V. (1979). The neuronal basis of rhythmic pumping in *Balanus hameri*. *J. Comp. Physiol.* **130,** 183.

Clark, J. V., and Dorsett, D. A. (1978). Anatomy and physiology of proprioceptors in the cirri of *Balanus hameri*. *J. Comp. Physiol. A* **123A,** 229-237.

Davis, W. J. (1968a). The neuromuscular basis of lobster swimmeret beating. *J. Exp. Zool.* **168,** 363-378.

Davis, W. J. (1968b). Lobster righting responses and their neural control. *Proc. R. Soc. London, Ser. B* **70,** 435-456.

Davis, W. J. (1968c). Quantitative analysis of swimmeret beating in the lobster. *J. Exp. Biol.* **48,** 643-662.

Davis, W. J. (1969a). Reflex organization in the swimmeret system of the lobster. I. Intrasegmental reflexes. *J. Exp. Biol.* **51,** 547-563.

Davis, W. J. (1969b). Reflex organization in the swimmeret system of the lobster. II. Reflex dynamics. *J. Exp. Biol.* **51,** 565-573.

Davis, W. J. (1971). Functional significance of motoneuron size and soma position in the swimmeret system of the lobster. *J. Neurophysiol.* **34,** 274-288.

Davis, W. J. (1973). Neuronal organization and ontogeny in the lobster swimmeret system. In "Control of Posture and Locomotion" (R. B. Stein, K. G. Pearson, R. G. Smith, and J. B. Redford, eds.), pp. 437-455. Plenum, New York.

Davis, W. J., and Ayers, J. L. (1972). Locomotion: Control by positive-feedback optokinetic responses. *Science* **177,** 183-185.

Davis, W. J., and Kennedy, D. (1972a). Command neurons controlling swimmeret movements in the lobster. I. Types of effects on motor neurons. *J. Neurophysiol.* **35,** 1-12.

Davis, W. J., and Kennedy, D. (1972b). Command neurons controlling swimmeret movements in the lobster. II. Interaction of effects on motor neurons. *J. Neurophysiol.* **35,** 13-19.

Davis, W. J., and Kennedy, D. (1972c). III. Command neurons controlling swimmeret movements in the lobster. Temporal relationships among bursts in different motor neurons. *J. Neurophysiol.* **35,** 20-29.

DiCaprio, R. A., and Clarac, F. (1981). Reversal of a walking leg reflex elicited by a muscle receptor. *J. Exp. Biol.* **90,** 197-203.

Evoy, W. H. (1977). Crustacean motor neurons. In "Identified Neurons and Behavior of Arthropods" (G. Hoyle, ed.), pp. 67-86. Plenum, New York.

Evoy, W. H., and Cohen, M. J. (1969). Sensory and motor interaction in the locomotor reflexes of crabs. *J. Exp. Biol.* **51,** 151-169.

Evoy, W. H., and Crabtree, R. L. (1978). Synaptic integration by crayfish walking leg motor neurons. *Neurosci. Abstr.* **4,** 192.

Evoy, W. H., and Fourtner, C. R. (1973a). Nervous control of walking in the crab *Cardisoma granhumi*. III. Proprioceptive influences on intra- and intersegmental coordination. *J. Comp. Physiol.* **83,** 303-318.

Evoy, W. H., and Fourtner, C. R. (1973b). Crustacean walking. In "Control of Posture and Locomotion" (R. B. Stein, K. G. Pearson, R. G. Smith, and J. B. Redford, eds.), pp. 477-493. Plenum, New York.

Field, L. H. (1974a). Sensory and reflex physiology underlying cheliped flexion behavior in hermit crabs. *J. Comp. Physiol.* **92,** 397-414.

Field, L. H. (1974b). Neuromuscular correlates of rhythmical cheliped flexion behavior in hermit crabs. *J. Comp. Physiol.* **92,** 415-441.

Field, L. H. (1976). Effects of proprioceptive disruption on the motor program for cheliped flexion behavior in hermit crabs. *J. Comp. Physiol. A* **105A,** 313-338.

Field, L. H. (1977). A description and experimental analysis of a stereotyped cheliped flexion behavior in hermit crabs. *Behaviour* **61,** 147-179.

Findlay, I. (1978). The role of the cuticular stress detector (CSD1) in locomotion and autotomy in the crab, Carcinus maenas. *J. Comp. Physiol. A* **125A,** 79-90.

Forssberg, H., Grillner, S., and Rossignol, S. (1975). Phase-dependent reflex reversal during walking in chronic spinal cats. *Brain Res.* **85,** 103-107.

Fourtner, C. R., and Evoy, W. H. (1973). Nervous control of walking in the crab, Cardisoma granhumi. IV. Effects of myochordotonal organ ablation. *J. Comp. Physiol.* **83,** 319-329.

Govind, C. K., Meiss, D. E., and Lang, F. (1979). Lobster claw motor neurons respond to contralateral sensory stimuli. *J. Neurobiol.* **10,** 513-517.

Grillner, S. (1975). Locomotion in vertebrates: Central mechanisms and reflex interaction. *Physiol. Rev.* **55,** 247-304.

Grote, J. R. (1981). The effect of load on locomotion in crayfish. *J. Exp. Biol.* **92,** 277-288.

Gwilliam, G. F. (1976). The mechanism of the shadow reflex in Cirripedia. III. Rhythmical and patterned activity in central neurons and its modulation by shadows. *Biol. Bull. Woods Hole, Mass.* **151,** 141-160.

Gwilliam, G. F., and Bradbury, J. C. (1971). Activity patterns in the isolated central nervous system of the barnacle and their relation to behavior. *Biol. Bull. (Woods Hole, Mass.)* **141,** 502-513.

Hafemann, D. R., and Hubbard, J. I. (1969). On the rapid running of ghost crabs (Ocypode ceratophthalma). *J. Exp. Zool.* **170,** 25-32.

Hartnoll, R. G. (1970). Swimming in the dromid crab (Homola barbata). *Anim. Behav.* **18,** 588-591.

Hartnoll, R. G. (1971). The occurrence, methods and significance of swimming in the Brachyura. *Anim. Behav.* **19,** 34-50.

Heitler, W. J. (1978). Coupled motor neurons are part of the crayfish swimmeret central oscillator. *Nature (London)* **275,** 231-234.

Heitler, W. J., and Mulloney, B. (1978). Crayfish motor neurons are an integral part of the swimmeret central oscillator. *Neurosci. Abstr.* **4,** 381.

Heitler, W. J., and Pearson, K. G. (1980). Non-spiking interactions and local interneurones in the central pattern generator of the crayfish swimmeret system. *Brain Res.* **187,** 206-211.

Hoyle, G. (1976). Arthropod walking. In "Neural Control of Locomotion" (R. M. Herman, S. Grillner, P. S. G. Stein, and D. G. Stuart, eds.), pp. 137-179. Plenum, New York.

Hoyle, G., and Burrows, M. (1973). Correlated physiological and ultrastructural studies on specialized muscles. III. Neuromuscular physiology of the power stroke muscles of the swimming leg of Portunus sanguinolentus. *J. Exp. Zool.* **185,** 83-96.

Hughes, G. M., and Wiersma, C. A. G. (1960). The co-ordination of swimmeret movements in the crayfish, Procambarus clarkii (Girard). *J. Exp. Biol.* **37,** 657-670.

Ikeda, K., and Wiersma, C. A. G. (1964). Autogenic rhythmicity in the abdominal ganglia of the crayfish. Control of swimmeret movements. *Comp. Biochem. Physiol.* **12,** 107-115.

Kennedy, D. (1969). The control of output by central neurons. In "The Interneuron" (M. A. B. Brazier, ed.), pp. 21-36. Univ. of California Press, Berkeley.

Kennedy, D., and Davis, W. J. (1977). The organization of invertebrate motor systems. In "Handbook of Physiology" (E. R. Kandel, ed.), Sect. 1, Vol. I, pp. 1033-1087. Am. Physiol. Soc., Bethesda, Maryland.

Krasne, F. B., and Bryan, J. S. (1973). Habituation: Regulation through presynaptic inhibition. *Science* **182,** 590-592.

Krasne, F. B., and Wine, J. J. (1977). Control of crayfish escape behavior. In "Identified Neurons and Behavior of Arthropods" (G. Hoyle, ed.), pp. 275-292. Plenum, New York.

Larimer, J. L., and Kennedy, D. (1969a). Innervation patterns of fast and slow muscles in the uropods of crayfish. *J. Exp. Biol.* **51**, 119-133.

Larimer, J. L., and Kennedy, D. (1969b). The central nervous control of complex movements in the uropods of crayfish. *J. Exp. Biol.* **51**, 135-150.

Laverack, M. S. (1976). External proprioceptors. In "Structure and Function of Proprioceptors in the Invertebrates" (P. J. Mill, ed.), pp. 1-63. Chapman & Hall, London.

Laverack, M. S., MacMillan, D. L., and Neil, D. M. (1976). A comparison of beating parameters in larval and post-larval locomotor systems of the lobster, *Homarus gammarus*. *Philos. Trans. R. Soc. London, Ser. B* **274**, 87-99.

Lent, C. M. (1971). Metachronal limb movements by *Artemia salina:* Synchrony of male and female during copulation. *Science* **173**, 1247-1248.

Lent, C. M. (1977). The mechanism for coordinating metachronal movements between male and female *Artemia salina* during precopulatory behavior. *J. Exp. Biol.* **66**, 127-140.

Lindsey, B. G., and Gerstein, G. L. (1977). Reflex control of a crayfish claw motor neuron during imposed dactylopodite movements. *Brain Res.* **130**, 348-353.

Lochhead, J. N. (1961). Locomotion. In "Physiology of Crustacea" (T. H. Waterman, ed.), Vol. 2, pp. 313-364. Academic Press, New York.

MacMillan, D. L. (1975). A physiological analysis of walking in the American lobster, *Homarus americanus*. *Philos. Trans. R. Soc. London, Ser. B* **270**, 1-59.

MacMillan, D. L. (1976). Arthropod apodeme tension receptors. In "Structure and Function of Proprioceptors in the Invertebrates" (P. J. Mill, ed.), pp. 427-442. Chapman & Hall, London.

MacMillan, D. L., and Dando, M. R. (1972). Tension receptors on the apodemes of muscles in the walking legs of the crab, *Cancer magister*. *Mar. Behav. Physiol.* **1**, 185-208.

MacMillan, D. L., Neil, D. M., and Laverack, M. S. (1976). A quantitative analysis of exopodite beating in the larvae of the lobster, *Homarus gammarus*. *Philos. Trans. R. Soc. London, Ser. B* **274**, 69-85.

Marrelli, J. D., and Evoy, W. H. (1976). Load compensation through CNS control of proprioceptive interaction with centrally generated motor activity. *Neurosci. Abstr.* **2**, 525.

Marrelli, J. D., and Larimer, J. L. (1978). Proprioceptive shaping of a centrally initiated motor program during movement of the crayfish claw. *Neurosci. Abstr.* **4**, 300.

Mendelson, M. (1971). Oscillator neurons in crustacean ganglia. *Science* **171**, 1170-1173.

Neil, D. M. (1977). Interaction of reflexes in the antenna of the spiny lobster, *Palinurus vulgarus*. *J. Physiol. (London)* **273**, 93P.

Neil, D. M., MacMillan, D. L., and Laverack. M. S. (1976). The structure and function of thoracic exopodites in the larvae of the lobster, *Homarus gammarus*. *Philos. Trans. R. Soc. London, Ser. B* **274**, 53-68.

Neil, D. M., Schöne, H., and Scapini, F. (1979). Leg resistance reaction as an output and an input. Reactions of the rock lobster, *Palinurus vulgaris,* to substrate tilt. VI. *J. Comp. Physiol. A* **129A**, 217-221.

Page, C. H. (1981). Thoracic leg control of abdominal extension in the crayfish, *Procambarus clarkii*. *J. Exp. Biol.* **90**, 85-100.

Parrack, D. W. (1964). Stepping sequences in the crayfish. Ph.D. Thesis, University of Illinois, Chicago.

Paul, D. H. (1971a). Swimming behavior of the sand crab, *Emerita analoga*. I. Analysis of the uropod stroke. *Z. Vergl. Physiol.* **75**, 233-258.

Paul, D. H. (1971b). Swimming behavior of the sand crab, *Emerita analoga*. II. Morphology and physiology of the uropod neuromuscular system. *Z. Vergl. Physiol.* **75**, 259-287.

Paul, D. H. (1971c). Swimming behavior of the sand crab, Emerita analoga. III. Neuronal organization of uropod beating. Z. Vergl. Physiol. **75,** 286-302.
Paul, D. H. (1972). Decremental conduction over "giant" afferent processes in an arthropod. Science **176,** 680-682.
Paul, D. H. (1976). Role of proprioceptive feedback from non-spiking mechanosensory cells in the sand crab, Emerita analoga. J. Exp. Biol. **65,** 243-258.
Paul, D. H. (1979). An endogenous motor program for sand crab uropods. J. Neurobiol. **10,** 273-289.
Pond, C. M. (1975). The role of the walking legs in aquatic and terrestrial locomotion in the crayfish Austropotamobius pallipes (Lereboullet). J. Exp. Biol. **62,** 447-454.
Ripley, S. H., Bush, B. M. H., and Roberts, A. L. (1968). Crab muscle receptor which responds without impulses. Nature (London) **218,** 1170-1171.
Ritzmann, R. E. (1973). Snapping behavior of the shrimp, Alpheus californiensis. Science **181,** 459-460.
Ritzmann, R. E. (1974). Mechanisms for snapping behavior of two alpheid shrimp, Alpheus californiensis and Alpheus heterochelis. J. Comp. Physiol. A **95A,** 217-236.
Russell, D. F. (1976). Rhythmic excitatory inputs to the lobster stomatogastric ganglion. Brain Res. **101,** 582-588.
Russell, D. F., and Hartline, D. K. (1978). Bursting neural networks: A reexamination. Science **200,** 453-456.
Scapini, F., Neil, D. M., and Schöne, H. (1978). I. Leg to body geometry determines eyestalk reactions to substrate tilt and orientation in rock lobsters. J. Comp. Physiol. A **126A,** 287-291.
Schmidt, W. (1915). Die Muskulatur von Astacus fluviatilis (Potamobius astacus L.) Ein Beitrag zur Morphologie der Decapoden. Z. Wiss. Zool. **113,** 165-251.
Schöne, H., Neil, D. M., Stein, A., and Carlstead, M. K. (1976). Reaction of the spiny lobster, Panulirus vulgaris to substrate tilt. I. J. Comp. Physiol. **107,** 113-128.
Sigvardt, K. A. (1977). Sensory-motor interactions in antennal reflexes of the American lobster. J. Comp. Physiol. A **118A,** 195-214.
Simmers, A. J., and Bush, B. M. H. (1980). Non-spiking neurons controlling ventilation in crabs. Brain Res. **197,** 247-252.
Smith, D. O. (1974). Central nervous control of excitatory and inhibitory neurons of the opener muscle of the crayfish claw. J. Neurophysiol. **37,** 108-118.
Spirito, C. P. (1972). An analysis of swimming behavior in the portunid crab, Callinectes sapidus. Mar. Behav. Physiol. **1,** 261-276.
Spirito, C. P., Evoy, W. H., and Barnes, W. J. P. (1972). Nervous control of walking in the crab, Cardisoma guanhumi. I. Characteristics of resistance reflexes. Z. Vergl. Physiol. **76,** 1-15.
Stein, P. S. G. (1971). Intersegmental coordination of swimmeret motor neuron activity in crayfish. J. Neurophysiol. **34,** 310-318.
Stein, P. S. G. (1974). Neural control of interappendage phase during locomotion. Am. Zool. **14,** 1003-1016.
Stein, P. S. G. (1977). A comparative approach to the neural control of locomotion. In "Identified Neurons and Behavior of Arthropods" (G. Hoyle, ed.), pp. 227-329. Plenum, New York.
Stein, P. S. G. (1978). Motor systems with specific reference to the control of locomotion. Ann. Rev. Neurosci. **1,** 61-81.
Stent, G. S., Kristan, W. B., Friesen,W. O., Ort, C. A., Poon, M., and Calabrese, R. L. (1979). Neuronal generation of the leech swimming movement. Science **200,** 1348-1357.
Storch, O. (1929). Die Schwimmbewegung der Copepoden, auf Grund von Mikrozeitlupenaufnehmen Analysiert. Verh. Dtsch. Zool. Ges. **33,** 118-129.

Vedel, J. P. (1980). The antennal motor system of the rock lobster: competitive occurrence of resistance and assistance reflex patterns originating from the same proprioceptor. *J. Exp. Biol.* **87,** 1-22.

Vedel, J. P., and Clarac, F. (1974). Etude neurophysiologique de la motricité antennaire chez la languste, *Palinurus vulgaris. C. R. Hebd. Seances Acad. Sci., Ser D* **278,** 927-930.

Vedel, J. P., and Clarac, F. (1975). Neurophysiological study of antennal motor patterns in the rock lobster, *Palinurus vulgaris.* II. Motor neuronal discharge patterns during passive and active flagellum movements. *J. Comp. Physiol. A* **102A,** 223-235.

Vedel, J. P., and Clarac, F. (1979). Combined reflex actions by several proprioceptive inputs in the rock lobster walking legs. *J. Comp. Physiol. A* **130A,** 251-258.

Vedel, J. P., Angaut-Petit, D., and Clarac, F. (1975a). Reflex modulation of motor neuron activity in the cheliped of the crayfish, *Astacus leptodactylus. J. Exp. Biol.* **63,** 551-567.

Vedel, J. P., Clarac, F., and Bush, B. M. H. (1975b). Coordination motrice proximo-distale au niveau des appendices locomoteurs de la langouste. *C. R. Hebd. Seances Acad. Sci., Ser. D* **281,** 723-726.

Wales, W., Clarac, F., Dando, M. R., and Laverack, M. S. (1970). Innervation of the receptors present at the various joints of the pereiopods and third maxilliped of *Homarus gammarus* and other macruran decapods. *Z. Vergl. Physiol.* **68,** 345-384.

Ward, J. (1879). Some notes on the physiology of the nervous system of the freshwater crayfish, *Astacus fluviatilis. J. Physiol. (London)* **2,** 214-227.

West, L., Jacobs, G., and Mulloney, B. (1979). Intrasegmental proprioceptive influences on the period of the swimmeret rhythm in crayfish. *J. Exp. Biol.* **82,** 281-288.

White, A. Q., and Spirito, C. P. (1973). Anatomy and physiology of the swimming leg musculature in the blue crab, *Callinectes sapidus. Mar. Behav. Physiol.* **2,** 141-153.

Wiens, T. J., and Gerstein, G. L. (1976). Reflex pathways of the crayfish claw. *J. Comp. Physiol. A* **107A,** 309-326.

Wiersma, C. A. G. (1938). Function of giant fibers of the central nervous system of the crayfish. *Proc. Soc. Exp. Biol. Med.* **38,** 661-662.

Wiersma, C. A. G. (1952). Neurons of arthropods. *Cold Spring Harbor Symp. Quant. Biol.* **17,** 155-163.

Wiersma, C. A. G. (1958). On the functional connections of single units in the central nervous system of the crayfish, *Procambarus clarkii* (Girard). *J. Comp. Neurol.* **110,** 421-471.

Wiersma, C. A. G., and Bush, B. M. H. (1963). Functional neuronal connections between the thoracic and abdominal cords of the crayfish, *Procambarus clarkii* (Girard). *J. Comp. Neurol.* **121,** 207-235.

Wiersma, C. A. G., and Ikeda, K. (1964). Interneurons commanding swimmeret movements in the crayfish, *Procambarus clarkii* (Girard). *Comp. Biochem. Physiol.* **12,** 509-525.

Wilson, D. M. (1966). Insect Walking. *Annu. Rev. Entomol.* **11,** 103-122.

Wilson, D. M., and Davis, W. J. (1965). Nerve impulse patterns and reflex control in the motor system of the crayfish claw. *J. Exp. Biol.* **43,** 193-210.

Wilson, D. M., and Wyman, R. J. (1965). Motor output patterns during random and rhythmic stimulation of locust thoracic ganglia. *Biophys. J.* **5,** 121-143.

Wine, J. J., and Krasne, F. B. (1972). The organization of escape behavior in the crayfish. *J. Exp. Biol.* **56,** 1-18.

Young, J. H. (1959). Morphology of the white shrimp, *Penaeus setiferus,* (Linneaus). *Fish. Bull.* **59,** 1-168.

# 4

Autotomy

A. McVEAN

I.	Introduction	107
II.	The Incidence and Consequences of Autotomy	109
	A. The Natural Incidence of Autotomy	109
	B. Forced Autotomy	111
	C. Physiological Consequences of Autotomy	112
III.	Physical Factors in Autotomy	112
	A. The Problem	112
	B. Structure of the Cuticle at the Fracture Plane	113
	C. Mechanical Properties of Rapidly Stressed Materials	115
	D. Anatomical Arrangement of Muscles, Tendons, and Cuticular Structures of the Basi-ischiopodite	117
IV.	Function of the B-I Levator Muscles in Normal Locomotion and Autotomy	121
	A. Motor Activity Not Associated with Autotomy	121
	B. Motor Activity during Autotomy	122
	C. The Autotomy Mechanism	123
	D. Central Response to Limb Damage	126
V.	Modifying Sensory Influences on Autotomy	126
	A. Sensory Reflexes Operating through CSD_1	127
	B. Role of CSD_1 in Autotomy	127
VI.	The Behavioral Roles of Autotomy	128
VII.	Perspectives	129
	References	130

I. INTRODUCTION

The term autotomy was coined by Fredericq (1883) to describe the act by which some animals shed a part of their body, and it is characterized by its speed, coupled with a local provision for restricting material damage. It may or may not be controlled by local musculature. The ability to discard various parts of the body has appeared, by convergent evolution, several times in disparate sections of the animal kingdom.

All animals that exhibit the ability to autotomize a part of their body have two main problems to solve. The first is how to prevent autotomy from occurring when it is not required and the second is how best to exploit the facility once it has evolved. This suggests that autotomy should be considered in two or more complementary ways. First of all, the mechanism by which autotomy is achieved in each animal may be different. Such differences will relate to the genetic distance between the possessors, so that we should understand afresh the mechanism in each species in which we are interested. Secondly, we should study the behavioral strategies that the appearance of this facility permits, remembering that the evolutionary pressures that led to the development of autotomy may be different from those that shaped its subsequent evolution. In attempting to appreciate the interaction of behavioral, ecological, and physiological pressures, we may realize the evolutionary pathways open to discrete parts of simple nervous systems. The variety and distribution of autotomy mechanisms in the Crustacea has been reviewed by Wood and Wood (1932) and Bliss (1960), and although our view of autotomy in the decapods has changed considerably in the last 10 years, there has been no further work done in any other Crustacea.

In the decapod Crustacea, the preformed breakage plane has close similarities to that of the insects (Brousse-Gaury, 1958; McVean, 1975). As in the insects, there is only one fracture plane for each appendage that can be autotomized; this plane is situated near the base of each limb. Where the basipodite and ischiopodite are fused, as in all the pereiopods of the Brachyura (Fig. 3), Scyllardae, Anomura (Fig. 4A), and the chelipeds of the Nephropidae (Fig. 1C), the breakage plane runs close to the line of fusion of these two segments (Wales et al., 1971). This condition is associated with good autotomy capability. The walking legs of the Nephropidae (Fig. 1A,B), however, retain between the basipodite and ischiopodite a functional joint, with which is associated an incomplete breakage plane. In *Homarus gammarus* the plane runs around the dorsal surface of the ischiopodite (Fig. 1), but merges with the articular membrane of the B-I joint near the insertion of the retractor muscle tendon (Paul, 1915; Wales et al., 1971). This condition is associated with a reduced facility for autotomy, since final separation of the limb is only achieved by tearing this membrane. As in insects (McVean, 1975), the exposed faces of the fracture plane show that the cuticle is not noticeably thinner than in adjacent regions (Fig. 2), though there are abrupt changes of angle. Detailed description of the region can be found in Section III,B.

4. Autotomy

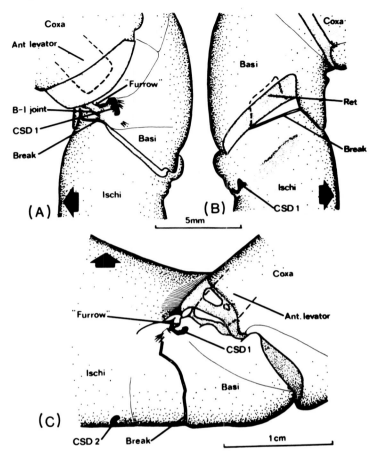

Fig. 1. The external anatomy of the basi-ischiopodite region of the pereiopods in *Homarus gammarus*. In the first pereiopod, the basipodite and ischiopodite are fused, as in *Carcinus maenas*; but in the second and fifth pereiopods, there is a functional joint between the basipodite and ischiopodite. (A) The anterior surface of the second right pereiopod. (B) The posterior surface of the second right pereiopod. (C) The anterior surface of the first right pereiopod. The arrows indicate the dorsal surface of the limbs. Coxa, coxopodite; Basi, basipodite; Ischi, ischiopodite; Break, preformed breakage plane; Ant levator, anterior levator muscle tendon; Ret, retractor muscle tendon; Furrow, Paul's furrow. (From Wales et al., 1971.)

II. THE INCIDENCE AND CONSEQUENCES OF AUTOTOMY

A. The Natural Incidence of Autotomy

Data on the incidence of autotomy in crustacean populations sampled in the field are presented in Table I, and while this provides substantial evi-

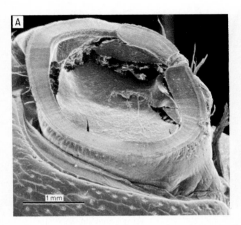 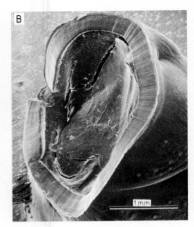

Fig. 2. (A) View of the complete fracture plane in the retained portion of the limb in *Carcinus maenus*, exposed by autotomy showing the sealing membrane, surface planes, and plug. (B) View of the matching fracture plane on the rejected portion of the limb.

dence that autotomy is used extensively by crustaceans in the field, a slightly more detailed analysis shows that the incidence of autotomy may vary within the species. One of the more important biological aspects of autotomy in the decapod Crustacea is the demonstration (McVean, 1975; McVean and Findlay, 1979) that the chelipeds in *Carcinus maenas* are more susceptible than are walking legs; of the total number of limbs lost, chelipeds comprised about 30%, while first, second, third, and fourth walking legs each comprised from 14 to 23.50% (Needham, 1953; McVean, 1976; McVean and Findlay, 1979). There is no differential susceptibility among walking legs or between left and right sides in *Carcinus*.

In both of the extensive population studies done on autotomy in *C. maenas* (McVean, 1976; McVean and Findlay, 1979), it was found that significantly more crabs exhibited multiple autotomy than would be expected if all autotomies occurred by chance; this suggests that while autotomy has immediate and valuable benefits, in the long term it may render the animal more vulnerable. McVean and Findlay (1979) found that there were only 223 regenerating limbs against 556 autotomized limbs for the same period in a discrete population of *C. maenas*, a difference that may result from differential mortality following autotomy. Easton (1972) has shown that multiple autotomy is resisted by increase in the threshold damage that elicits autotomy in the Pacific shore crab *Hemigrapsus oregonensis*, although this effect disappears after 5 min. It is estimated that adult *C. maenas* may, on average, autotomize a limb between once every 4 years

4. Autotomy

TABLE I

The Natural Incidence of Autotomy in Wild Populations of Crustaceans

Animal	% Individuals with losses	Sample size	Authority
Homarus (chelipeds)	7	725	Herrick (1895)
Pagurus (pereiopods)	11	188	Morgan (1901)
Cancer (chelipeds in ♀)	14	776	Pearson (1909)
Porcellana platycheles	28.8	115	Needham (1953)
Leander serratus	34.5	206	Needham (1953)
Carcinus maenas	26.5	234	Needham (1953)
Carcinus maenas	13.2	3342	McVean and Findlay (1979)
Carcinus maenas	42.6	1023	McVean (1976)

and once every 16 months. Autotomy must be seen as a rare event for sexually mature *C. maenas*. In immature *C. maenas*, 12-13% of any one sample exhibit autotomy; and since 10-12 molts occur in their first postlarval year, each crab probably undergoes autotomy at least once before reaching sexual maturity. If autotomy saves the life of the individual, it can be seen as a device that permits these crabs to reach reproductive age, or alternatively allows the crabs to delay the onset of sexual maturity. In both sexes there is a positive but unexplained correlation between carapace width, which increases with age, and the incidence of autotomy (McVean and Findlay, 1979).

B. Forced Autotomy

In most animals autotomy is caused by damage inflicted to the appendage (Bliss, 1960; McVean and Findlay 1976), and there is some evidence that cutting the limb nerve in *Carcinus maenas* and *Pagurus bernhardus* stimulates autotomy. The situation is complicated by the common observation that autotomy does not always follow immediately from inflicted damage but may be delayed. There are reports suggesting that the adequate stimulus for autotomy may be quantitatively and qualitatively different in different species. Needham (1953) reports that it is quite difficult to make *Carcinus* autotomize; *Uca rapax* is said to autotomize readily, whereas the opposite is true for *Uca thayeri* (Weis, 1977). *Porcellana platycheles* is renowned for the ease with which it sheds its limbs. Autotomy may be initiated without mechanical damage. Robinson et al. (1970) describe how two land crabs, the freshwater *Potamocarcinus richmondi* and the terrestrial *Gecarcinus quadratus,* will sometimes detach chelipeds when induced to attack an inert object.

C. Physiological Consequences of Autotomy

Although these are discussed fully in a later volume, a brief mention should be made of the adaptive significance of physiological responses to autotomy. Functional limb replacement in Crustacea, as in other arthropods, has at best to wait until the next ecdysis, though even then the regenerate may be much below its proper size. For instance, Miller and Watson (1976) have shown that the meropodites of autotomized limbs of the spider crab *Chionoecetes opilio* regenerated to 48 and 73% of their full length, irrespective of crab size, on the first and second molt, respectively, whereas in the king crab *Paralithodes camtschatica,* total limb length was still only 91% after 7 molts (Kurata, 1963). Miller and Watson (1976) calculated that the difference between the regenerating and fully grown limb was nearly constant at 35% per molt for juvenile and 30% for adult *P. camtschatica.* In such animals autotomy might appear a drastic step, reaching far into the future life history of the animal, particularly for animals such as adult *Cancer pagurus* and *Carcinus maenas,* which only molt once or twice a year in temperate waters (Bennet, 1973; Broekhuysen, 1936). However, the disadvantage that accrues from limb loss is partially offset by a well-established interaction between autotomy and molting (Bliss, 1956; Skinner and Graham, 1970, 1972; Holland and Skinner, 1976; McCarthy and Skinner, 1977), which has two main effects. The first is to reduce the expected period before the next molt; and the second, apparently paradoxical effect, is to delay imminent molting until new regenerate limb buds have formed. Both are facets of the same strategy, which ensures that the animal is returned to a full limb complement as quickly and economically as possible while minimizing the period spent in the vunerable, newly molted condition.

III. PHYSICAL FACTORS IN AUTOTOMY

A. The Problem

Autotomy, as considered here, involves breaking part of the supporting skeleton, whether external or internal. In the decapod Crustacea, it has been known for a long time (Fredercq, 1883) that the fracture plane is not the weakest point in the limb. Though not as impressive as some insect cuticles, crustacean cuticle appears to be reasonably strong, with a tensile strength of

4. Autotomy

between 18 and 31 N/mm in *Penaeus monodon* (Joffe et al., 1975). The exposed area of the breakage plane in limbs of large *Carcinus maenas* is about 14 mm (McVean, 1973); hence, with use of the lowest estimate for *P. monodon*, the minimum force that would break the intact plane would be in the order of 252 N. Almost all this force would have to be supplied by the basi-ischiopodite levator muscle (see Section D,1), the wet weight of which is less than 1 g (McVean and Findlay, 1976). Although this muscle (anterior levator muscle, ALM) is unlikely to be composed entirely of slow muscle fibers, we can, for the purpose of argument, assume that it might develop as much force as the lobster slow remotor muscle, or 0.255 n/mm^2 (Prosser, 1973). This would mean that the ALM would have to have a minimal cross-sectional area of 9.9 cm^2, which is clearly absurd. Even if we attribute this muscle with a force capability of 1 N/mm^2, as in the locust extensor tibiae (Bennet-Clarke, 1975), the required cross-sectional area is unreasonably large. This suggests that either the preformed breakage plane is so fragile that it is in danger of breaking in normal use or else the ALM is enormously powerful. Experimental evidence has shown that neither is the ALM as strong as a completely intact plane would imply (Ellam, 1978), nor is the fracture plane particularly weak (Fredericq, 1883; McVean, 1973). Autotomy in the decapod Crustacea must therefore be achieved through application of force across the fracture plane in a particular way, especially since normal levatory forces can be larger than the levatory force at autotomy (McVean, 1973).

B. Structure of the Cuticle at the Fracture Plane

The fracture plane can be detected in the cuticle of decapod Crustacea as a line encircling or partially encircling the second or third proximal limb segment (Figs. 3, 4A, and 6). The course taken by this line contains sharp deviations and corners. The most notable of these is found immediately distal to the insertion of the large basi-ischiopodite levator muscle, where the fracture line delimits a pluglike projection (Fig. 2). When exposed by autotomy, the greater part of the opposed faces have a smooth unbroken surface with quite distinct planes (Fig. 2A, B), except where the plug was situated. For the most part the separate lamina of the cuticle are clearly visible. In *Pagurus bernhardus,* the opposing faces are slightly ridged and carry projecting globules of cuticular material, which can be seen to bridge the intact plane. The function of these is uncertain. It has been postulated (McVean, 1973), though not proven, that the surfaces of the plug form the only mechanically secure link across the breakage plane. It is envisaged that weak links across the major portion of the plane are normally protected from

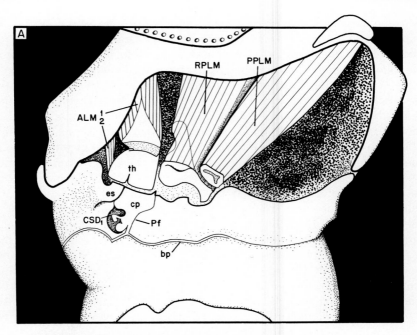
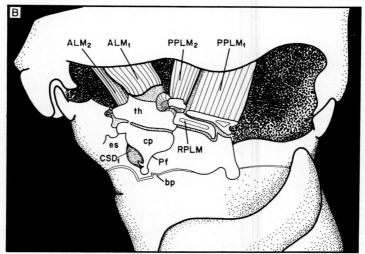

Fig. 3. The disposition of the B-I levator muscles in *Carcinus maenas* as seen when looking down onto the articular membrane that spans the dorsal surface of the C-B joint. Limb extends downward out of the picture. (A) In the walking legs the anterior levator group is divided into two parts, as is the posterior levator group. ALM_1 tendon head makes contact with the dorsal rim of the B-I at two locations: the cuticular projection, which forms the energy store (es), and

excessive shear forces by the angular nature of the plane surfaces. The unbroken fracture plane is pitted in *Cardisoma guanhumi,* with cavities that Moffett (1975) has likened to perforations around a postage stamp.

C. Mechanical Properties of Rapidly Stressed Materials

Arthropod cuticle is a complex composite material, the mechanical properties of which are derived from the individual components and their interaction with one another. When autotomy is considered, it is clear that some fraction of the cuticle has to be broken.

In many materials the application of a small stress produces a correspondingly small strain, which is exactly reversed when the applied stress is removed. Such material are described as elastic. If the load on an elastic material is continuously increased, it will eventually fracture. At fracture, new surfaces are created, a process that requires a supply of energy. Much of this energy may derive from the initial elastic strain energy (Wainwright et al., 1976). Fracture starts by the formation of a crack. There are likely to be many small cracks present in any material other than perfect crystals. Short cracks, however, are resistant to propagation, since more energy is required to increase the surface area of the crack than is available from the corresponding release of strain energy. At a critical crack length, the rate of release of strain energy by extension of the crack equals the rate at which work is done in creating a new surface. Beyond this length, the crack must continue to grow, since an increase in length now decreases the stress needed to propagate the crack; thus, not only does the crack grow, but its rate of growth accelerates (Wainwright et al., 1976).

Bone and arthropod cuticle are complex elastic polymers, whose fracture properties alter with loading time. Figure 5 shows what happens when different strain rates are imposed on a biological polymer (Goldspink, 1977). As the strain rate begins to increase, so does the breaking force: but beyond an extension rate of 80 mm/sec, the force required to break the preparation decreases. Moreover, the curve is asymmetric, so that the breaking force at the highest strain rate is less than one tenth of the force required at the

---

posteriorly, onto the cuticular plug (cp), which forms, at its distal end, the intact bridge across the breakage plane. The plug is bounded anteriorly by the membrane overlying CSD_1 and posteriorly, onto the cuticular plug (cp), which forms, at its distal end, the intact bridge across comprising the energy store is reduced and an extra muscle, PPLM, helps to keep the cuticular catch engaged. Details of the muscles are given in Tables II and III. ALM, anterior levator muscle; bp, limb breakage plane; CDS, cuticular strain detector; PPLM, posterior branch of the posterior levator muscle; RPLM, rotatory portion of the posterior levator muscle; th, cuticular block that functions as a tendon head. (From McVean and Findlay, 1976.)

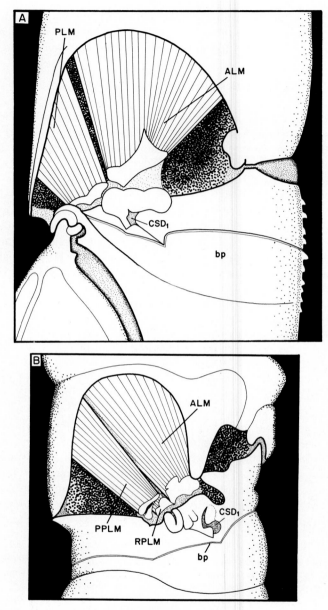

Fig. 4. In *Pagurus bernhardus* PPLM is absent, even in the cheliped (A), though the most dorsal fibers may fulfill its function. In *Màja squinado*, with its immensely long legs, PPLM is exceptionally well developed, even in the fifth pereiopod (B). Aspect and legend as for the previous figure. (From McVean and Findlay, 1976.)

4. Autotomy

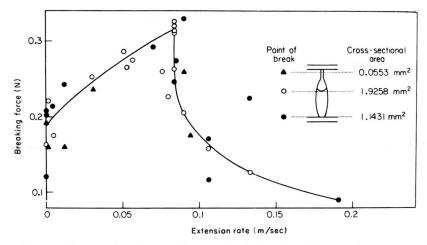

Fig. 5. This shows the variation in force required to break a polymer, in this case a muscle, at different rates of extension. The muscle preparation was the biceps brachii of the mouse. Note the low breaking force at high strain rates. (From Goldspink, 1977.)

slowest strain rates. It is relevent that bones are nearly always broken by impact loading (Wainwright et al., 1976). It could be that the intact cuticle spanning the breakage plane in crustacean limbs is also ruptured by impact loading, since this would require a smaller force and thus a smaller driver muscle. However, strain rates of the required value are unknown in single crustacean muscles (Prosser, 1973), suggesting that if impact loading is the means by which autotomy is achieved, then such loading must be achieved by the sudden release of force in a pretensioned muscle.

D. Anatomical Arrangement of Muscles, Tendons, and Cuticular Structures of the Basi-ischiopodite

1. THE MUSCLES

In the following account, most attention will be given to *Carcinus maenas, Pagurus bernhardus,* and *Cardisoma guanhumi*. The arrangement of muscles inserting upon the first two limb segments in crabs is complicated. Their form has been described in the blue crab, *Callinectes sapidus* (Cochran, 1935), and in great detail and with specific reference to function by Ellam (1978). Both form and function are complicated by the fact that the majority of these muscles have multiple origins within the thorax. In order to simplify matters, this account will concentrate exclusively on muscles and tendons immediately involved in autotomy; these muscles are those inserting onto the proximal rim of the basi-ischiopodite. The basi-ischiopodite receives a single depressor muscle (DM), which has its origin within the thorax and traverses the coxopodite. This muscle also influences the attitude of the coxopodite, as does the large basi-ischiopodite levator muscle (ALM), but this does not alter the role of these muscles in autotomy.

The critical automizer muscles are the B-I levator muscles. In *C. maenas*, these are split into two main groups, the anterior levator muscles (ALM) and the posterior levator muscles (PLM). Each of these two groups is further subdivided (Table II and Fig. 3A). In *Cardisoma guanhumi*, in addition to the two subdivisions of ALM that arise in the thorax, there is a small branch of this muscle that arises within the coxa and runs parallel to PLM (Moffett, 1975). It shares innervation with ALM and is not innervated by PLM motoneurons. In the cheliped of *C. maenas*, the posterior levator muscles are subdivided into three distinct muscles (Fig. 3B). The extra muscle (PPLM) originates in the coxopodite and inserts, via the arthrodial membrane, onto a sclerite that forms the ALM tendon head. The PPLM is absent in *Pagurus bernhardus* (Fig. 4A) and highly developed in *Maja squinado* (Fig. 4B). The tendons of all the muscles, except one, lie in the same axis as their muscle fibers. The one important exception is the tendon of the rotating PLM (RPLM), which projects inward (Fig. 6). Contraction of RPLM causes this tendon to rotate about its insertion so as to displace the ALM tendon dorsally at its insertion (McVean 1973; Moffett, 1975; McVean and Findlay, 1976).

TABLE II

The Origin, Insertion, and Action of each B-I Levatory Muscle in the Walking Limbs of *Carcinus maenas*[a]

Muscle group	Muscle	Origin	Insertion	Action
ALM	ALM_1	Thorax	Main tendon blade of ALM; attaches to dorsal rim of the B-I via independent tendon head	Limb levation and autotomy
	ALM_2	Thorax	Directly onto the tendon head	Closes the catch
PLM	PPLM	Dorsal surface of the coxa	Onto small tendon attached to arthrodial membrane in region of RPLM tendon insertion	Limb levation and closing the catch
	RPLM	Dorsal surface of the coxa	Onto tendon that rotates against tendon head of ALM	Opens the catch

[a] The cheliped contains another muscle in the PLM group

4. Autotomy

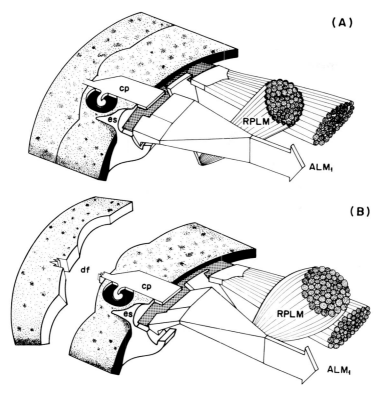

Fig. 6. Diagrammatic representation of the main elements in the walking legs of *Carcinus maenas* that contribute to limb autotomy. The function of the individual muscles is discussed in the text. (A). The main levator muscle ALM_1 connects to a cuticular projection (es), which is capable of acting as an energy store. Functional connection is maintained to the energy store, which is tolerant of considerable distortion, as long as the catch remains closed. (B) When the catch is disconnected by rotation of the RPLM tendon, tension accumulated in ALM_1 is transferred directly onto the cuticular plug (cp). This plug is loosely connected to the surrounding cuticle, so that the full force of ALM is concentrated on the distal face (df) of the plug, which breaks under the strain. The limb is now free to fall away since the preformed breakage plane parts with little resistance.

The RPLM tendon acts against the ALM tendon with a fivefold mechanical advantage.

2. INNERVATION

Details are available for only a few species of crabs. Moffett (1975) traced eight to ten axons in the nerve serving the anterior and posterior branches of ALM in *Cardisoma guanhumi* and could distinguish at least six distinct physiological units on the basis of spike size. McVean (1974) also found an

TABLE III

Details of the Innervation and Properties of the Basi-ischiopodite Levator Muscles in Three Crabs

Animal	Muscle	Number of axons	Number of discernable units	EPSP[a] (mV)	Tension response	Authority
Carcinus maenas	ALM	13	3	1	Tonic, facilitating	McVean (1974)
				5	Phasic	
	RPLM[b]			25–30	Fast phasic	
			2	4	Tonic	Findlay (1978)
				8	Phasic	
	PPLM[b]		2	1–2	Tonic	Findlay (1978)
				3–4	Phasic	
Cardisoma guanhumi	ALM	8–10	3–4			Moffett (1975)
	PLM	2	2			
Pagurus bernhardus	ALM	?	?			Findlay and McVean (1977)
	PLM	?	2	1–2	Tonic	
				3–4	Phasic	

[a] Excitatory postsynaptic potential.
[b] Receives common innervation.

4. Autotomy

excess of axons over the three units serving ALM in *Carcinus maenas*. Such details as are known can be found in Table III. Where PLM is split into RPLM and PPLM, as in *C. maenas*, the two subdivisions receive common innervation (Findlay, 1978).

3. CUTICULAR STRUCTURES INFLUENCED BY THE B-I LEVATOR MUSCLES

Interposed between ALM and the basi-ischiopodite is a small cuticular block (th, Figs. 3, 4, and 6), which is flexibly linked to both. Its connection with the B-I is broad and divided into two regions. The anterior region consists of a catch (Fig. 6) that engages with the cuticle of the B-I, raised in this area in *C. maenas* into a small but definite bump, which is less distinct in *P. bernardus* (McVean and Findlay, 1976). The posterior half of the connection leads directly to the cuticular plug (Section III,B). The posterior margin of the cuticular plug is bounded by a groove known as Paul's furrow and anteriorly by a membrane (Figs. 3 and 4). The plug is insecurely anchored into the surrounding cuticle.

The attitude of the ALM ensures that the faces of the cuticular catch are closed. The state of this catch is influenced by the posterior levator muscles. In the walking legs of *C. maenas*, rotation of the RPLM tendon opens the catch, an action that is opposed by tension in the PPLM. Tension in the cheliped PPLM aids the $PPLM_1$ (Fig. 3) by depressing the ALM tendon head directly (McVean and Findlay, 1976).

IV. FUNCTION OF THE B-I LEVATOR MUSCLES IN NORMAL LOCOMOTION AND AUTOTOMY

A. Motor Activity Not Associated with Autotomy

1. ANTERIOR LEVATOR MUSCLE

Resistance reflexes can be elicited from the ALM tonic and smaller phasic motoneurons in response to imposed depression of the limb about the coxobasi-ischiopodite (C-B) articulation in *C. maenas* (Bush, 1965a; Mc-Vean, 1974; Findlay, 1978) and in *Cardisoma guanhumi* (Moffett, 1975). Activity in ALM was recorded during walking in *Cardisoma* and paddling in *C. maenas* (Findlay, 1978). The motor activity recorded during these activities consisted of one to two motor units firing during active limb levation and imposed depression.

2. POSTERIOR LEVATOR MUSCLES

A single PLM motoneuron fires synergistically with ALM motoneurons, both in the resistance reflex and active paddling. This tonic motoneuron serves both RPLM and PPLM in C. *maenas* where it preferentially innervates the dorsal muscle fibers with least mechanical advantage for rotation (Findlay, 1978): it is incapable of eliciting sufficient tension from RPLM to rotate its tendon (Fig. 7). This tonic motoneuron evokes small excitatory postsynaptic potentials (EPSP's) in both C. *maenas* (Findlay, 1978) and *P. bernhardus* (Findlay and McVean, 1977). The innervation and electrical responses of ALM_2 and $PPLM_2$ in C. *maenas* are not well known. Initial results suggest that $PPLM_1$ and $PPLM_2$ share common innervation.

B. Motor Activity during Autotomy

1. ANTERIOR LEVATOR MUSCLE

McVean (1973) showed that the larger phasic motoneuron in C. *maenas* could only be invoked by damage inflicted to the limb; during this time both the tonic and smaller phasic motoneurons were firing at high frequency. The larger phasic unit can be activated by injury in intact C. *maenas* (McVean, 1974), and in C. *guanhumi* three large units not active in resistance reflexes are activated by injury (Moffett, 1975). In *P. bernhardus*, the ALM motoneurons also fire at high frequency during autotomy (Findlay and

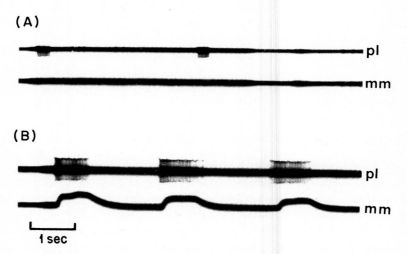

Fig. 7. The position of the rotatory posterior levator muscle (RPLM) tendon in C. *maenas* was monitored during excitation of posterior levator motoneurons, recorded *en-passant*. (A) Firing of the tonic posterior levator motoneurons, elicited by stretching the basi-ischium chordotonal organ, has no effect upon the orientation of the RPLM tendon. (B) Activity of the phasic PL motoneuron, caused by mechanical stimulation of CSD_1, causes this tendon to rotate. pl, action potentials recorded from the posterior levator nerve. mm, movement monitor attached to the RPLM tendon. (From Findlay, 1978.)

McVean, 1977). On the evidence of tension records (McVean, 1974; Ellam, 1978), the introduction of the larger phasic motoneuron in C. maenus signals a massive and rapid increase in muscle tension.

2. POSTERIOR LEVATOR MUSCLES

As shown in Table III, RPLM and PPLM in C. maenas, the only brachyuran in which these muscles have been studied in detail, receive common innervation, from both the phasic and tonic PL motoneurons (Findlay, 1978). This precludes the possibility that PPLM can operate with separate functions on the basis of motoneuron activity alone. The phasic PL motoneuron preferentially innervates the mechanically advantageous muscle fibers in RPLM, where its large EPSP's are capable of eliciting sufficient tension to rotate the tendon (Fig. 7). In P. bernhardus, where regional innervation of the single PL muscle might achieve separate mechanical effects, with the tonic motoneuron preferentially innervating the dorsal and the phasic unit innervating the ventral fibers, no such segregation is found (Findlay and McVean, 1977). Damage inflicted to the limb recruits not only the AL motoneurons, but also, in C. maenas, the PL tonic unit. Shortly before autotomy, however, the tonic PL unit is silenced and the phasic PL unit fires (Fig. 8A). In P. bernhardus, autotomy is achieved by a more coincidental relationship between ALM and the PLM phasic units (Fig. 8B). As in C. maenas, the PLM tonic motoneuron is inhibited at autotomy (Findlay, 1977).

C. The Autotomy Mechanism

Two mechanisms of autotomy have been proposed within the last decade. Both agree that the role played by the RPLM and its tendon is crucial in determining whether the limb is levated or autotomized by tension developed in ALM, but they differ as to whether autotomy is achieved when rotation of the RPLM tendon displaces the AL tendon dorsally so that the cuticular catch is opened (McVean and Findlay, 1976, on *Carcinus* and *Pagurus*) or whether it occurs when RPLM relaxes and allows the ALM to use the cuticular projection arising from the interior rim of the basipodite as a fulcrum (Moffett, 1975, on *Cardisoma*). Moffett (1975), however, was not aware that the PLM in Brachyura consists of two functionally distinct muscles and may have confused the role played by PPLM and RPLM.

In C. maenas, autotomy is distinguished by the introduction of the largest phasic ALM motoneuron, which responds to injury with a high frequency barrage. This generates a high tension in this muscle. A delayed burst from the PLM phasic unit begins shortly before autotomy.

The conclusion is that autotomy in C. maenas occurs when the RPLM contracts and in doing so rotates its tendon. The accumulating tension in the

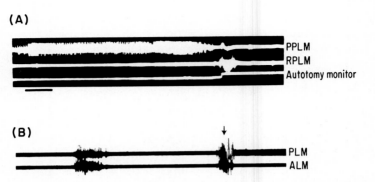

Fig. 8. Electrical activity of B-I levator muscles during autotomy. (A) Responses of the PPLM and RPLM of Carcinus maenas to an injury applied to the merus, leading to limb autotomy. The upward shift of the autotomy monitor beam (am) indicates the moment of autotomy (from McVean and Findlay, 1976). (B) Electromyogram recording from the ALM and PLM in Pagurus bernhardus during an ineffectual and a successful attempt at limb autotomy. The wire electrodes were displaced from the PLM at the moment shown by the arrow. During both responses, the PLM fired simultaneously with the ALM (from Findlay and McVean, 1977).

ALM progressively strains the energy store until the moment when the cuticular catch (Fig. 7) is disengaged by rotation of the RPLM tendon. The cuticular plug is thus subjected to impact loading, which, in conjunction with the elastic recoil of the energy store in the opposite direction, severs the connection of this plug across the limb breakage plane, allowing the preformed breakage plane to separate.

In Pagurus bernhardus the PLM phasic unit is not delayed, as in Carcinus, but fires in conjunction with ALM motoneurons as the limb is injured. The growing tension in the ALM is thus directed onto the cuticular plug from the start. In this case, there can be no impact loading, so that limb separation follows from concentration of a large force upon the small area of the plug connection. It seems probable that fracture in C. maenas is achieved in the fashion of P. bernhardus when the unrestricted limb is fully elevated, since in this position the cuticular catch is open (see above). This would happen if the limb were severed distally and free to elevate, while impact load autotomy would occur if the limb were held.

By the appropriate intervention of the associated PL muscles, the single ALM in C. maenus and P. bernhardus can fulfil two distinct roles; that of normal limb levation and autotomy. Findlay (1977) reports that the RPLM tendon in Homarus gammarus does not interact with the ALM tendon, although the limbs are capable of autotomy (Wales et al., 1971). The more sophisticated solution to limb autotomy prevalent in the Brachyura, with the introduction of impact loading, has probably derived from a more straightforward system, as in P. bernhardus, in which tension in the RPLM

4. Autotomy

disengages the cuticular catch immediately in response to limb injury. One attraction of impact loading as practiced by *C. maenas* is that the elasticity of the cuticle comprising the energy store will be matched to the elasticity of the cuticle across the fracture plane. The mechanism in *P. bernhardus* may in turn have evolved from that in *H. gammarus,* where the RPLM, though not influencing ALM tendon, serves to strain the cuticle along Paul's furrow, enhancing strain imposed by the ALM (Paul, 1915). It is not difficult to envisage the basic mechanism in *H. gammarus* evolving from slight modification in angle of insertion of the accessory levator muscle tendon. Since *C. maenas* can still autotomize limbs after the PLM tendons are cut (Moffett, 1975), the assumption must be that impact loading is a sophistication which, though not absolutely necessary, improves the reliability and timing of autotomy.

It is surprising that the considerable modifications that have taken place in the B-I muscles, their tendons, and associated cuticular structures have been accompanied by only slight modification in the attendant motoneurons. Two motor units serve the single PL muscle in *P. bernhardus* as they serve the anatomically separate PPLM and RPLM in *C. maenas;* in the latter case, functional separation of the muscles has been achieved not by the separation or addition of motoneurons, but by the inability of the tonic PL motoneurons to create sufficient tension in the RPLM to rotate the tendon. We do not yet know if this tonic motoneuron is uncoupled from the tension-generating mechanism in RPLM, or if the tension that it generates in either PL muscle is immoderately low. Whichever may be the case, the continued innervation of RPLM in *Carcinus* by the tonic motoneuron is intriguing.

According to Moffett (1975), autotomy occurs in *Cardisoma guanhumi* when RPLM relaxes at the same time that the ALM strongly contracts. For evidence, she presents three records obtained during autotomy, in which the PLM nerve was monitored in conjunction with a myogram of ALM. She proposes that the plug crossing the breakage plane is protected from excessive tension generated in ALM, by coincidental tension in RPLM, whose action prevents the ALM tendon from using the cuticular block as a fulcrum. However, since the ALM tendon is connected to the cuticular block by a flexible linkage, such a fulcrum would diminish, not magnify, the force generated by ALM. Moffett's (1975) evidence indicated that large motoneurons supplying ALM are recruited by injury, and that this muscle, acting alone, is capable of producing the fracture. By combining cinematography with electrophysiological recordings, Moffett demonstrated that a long PLM motoneuron barrage follows rather than precedes production of the fracture. In contrast, in locomotion and resistance reflexes, PLM and ALM act synergistically. Thus, according to Moffett's hypothesis, autotomy is effected by activation of ALM alone. In the absence of injury stimulus, both levators are postulated to share the levatory load and, in addition, PLM contraction forces the ALM tendon upward, preventing it from contacting the projection arising from the interior rim of the basipodite. This projection, described by Moffett as a fulcrum rather than a catch [compare Moffett's (1975) Fig. 4A and Fig. 5 with Figs. 6A and B of this chapter] is viewed as important in the transfer of force to the fracture plane. Moffett, however, did not monitor the position of the PL tendon during locomotion, and as described above for *C. maenas,* electrical activity in RPLM is no guarantee that the RPLM tendon is rotated (Fig. 7). Moffett suggests that reflex activation of PLM by the cuticular stress detector distal to the ALM tendon insertion (CSD_1, discussed below) would serve to prevent unintentional limb fracture by forcing PLM to assume a greater share of the levatory load. However, she does not explain why PLM motoneurons are not active prior to autotomy, when strain in the cuticle distal to the ALM tendon would be maximal, and her use of the term "fulcrum" is misleading, since the flexible linkage between the ALM tendon and cuticular block would prevent the lever action this term implies. Until an analysis of PPLM and RPLM innervation and role in autotomy is done for *C. guanhumi* as has been done for *C. maenas* and *P. bernhardus,* it is difficult to decide whether *C. guanhumi* autotomizes in the same way or not.

D. Central Response to Limb Damage

If the B-I levator tendons are severed or their motor nerves cut, the motor response that emanates from the CNS to these muscles is isolated from any feedback that might derive from associated sense organs. When sensory feedback to ALM motoneurons is prevented in *C. maenas,* the response differs from that during autotomy in the intact animal, in that the PL phasic unit remains silent. In *P. bernhardus,* both PL units respond to inflicted damage but fall silent during the intensive barrage seen in ALM, although in the intact animal both AL and PL phasic units operate simultaneously to produce autotomy.

V. MODIFYING SENSORY INFLUENCES ON AUTOTOMY

Although apodeme tension receptors (Macmillan, 1976) might be associated with the autotomy musculature, none has yet been demonstrated in this region. Even if present, such receptors may not play a direct role in autotomy, since it is cuticular strain, not apodemal tension, that should be monitored.

A change in tension in the large anterior levator muscle of the basi-ischiopodite will either alter the angle of the C-B joint or alter the strain on the cuticle in the region of the insertion of this muscle. Joint rotation is monitored by the single CB chordotonal organ that spans this articulation (Whitear, 1962; Bush, 1965b).

Inspection of the cuticle distal to the ALM insertion (Figs. 3, 4, and 6) reveals that the cuticular plug is bounded by a pliable membrane along its anterior edge, which in *C. maenas* carries an island of cuticle attached to and projecting from the cuticular plug. Any differential strain experienced by this plug in relation to the surrounding cuticle will distort this membrane. Wales et al. (1971) have shown that a strand of connective tissue extends from the inner surface of this membrane to an adjacent peg (*C. maenas*) or area of cuticle (*H. gammarus*). Bipolar neurons and their dendrites, embedded in the strand, respond to regional strain imposed upon this membrane (Clarac et al., 1971). These sense organs were termed cuticular stress detectors or CSD_1 (Wales et al., 1971), since they appeared to respond to cuticle stress; but it is probably more accurate to say that they respond to cuticular strain, which will be a variable function of applied stress, dependent on the elastic state of the cuticle at the time. CSD_1 responds to strains imposed upon this region by tension in the ALM (Clarac et al., 1971; Moffett, 1975; Findlay, 1978).

A. Sensory Reflexes Operating through CSD_1

Tension generated in ALM in *Carcinus maenas* and *Pagurus bernhardus* operates a dual influence, via CSD_1, upon the B-I levator motoneurons. In both animals, such tension inhibits active ALM motoneurons and the PLM tonic motoneuron, causing simultaneous recruitment of the phasic PLM motoneuron. The response to tension in the ALM tendon is diminished when the cuticular catch is closed (Findlay and McVean, 1977), and it is enhanced when the limb encounters mechanical resistance to elevation (Findlay, 1978). In normal limb elevation, therefore, excessive strain imposed by the ALM upon the cuticular plug is prevented by reflex inhibition of the ALM motoneurons.

In autotomy, tension in ALM is not diminished by the CSD_1 reflex, because one or more ALM motoneurons are exempt from an inhibitory influence from CSD_1 (Findlay, 1978). How this is achieved is not yet known.

B. Role of CSD_1 in Autotomy

In *P. bernhardus* the cuticular catch, opened by the initial response to injury, is held in that state by the CSD_1-PLM phasic unit reflex, which constitutes a positive feedback loop, so that ALM tension is directed from the start upon the cuticular plug across the fracture plane. In *C. maenas,* tension accumulates with the catch engaged, since the initial response to injury does not include the PLM phasic unit. The CSD_1-PLM phasic motoneuron reflex is delayed by the energy store, so that when RPLM tendon finally rotates, the ALM has achieved high tension; this leads to impact loading when this tension is released onto the cuticular plug.

VI. THE BEHAVIORAL ROLES OF AUTOTOMY

At the most basic level, autotomy converts all random injuries that are suffered by a particular appendage and that cross a particular damage threshold into one predictable injury. All regenerating and wound healing resources can then be invested in this region. It seems reasonable to suppose that autotomy initially appeared as an injury-limiting strategy. Evidence for this assertion is found in the limbs of the shrimp *Palaemon serratus,* which have a preformed fracture plane but no muscular autotomy mechanism (Bliss, 1960).

It is harder to understand how autotomy, once developed, was exploited. The Anomura and Brachyura, particularly some land crabs (Robinson *et al.,* 1970), have developed different levels of muscular and cuticular adaptations for autotomy. The further development of attack autotomy must have required adaptation of interneurons to permit autotomy without prior limb damage.

Autotomy, in the decapod Crustacea, almost certainly accrues from fighting. Recent commentaries on fighting behavior are worth a short review, if we are to construct hypotheses about the evolution of autotomy. Animals may fight members of their own species or other species. The contests may take two forms: symmetric contests, in which combatants are equally matched in all ways that could influence the outcome of a contest; and asymmetric fights, in which one combatant possesses some quality that is construed by the other as influencing the result of the contest (Maynard-Smith and Price, 1973). Asymmetric contests are worth particular attention, not only because the fighting is often carried out in defence of accrued gains, which produces asymmetry, but also because fights between prey and predator might be considered examples of extreme asymmetry. Prey–predator conflicts may be thought of as competition for resource-holding potential between the genes of the prey and the genes of the predator. The genes of the prey will organize the behavior of their phenotype so as to conserve their investment in it; those of the predator will produce behavior appropriate to the assimilation of foreign protein. The main difference between intra- and interspecific asymmetrical fighting is that in the latter case, the prey has considerably more at stake. From this it follows that if a sacrifice can be made that will reduce loss of resource-holding potential, then such a sacrifice should more readily be made when the prey is challenged by a predator than when fighting occurs intraspecifically, with the proviso that losing an intraspecific fight does not reduce the probability of subsequent reproduction to zero.

Autotomy fits the economics of asymmetric fighting well; it allows the possessor, once committed to fighting with a predator, the possibility of making effective sacrifices, not only because each sacrifice is a limited one, but also the rejected portion of the body is, or gives the illusion of being, a transfer of resource-holding potential. Such an exchange may serve to reduce the willingness of a predator to prosecute the contest further, especially if the potential prey can underwrite the exchange with offensive weaponry. Two predictions follow from this argument. Where the individual is prone to attacks from predators, autotomy should not be used by the animal in fighting conspecifics, since by doing so the animal immediately reduces its future viability. Secondly, the greater the offensive capability of the prey, the smaller need be its sacrifice.

Intraspecific aggression between decapod Crustacea, where both ritualized and escalated fighting occur, is well documented (Crane, 1966, 1967; Hazlett, 1968, 1969, 1972; Heckenlively, 1970; Hyatt and Salmon, 1978). Significantly, in such combats, autotomy almost never occurs (although there are reports that autotomy in *Gecarcinus lateralis* does increase during the breeding season). In at least two documented cases, the position considered to represent the point of greatest escalation consists of the combatants interlocking their chelipeds. At this point either combatant is in a position to damage, and thus induce autotomy of, its competitor's cheliped, thereby placing the competitor at a considerable disadvantage both in mate selection and defence.

VII. PERSPECTIVES

Autotomy, widespread through the decapod Crustacea, has been investigated thoroughly in only two animals, *Carcinus maenas* and *Pagurus bernhardus;* even in these animals some details of the autotomy mechanism are based on surmise. In these and other decapods, a description of the role of interneurons in autotomy has not yet been, but should be, attempted. Certain useful hypotheses concerning the role of autotomy in vertebrates are available in the literature. Little is known about the uses and consequences of autotomy for decapod Crustacea in the field.

REFERENCES

Bennet, D.B. (1973). The effect of limb loss and regeneration on the growth of the edible crab, *Cancer pagurus*, L. *J. Exp. Mar. Biol. Ecol.* **13,** 45-53.

Bennet-Clarke, H. (1975). The energetics of the jump of the locust, *Schistocerca gregaria*. *J. Exp. Biol.* **63,** 53-83.

Bliss, D.B. (1956). Neurosecretion and the control of growth in a decapod crustacean. In "Bertil Hanstrom. Zoological Papers in Honour of his Sixty Fifth Birthday, November 20th, 1956" (K. G. Wingstrand, ed.), pp. 55-75. Zool. Inst., Lund, Sweden.

Bliss, D.E. (1960). Autotomy and regeneration. In "The Physiology of Crustacea" (T.H. Waterman, ed.), Vol. 1, pp. 561-589. Academic Press, New York.

Broekhuysen, G.J. (1936). On development, growth and distribution of *Carcinides maenas* (L.). *Arch. Neerl. Zool.* **2,** 257-399.

Brousse-Gaury, P. (1958). Contribution à l'étude de l'autotomiè chez *Acheta domestica* L. *Bull.Biol.Fr.Belg.* **92,** 55-85.

Bush, B.M.H. (1965a). Leg reflexes from chordotonal organs in the crab, *Carcinus maenas*. *Comp.Biochem.Physiol.* **15,** 567-587.

Bush, B.M.H. (1965b). Proprioception by the coxo-basal chordotonal organ, CB, in legs of the crab, *Carcinus maenas*. *J.Exp.Biol.* **42,** 285-297.

Clarac, F., Wales, W., and Laverack, M.S. (1971). Stress detection at the autotomy plane in the decapod Crustacea. II. The function of receptors associated with the cuticle of the basi-ischiopodite. *Z.Vergl.Physiol.* **73,** 357-382.

Cochran, D.M. (1935). The skeletal Musculature of the Blue Crab. *Smithson.Misc.Collect.* **92,** 1-76.

Crane, J. (1966). Combat, display and rutualisation in Fiddler crabs (Ocypodidae, genus *Uca*). *Philos.Trans.R.Soc. London, Ser.B* **251,** 459-472.

Crane, J. (1967). Combat and its ritualisation in Fiddler crabs (ocypodidae), with special reference to *Uca rapax*. *Zoologica (N.Y.)* **52,** 49-76.

Easton, D.M. (1972). Autotomy of walking legs in the pacific shore crab *Hemigrapsus oregonensis*. *Mar.Behav.Physiol.* **1,** 209-217.

Ellam, L. (1978). A neurophysiological analysis of aggressive behaviour in *Carcinus maenas*. Ph.D. Thesis, University of London.

Findlay, I. (1977). A study of autotomy in decapod Crustacea. Ph.D. Thesis, University of London.

Findlay, I. (1978). The role of the cuticular stress detector, CSD_1, in locomotion and limb autotomy in the crab, *Carcinus maenas*. *J.Comp.Physiol.* **125,** 79-90.

Findlay, I., and McVean, A.R. (1977). The nervous control of limb autotomy in the hermit crab *Pagurus bernhardus* (L.) and the role of the cuticular stress detector, CSD_1. *J.Exp.Biol.* **70,** 93-104.

Fredericq, L. (1883). Sur l'autotomie ou mutilation par voie réflexe comme moyen de défense chez les animaux. *Arch.Zool.Exp.Gen.* [2^B Ser.] **1,** 413-426.

Goldspink, G. (1977). Mechanics and energetics of muscle in animals of different sizes, with particular reference to the muscle fibre composition of vertebrate muscle. In "Scale Effects in Animal Locomotion" (T.J. Pedley, ed.), pp. 37-55. Academic Press, New York.

Hazlett, B.A. (1968). Size relationships and aggressive behaviour in the hermit crab *Clibanarius vittatus*. *Z.Tierpsychol.* **25,** 608-614.

Hazlett, B.A. (1969). Further investigations of the cheliped presentation display in *Pagurus bernhardus* (Decapoda, Anomura). *Crustaceana* **17,** 31-34.

Hazlett, B.A. (1972). Responses to agonistic postures by the spider crab *Microphrys bicornutus*. *Mar.Behav.Physiol.* **1,** 85-92.

4. Autotomy

Heckenlively, D.B. (1970). Intensity of aggression in crayfish, *Orconectes virilis* (Hagen). *Nature (London)* **225,** 180-181.

Holland, C.A., and Skinner, D.M. (1976). Interations between moulting and regeneration in the land crab. *Biol.Bull. (Woods Hole, Mass.)* **150,** 222-240.

Hyatt, G.W., and Salmon, M. (1978). Combat in the Fiddler crabs *Uca pugilator* and *Uca pugnax:* A quantitative analysis. *Behaviour* **65,** 182-211.

Joffe, I., Hepburn, H.R., Nelson, K.J., and Green, N. (1975). Mechanical properties of a crustacean exoskeleton. *Comp.Biochem.Physiol. A* **50A,** 545-549.

Kurata, W. (1963). Limb loss and recovery in the young king crab. *Paralithodes camtschatica. Bull.Hokkaido Reg.Fish.Res.Lab.* **26,** 75-80.

McCarthy, J.F., and Skinner, D.M. (1977). Interruption of proecdysis by autotomy of partially regenerated limbs in the land crab, *Gecarcinus lateralis. Dev.Biol.* **61,** 299-310.

Macmillan, D.L. (1976). Arthropod apodeme tension receptors. In "Structure and Function of Proprioceptors in the Invertebrates" (P.J.Mill ed.), pp. 427-442. Chapman & Hall, London.

McVean, A.R. (1973). Autotomy in *Carcinus maenas* (Decapoda:Crustacea). *J.Zool.* **169,** 349-364.

McVean, A.R. (1974). The nervous control of autotomy in *Carcinus maenas. J.Exp.Biol.* **60,** 423-436.

McVean, A.R. (1975). Autotomy: Mini-Review. *Comp. Biochem. Physiol.* **51**(3A), 497-505.

McVean, A.R. (1976). The incidence of autotomy in *Carcinus maenas* L. *J.Exp.Mar.Biol.Ecol.* **24,** 177-187.

McVean, A.R., and Findlay, I. (1976). Autotomy in *Carcinus maenas:* The role of the basi-ischiopodite posterior levator muscles. *J.Comp.Physiol.* **110,** 367-381.

McVean, A.R., and Findlay, I. (1979). The incidence of autotomy in an estuarine population of the crab *Carcinus maenas. J.Mar.Biol.Assoc.U.K.* **59,** 341-351.

Maynard-Smith, J., and Price, G.R. (1973). The logic of animal conflict. *Nature (London)* **246,** 15-18.

Miller, R.J., and Watson, J. (1976). Growth per moult and limb regeneration in the spider crab, *Chionecetes opilio. J.Fish.Res.Board Can.* **33,** 1644-1649.

Moffett, S.B. (1975). Motor patterns and structural interactions of basi-ischiopodite levator muscles in routine limb elevation and production of autotomy in the land crab, *Cardisoma guanhumi. J.Comp.Physiol.* **96,** 285-305.

Needham, A.E. (1953). The incidence and adaptive value of autotomy and of regeneration in Crustacea. *Proc.Zool.Soc.London* **123,** 111-122.

Paul, J.H. (1915). A comparative study of the reflexes of autotomy in decapod Crustacea. *Proc.R.Soc.Edinburgh* **35,** 232-262.

Prosser, C.L. (1973). "Comparative Animal Physiology." Saunders, Philadelphia, Pennsylvania.

Robinson, M.H., Abele, L.G., and Robinson, B. (1970). Attack autotomy: A defense against predators. *Science* **169,** 300-301.

Skinner, D.M., and Graham, D.E. (1970). Moulting in land crabs: Stimulation by leg removal. *Science* **169,** 383-384.

Skinner, D.M., and Graham, D.E. (1972). Loss of limb as a stimulus to ecdysis in Brachyura (true crabs). *Biol.Bull. (Woods Hole, Mass.)* **143,** 222-233.

Wainwright, S.A., Biggs, W.D., Currey, J.D., and Gosline, J.M. (1976). "Mechanical Design in Organisms." Arnold, London.

Wales, W., Clarac, F., and Laverack, M.S. (1971). Stress detection at the autotomy plane in the decapod Crustacea. 1. Comparative anatomy of the receptors of the basi-ischiopodite region. *Z.Vergl.Physiol.* **73,** 357-382.

Weis, J.S. (1977). Limb regeneration in Fiddler crabs; species differences and effects of methyl mercury. *Biol.Bull. (Woods Hole, Mass.)* **52**, 263-274.

Whitear, M. (1962). The fine structure of crustacean proprioceptors. 1. The chordotonal organs in the legs of the shore crab, *Carcinus maenas*. *Philos.Trans.R.Soc. London, Ser. B* **245**, 291-324.

Wood, F.D., and Wood, H.E. (1932). Autotomy in decapod Crustacea. *J.Exp.Zool.* **62**, 1-55.

5

Compensatory Eye Movements

DOUGLAS M. NEIL

I.	Introduction	133
II.	The Eye as a Motor System	134
	A. Functional Anatomy of the Eye Assembly	134
	B. Neuromuscular Organization	136
III.	Visual Responses	138
	A. Optokinetic Nystagmus	138
	B. Optokinetic Memory	143
	C. Field Effect of Illumination	144
IV.	Responses Induced by the Statocysts	144
	A. Compensation for Tilt	145
	B. Rotational Nystagmus	149
V.	Responses to Substrate	152
VI.	Multimodal Interaction	154
	References	157

I. INTRODUCTION

In stalk-eyed Crustacea, as in other visual animals, control systems exist that stabilize the position of the eye relative to a spatial frame of reference. Movements of the body relative to the earth are detected by various sense organs (eyes, statocysts, and limb proprioceptors), inputs from which initiate appropriate compensatory movements of the eyes, returning them to their original positions. In addition, objects moving in the visual field are tracked by an optokinetic pursuit system, which initiates a sequence of slow and fast eye movements. Image stabilization on the eye may serve to improve the acuity of vision and the recognition of moving objects. Furthermore, by maintaining the ommatidial axes and corresponding receptor structures in a particular spatial orientation, compensatory eye movements facilitate processes of perception that depend on vector abstraction, e.g., that for polarized light.

As a reflex system in which to study the relationship between an initial stimulus and a final behavioral response, eye movements offer several practical advantages. Movements of the animal or of the surroundings provide controllable stimuli, while input and output can be readily quantified as angles or velocities. Data on overall transfer characteristics have been used to construct testable hypotheses about the underlying control systems. The development of particular electrophysiological techniques and viable dissected preparations has made it possible to combine behavioral measures with sampling of the nervous system at various levels between the receptor cells and the eye muscles. It has thus been possible to identify integrative steps with particular mechanical and neuronal events in the reflex pathway. Studies of crustacean eye movements show that the principles governing their control are essentially similar to those for the control system of the vertebrate eye. Common features can be recognized at many levels. Crustacean eye movements provide a model system, in which processes such as sensory coding and neuronal integration are readily accessible to experimental analysis.

II. THE EYE AS A MOTOR SYSTEM

A. Functional Anatomy of the Eye Assembly

In decapods and mysids, the complete eye assembly consists of five main skeletal elements (Parker and Rich, 1893; Schmidt, 1915; Baumann, 1921; Berkeley, 1928; Cochran, 1935; Mayrat, 1956; Elofsson, 1964; Paterson, 1968). A fused middle cylinder provides structural support and is clearly seen in palinurids as a transverse bar (Parker and Rich, 1893). In carid shrimps and astacids it is covered by a rostrum, and in brachyurans it lies beneath an anterior fold of the carapace (Fig. 1A).

Each eye is composed of two segments: a proximal subcylindrical eyestalk, and a distal eyecup bearing ommatidia and containing the optic ganglia. These segments are held out to the side, forward, or upward, and they may be flanked by protective spines above and below (e.g., palinurids) or may project from a socket formed by the lateral cephalothorax (e.g., as-

5. Compensatory Eye Movements

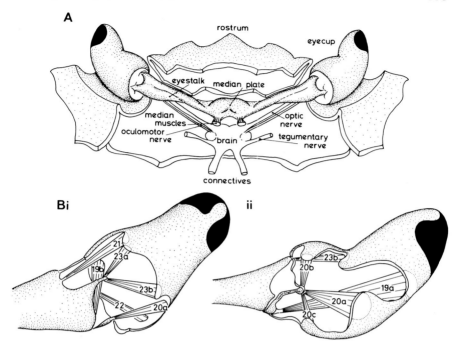

Fig. 1. (A) Assembly of the eye of the crab *Carcinus maenas* as revealed by dissecting away part of the dorsal carapace. The eyestalks unite in the median plate, upon which the median muscles act to roll the eyestalks around their longitudinal axes. The eyecups are borne on the distal ends of the eyestalks and are free to move within the limits of the sockets (from Sandeman, 1977; after Horridge and Sandeman, 1964). (B) Eye muscles of *Carcinus:* right eye from (i) above, and (ii) the side. Muscles 19b, 21, and 20a are the prime movers for horizontal eye movements (from Sandeman, 1977; after Burrows and Horridge, 1968a).

tacids, brachyurans). In many species this resting position of the eye maintains ommatidial axes in a particular orientation relative to external spatial cues, thus providing a reference for visual perception (Horridge, 1966d; Waterman and Horch, 1966).

Protective withdrawal of the eye is a common feature in stalk-eyed crustaceans, and it is achieved by a medial movement beneath the rostrum or (in crabs) by a lateral movement into the socket. Such movements necessarily change the disposition of the ommatidia, although box crabs such as *Calappa hepatica* utilize a vertical guide tube for withdrawal, so that the periscope-like movements of the eye do not change ommatidial orientation (Wiersma and Bush, 1963).

Within the limits set by the exoskeleton, the eyes can move in all directions in space. Movements about the basal joints contribute to rotation of the eye about its own long axis, but major horizontal and vertical excursions occur primarily about the outer joint between the eyestalk and eyecup [although in certain crabs, e.g., *Podophthalmus vigil*, the proximal segment itself is elongated and mobile (Wiersma, 1966)]. The outer joint of the eye has no fixed condyles, but it is surrounded by a flexible arthrodial membrane, which may be stiffened by calcified sclerites. The eye is virtually suspended by the action of the various muscles spanning this joint. The muscles serve both to maintain posture of the eye and to bring about its movements (Burrows and Horridge, 1968a) (Fig. 1B).

B. Neuromuscular Organization

Up to thirteen muscles are involved in producing eye movements, and their arrangement has been determined anatomically in a number of species. Schmidt (1915) recognized antagonists for movements about the three major axes, but Robinson and Nunnemacher (1966) have pointed out the inadequacies of such anatomical schemes: homologous muscles in different species have differing actions, and a single muscle may have subdivisions, each with a separate function. Functional studies confirm these discrepancies. In the crab, *Carcinus maenas*, electrophysiological recordings from different muscles of the eyecup indicate that all muscles are tonically active when the eye is stationary, and that most change their firing rates whenever movement occurs (Burrows and Horridge, 1968a,b,c). Since, additionally, the pivot point of the eye shifts during rotation, even those muscles generating constant tension exert an altered turning moment about the joint (Burrows and Horridge, 1968a). Therefore, it seems incorrect to assign single, particular functions to individual muscles of the eyecup, but rather to regard them as acting in concert, with considerable functional overlap. Nevertheless, particular muscles do make major contributions to a given direction of movement (Fig. 2). Horizontal eye movements in *Carcinus* are dominated by the action of muscle 20a for movements away from the midline, and by the combined actions of muscles 19b and 21 for movements toward the midline (Burrows and Horridge, 1968a). A similar concept of prime movers acting within a framework of cooperative muscle action emerges from electrophysiological studies of the oculomotor system in the crayfish *Procambarus clarkii* (Wiersma and Oberjat, 1968; Hisada and Higuchi, 1973; Mellon, 1977a).

In crabs, the cell bodies of all motoneurons to eyecup muscles are clustered together in a nucleus on the dorsolateral side of the brain (Sandeman and Okajima, 1973b; Silvey and Sandeman, 1976b), and they send axons along both the optic nerve and the oculomotor nerve to muscles of the eyecup (Horridge and Sandeman, 1964; Sandeman, 1964; Burrows and Horridge, 1968a). Certain motoneurons supplying muscles 20a and 21

5. Compensatory Eye Movements

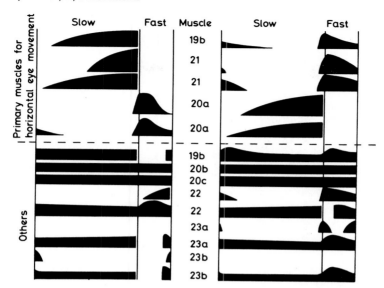

Fig. 2. Activity of eye muscles of *Carcinus* during nystagmus. The relative firing frequencies (represented by the thickness of the bars) during a slow phase toward the midline and a subsequent fast phase (left hand columns), and a nystagmus sequence away from the midline (right hand columns). The fast phases are shown on an expanded time scale. Note that different parts of individual muscle blocks have different patterns of activity. (From Sandeman, 1977; after Burrows and Horridge, 1968a.)

(horizontal prime movers) have been identified from their intracellular responses to antidromic stimulation of appropriate nerve endings (Silvey and Sandeman, 1976b).

In crayfish (*Procambarus clarkii*), a detailed analysis of the innervation patterns of eyecup muscles has been made by use of the cobalt backfilling technique. This study has revealed that the functional subdivision of these muscles is reflected in an anatomical separation of their motoneurons (Mellon, 1977a). Axons with cell bodies lying in three distinct clusters in the supraesophageal ganglion pass as separate bundles along the optic and oculomotor nerves to supply particular muscles involved in different movements of the eye (Fig. 3). Thus, cells from the lateral cluster (LC) exclusively supply four muscles that move the eye horizontally. Axons from the anterior motor cluster (AMC) pass along the optic nerve to emerge in the optic nerve motor bundle (ONMB) and distal motor bundle (DMB), which supply the levator and depressor divisions, respectively, of the subset of muscles responsible for vertical movements. Other fibers routed through ONMB include three giant axons that mediate the eye-withdrawal reflex, and two axons to a major levator muscle that have their cell bodies in the most proximal optic ganglion (the medulla terminalis). The destinations of

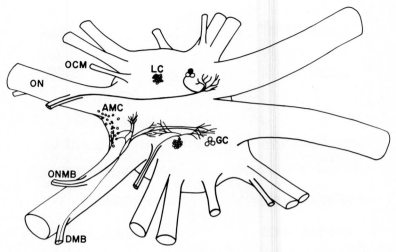

Fig. 3. Diagram of the brain of a crayfish (*Procambarus clarkii*), showing the relative disposition of the somata of the oculomotor neurons. The axons of representative neurons are shown leaving the ganglion, and major regions of the dendritic arborization are included. OCM, oculomotor nerve; ON, optic nerve; ONMB, optic nerve motor bundle; DMB, distal motor bundle; AMC, anterior motor cluster; LC, lateral motor cluster; GC, giant cell cluster. (From Mellon, 1977a.)

all identified ONMB axons have been traced. Muscles that are functional synergists receive both common and private motor innervation, but no peripheral inhibitory input. Within each muscle, there are a number of fiber types that, according to their various structural and physiological properties, may be classified as phasic, intermediate, or tonic (see Chapters 2 and 4 of Volume 3). There is some evidence that, as in *Carcinus* (Burrows and Horridge, 1968a,b,c), axonal supply to different fiber types is not the same, with phasic muscle fibers receiving a predominant fast-axon input and tonic muscle fibers a slow-axon or mixed input (Mellon, 1977a). Thus, for the crayfish we have a reasonably complete neuronal wiring diagram of the eye muscles, thereby providing an opportunity to analyze the mechanisms underlying compensatory responses at the cellular level.

III. VISUAL RESPONSES

A. Optokinetic Nystagmus

Optomotor reactions, in which movements of the body compensate for displacements of images over the eye, are well documented in higher crustaceans (Bethe, 1897; von Buddenbrock and Friedrich, 1933; Hassenstein, 1954). These movements stabilize the primary body orientation (Fraenkel and Gunn, 1961; Schöne, 1975a) and are most strongly developed in active swimming species (De Bruin, 1956). In stalk-eyed crustaceans, distinct compensatory reactions of the eyes that tend to stabilize images on the retinae also take place. When a freely moving decapod or mysid turns, both eyes, after an initial flick in the direction of intended motion, maintain their positional relationship to the environment by moving slowly against the direction of turn. Eventually, toward the end of the traverse, the eyes again flick in the direction of body movement before resuming compensatory rotation (Dijkgraaf, 1956b). This sequence of slow compensatory eye movements and fast return flicks constitutes optokinetic nystagmus; it is distinct from other small eye movements (tremor, saccades, waving), which are concerned with the prevention of visual adaptation (Horridge and Sandeman, 1964; Horridge, 1966d; Barnes and Horridge, 1969a; Sandeman, 1978b).

Optokinetic nystagmus can readily be evoked experimentally by rotating a large contrasting visual field, routinely a pattern of black and white stripes, around a stationary animal. Complete nystagmus occurs only in the horizontal plane, for although many species make compensatory eye movements in other planes, these are not accompanied by return flicks (Wiersma and Oberjat, 1968; Hisada et al., 1969; Neil, 1975b). This may be due to the influence of inputs from the statocysts: removal of these organs of balance from the crayfish *Procambarus* leads to the appearance of a full nystagmus in the vertical plane (Higuchi, 1973).

Both eyes perform horizontal optokinetic nystagmus, and their movements may be completely conjugate, as in the crab *Carcinus* (Horridge and Sandeman, 1964; Barnes and Horridge, 1969b), loosely coupled, as in the crayfish *Procambarus* (Wiersma and Oberjat, 1968), or predominantly independent, as in the spiny lobster, *Panulirus interruptus* Randall (York et al., 1972a). To elicit a full-sized optokinetic response, the visual stimulus presented to the eye must subtend a minimum field, both horizontally and vertically (Kunze, 1963, 1964; Waterman, 1961). In some crabs, e.g., *Carcinus* (Horridge and Sandeman, 1964) and *Leptograpsus variegatus* (Sandeman, 1978a), certain regions of the eye are specifically sensitive to optokinetic stimuli, whereas in species of *Uca* (Kunze, 1964), all regions of the eye show equal responsiveness.

Earlier workers utilized optokinetic nystagmus as a convenient measure of visual acuity (von Buddenbrock and Friedrich, 1933; Hassenstein, 1954). However, our knowledge of the performance of the optokinetic control system in crabs and its underlying neuronal circuitry has been extensively enlarged through the work of G. A. Horridge, D. C. Sandeman, and co-

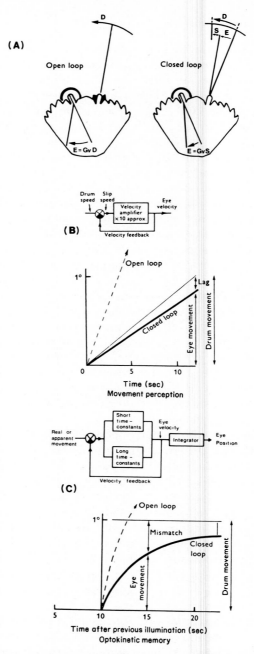

Fig. 4. Optokinetic performance of the crab *Carcinus*. (A) Experimental open-loop and closed-loop situations. In both, the right seeing eye drives the left shaded eye. The loop is opened by clamping the seeing eye to the carapace. Under closed-loop conditions, the eye

workers (reviews: Horridge, 1966a; Sandeman, 1977). In darkness, the eyes of *Carcinus* drift aimlessly (Barnes and Horridge, 1969a), but when presented with a moving striped drum or pinlight, they follow the movement, always lagging by a small amount, over a wide range of stimulus speeds (Horridge and Sandeman, 1964; Horridge, 1966d,e) (Fig. 4A). The responses of *Carcinus* to slow movements is impressive, and it has been directly demonstrated that their eyes can follow the path of the sun across the sky (Horridge, 1966c; Barnes and Horridge, 1969a).

The difference between the velocity of the visual stimulus and that of the eye (i.e., the "slip" of the pattern over the eye) effectively drives the optokinetic following response. This can be demonstrated under open-loop conditions by fixing one eye in its socket and measuring the movement of the other blinded, but freely moving eye (Fig. 4B). Under these conditions, the slip speed of the pattern over the seeing eye is equal to the drum speed, and the blinded eye, driven by the strong contralateral linkage, moves at a velocity determined by the forward gain of the control circuit. The open-loop response is nonlinear, and at very low velocities the gain approaches values of 25 (Fig. 4C). The control system inferred from both closed-loop and open-loop analyses incorporates an amplifier (of variable gain) and a visual negative feedback loop, which acts to minimize the error between input and output (Horridge, 1966d).

No proprioceptors have been detected at the eyestalk–eyecup joint in *Carcinus* (Sandeman, 1964; Horridge and Sandeman, 1964), *Procambarus* (Wiersma and Oberjat, 1968) or *Panulirus* (York et al., 1972a), and several lines of evidence suggest that proprioceptive inputs are not involved in the eye control system in crabs. Thus, an eye pushed to a new position in darkness "locks" onto a striped pattern on reillumination and does not return to its original position. All forced movements of a seeing eye are, in fact, interpreted as if they were equivalent movements of the visual field relative to the crab (Horridge and Sandeman, 1964; Sandeman, 1977). The only documented involvement of proprioceptive feedback in the control of eye position in crabs is for movements of the eyestalk about its own long axis; these take place about the basal joints (Steinacker, 1975). Slow phase of nystagmus is thus under the sole direction of the visual negative feedback loop, although this is apparently also disregarded under certain circum-

---

movement (E) lags behind the drum movement (D) by an amount (S) that measures the forward gain of the system (G_v). Under open-loop conditions D = S. (B) The optokinetic movement response plotted as a function of time. The block diagram shows the inferred control circuit. (C) The optokinetic "memory" response. Movement of the eye has the effect of reducing the mismatch. The block diagram incorporates amplifiers with different time constants. (A, D, E from Horridge, 1966d; B,C after Horridge, 1966a; W. J. P. Barnes, unpublished observations.)

stances, e.g., during extension of the eye after a protective withdrawal (Horridge and Burrows, 1968b).

For a given direction of optokinetic response, different groups of muscles are active in the two eyes: muscles 19b and 21 move one eye toward the midline, muscle 20a moves the other eye away from the midline. Discharges in other muscles cannot be clearly related to eye position (Burrows and Horridge, 1968a) (Fig. 2). Each of the principal muscles is made up of mixed motor units and generates a progressive turning force as a result of an orderly recruitment of an early firing "tonic" motoneuron and a later firing "phasic" motoneuron (Burrows and Horridge, 1968a; Sandeman et al., 1975a). At some point in the slow phase, often toward the end of the traverse, a rapid return flick is initiated by a control mechanism that is triggered neither by eye position nor by visual feedback (Horridge and Sandeman, 1964; Horridge and Burrows, 1968b; Barnes and Horridge, 1969b; Sandeman et al., 1975a). Initiation of fast phase is, however, conditional upon the recruitment of the later firing motor units. Sandeman et al. (1975a) propose that these motoneurons are coupled to an interneuronal network, the "fast phase generator," which at threshold briefly drives the motoneurons responsible for fast phase. It is consistently found that the eye traversing toward the midline begins its fast phase 30–40 msec before the other eye (Horridge and Burrows, 1968b; Barnes and Horridge, 1969b). Possibly single or paired central generators govern the fast phases of both eyes, while receiving stronger synaptic input from premotor elements linked to muscles 19b and 21 than from those linked to muscle 20a (Sandeman et al., 1975a). Muscle activity during the fast phase is not merely the mirror image of the slow-phase pattern, but represents the expression of a separate motor program (Burrows and Horridge, 1968a; Horridge and Burrows, 1968b), which involves central inhibition of both slow-phase motoneurons (Sandeman and Okajima, 1973b) and a number of premotor elements (Sandeman et al., 1975a,b). In this way, the system is essentially reset for another slow phase.

Despite a detailed knowledge of the properties of visual interneurons in crustaceans (see Chapter 1 of this volume), surprisingly few established optokinetic interneurons have been reported (Wiersma and Yanagisawa, 1971). This can be partly explained by the technical difficulties of recording such units in a moving eyecup, a problem that can be avoided by monitoring the response as the isometric force generated by a clamped eye. Using this arrangement, Sandeman et al. (1975b) have demonstrated three classes of neurons with axons running between the internal medulla and the medulla terminalis; these neurons are unidirectionally sensitive to moving stripes. Some units fire during the slow phase, others are excited prior to the fast phase. From their latencies to firing, time constants of decay, and the close match between their discharge patterns and torque development, a causal link between these interneurons and the optokinetic response is strongly indicated. In at least two classes of optokinetic interneuron, sensitivity to velocity spans the whole range of normal movement, suggesting that range fractionation does not occur at this level. Thus, contrary to an earlier suggestion that velocity detectors sensitive to slow and fast movements may be linked in parallel to tonic and phasic motor systems (Horridge and Burrows, 1968a), it seems more probable that sensory interneurons with wide velocity sensitivity are coupled to a motor system in which early and late components are recruited at different threshold levels (Sandeman et al., 1975a,b).

B. Optokinetic Memory

A phenomenon closely related to optokinetic eye movements is the so-called "optokinetic memory", first described in *Carcinus* (Horridge and Shepheard, 1966; Horridge, 1966a), but probably of widespread occurrence (Wiersma and Hirsh, 1974; Sandeman and Erber, 1976). In a typical experiment, a crab is allowed to view a stationary pattern of wide-field stripes, and then, in a period of darkness lasting for several minutes, the pattern is displaced by a small amount. On reillumination, the new stationary position is revealed. The crab, as if it were interpreting the change in position of the visual field as movement, moves its eyes in the direction of apparent motion. The final position of the eye is reached in 1–2 min and always falls short of the new position of the striped pattern (Horridge, 1966a,d).

If, on reillumination, the striped drum is oscillated about its new position, the eye moves in the direction of displacement with a superimposed oscillation. Thus, the optokinetic response is still operating and must interact with optokinetic memory as the eyes sweep across stationary stripes. By comparing closed-loop and open-loop responses, Horridge (1966a) concludes that the memory system incorporates the optokinetic velocity amplifier, and a model that fits all the data places the movement and memory systems in parallel (W. J. P. Barnes, unpublished). A concept central to this theory is that the two stationary drum positions are correlated by the mechanism that infers movement, and that this apparent motion is processed by the optokinetic system as if it were real motion (Horridge, 1966a; Sandeman and Erber, 1976).

Memory formation is a function of exposure time, being detectable within 0.5 sec and building up to maximum strength in 30 sec in *Carcinus* (Horridge, 1966a) and *Pachygrapsus crassipes* (Wiersma and Hirsh, 1974). In *Leptograpsus,* an optimum viewing time of 15 sec is reported, longer exposures producing weak responses, or no responses at all (Sandeman and Erber, 1976). Memory decays exponentially during the period of darkness, but measurable responses are still present after periods of 10 min and longer

(Horridge and Shepheard, 1966; Wiersma and Hirsh, 1974; Sandeman and Erber, 1976). It is evident that there is a mechanism in the visual system with long time constants that is able to retain, for a limited period, a pattern of excitation representing the original position of the visual stimulus.

C. Field Effect of Illumination

In addition to an optokinetic response, motor fibers to eye muscles of crayfish also exhibit distinct reactions to light falling on different parts of the eye (Wiersma and Oberjat, 1968; Hisada and Higuchi, 1973). Fibers designated "head-up" because they fire when the body is pitched in this direction are excited by light on the front rim of the eye and inhibited by light on the back rim. The reverse effect occurs in the antagonistic "head-down" fibers. This field-effect of illumination produces eye movements toward the source of illumination. Thus, this response of the eye bears a close relationship to the "dorsal light reaction," whereby a crustacean, in the absence of statocyst stimulation, orients its body with reference to the direction of illumination (Fraenkel and Gunn, 1961). There is some evidence that the field effect is mediated in part by a motor pathway separate from that for optokinetic responses. Motoneurons in the medulla terminalis of crayfish (see Section II,B) are activated by illumination of the eye and produce reflex effects that maintain the eye's horizontal position against the imposed roll of the body in overhead light (Mellon and Lorton, 1977).

IV. RESPONSES INDUCED BY THE STATOCYSTS

The statocysts of higher Crustacea provide powerful inputs both to equilibrium reflexes, which stabilize the position of the eye by restoring normal posture of the body (Kennedy and Davis, 1977), and to compensatory eye movements, which provide a more immediate and direct regulation of the spatial orientation of the eye (Schöne, 1975a). Two categories of eye movement can be distinguished: counterrolling responses, which compensate for body tilt about horizontal axes, and rotation nystagmus, which occurs when some species of crab and lobster are turned about a vertical axis. These distinct reflex effects closely resemble the static and dynamic vestibulo-ocular responses in vertebrates. They are driven by receptor subsystems responsive to linear and angular accelerations, respectively, that are analogous to those of the vertebrate vestibular apparatus (Carpenter, 1977; Precht, 1978).

A. Compensation for Tilt

Although descriptions were provided by several earlier workers (Clark, 1896; Bethe, 1897; Lyon, 1900; Prentiss, 1901; von Buddenbrock and Friedrich, 1933), the first reliable quantitative studies concerning the responses of the eyes of crustaceans to tilt of the body were those of Schöne (1951, 1954, 1957), who utilized them to investigate the mechanism of statocyst function (reviews: Cohen and Dijkgraaf, 1961; Schöne, 1971, 1975a). During a full roll of the body, the eye's position shows a sinosoidal relationship to the body's position that directly corresponds to the changing shearing force acting on the mass of the statolith. Thus, a linear relationship exists between the initial mechanical stimulation and the final behavioral response (Fig. 5).

Direct manipulation of statolith hairs demonstrates that a single statocyst elicits coordinated movements of the two eyes in both directions, although under normal conditions the sensory inputs from the two organs of balance summate centrally in a simple algebraic manner. This is particularly well demonstrated in those species (*Palemonetes, Crangon, Astacus*) in which the floor of each statocyst is tilted with reference to the major axes (Fig. 5A): unilateral removal of a statolith shifts the maxima and minima of the eye-response curves to positions corresponding to the vertical and horizontal positions of the intact statocyst, rather than of the body (Schöne, 1954; Fig. 5B,C).

The direct relationship between the input and output of the eye reflex applies not only to magnitude of tilt, but also to its direction. In the crayfish, the plane of the reflex eye movement closely matches the plane of imposed tilt about all horizontal axes between those of pure pitch and roll (Stein, 1975). Controlled manipulation of statolith hairs produces eye movements that are directed into a plane corresponding to the polarization plane of the stimulated hairs (Stein, 1975; Schöne and Steinbrecht, 1968) (Fig. 6). It thus appears that the statocysts are separately abstracting information about the direction and magnitude of tilt (Schöne, 1975b).

The directional component is embodied in the pattern of excitation around the "sensory crescent," which has a peak at those hair locations where polarization and shear directions coincide. We do not know how such a raster system defines stimulus direction: inputs from sensory cells may be fed separately into central pools for yaw, pitch, and roll, according to their polarization, or the ordered circular distribution of afference may be compared with a central circular reference pattern (Mittlestaedt, 1972, 1975). The magnitude of tilt in a particular direction is proportional to the overall excitation of sensory units, the largest contribution again being made

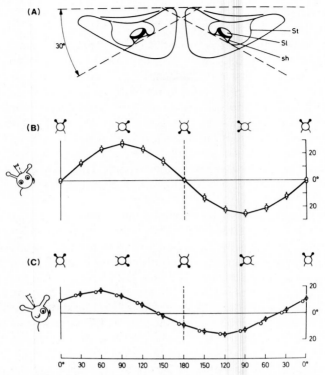

Fig. 5. (A) Section through the antennules of *Palaemonetes varians* at the level of the statocyst. The floor of each cyst is tilted by 30°: sh, sensory hair; Sl, statolith; St, statocyst. (B and C) Compensatory eye positions of *Palaemonetes* for various degrees of rotation of the body about the longitudinal axis. Ordinate, deviation of bisector of the angle between the eyes from the dorsoventral axis (see inset on left); abscissa, angle between dorsoventral body axis and gravity (see insets above); vertical lines, threefold standard deviation; small circles, best fitting sine curve. (B) Intact animals. (C) After removal of statolith on one side. The maximum reaction is here shifted by 30° to the position in which the floor of the intact cyst stands vertical, and shear is maximal on the sensory hairs. (From Markl, 1974; after Schöne, 1954.)

by those receptors polarized in the direction of shearing. For this reason, stimulation of small groups of hairs is sufficient to produce eye movements that are one-third as large as the full response to tilt of the body (Stein, 1975).

Together with results from other species (*Carcinus:* Horridge, 1966b; *Praunus:* Neil, 1975a,c), these findings provide a consistent picture of the static eye response, which is in good agreement with the observed properties of statolith hair receptors (Cohen, 1955, 1960; Patton, 1969; Ozeki et al., 1978). However, available evidence about the dynamic properties of the reflex suggests that eye movements reflect less clearly the neurally coded

5. Compensatory Eye Movements

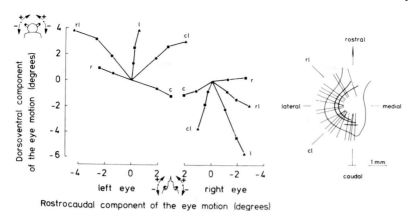

Fig. 6. The three-dimensional eye position of the crayfish *Astacus leptodactylus* resulting from stimulation of various groups of hairs in the statocyst (see inset). Circles, squares, and triangles on each curve mark eye positions corresponding to 40, 70, and 100 μm of hair displacement. Inset: scheme of statocyst indicating the arrangement of the polarization planes of sensory hairs. Thick lines delimit locations of the hairs; radial lines indicate the mean of the polarization plane of two to eight hairs within a 5.6° sector. (After Stein, 1975.)

output of statoreceptors. In crayfish, the eye-control system behaves as a low-pass filter: gain and phase-shift to sinusoidal body oscillations fall off with frequencies of oscillation above 0.2 Hz (Fay, 1975). Eye position phase-lags body position over the whole range tested (0.02–4.0 Hz), but discharge of the lith hair receptors phase-leads body position by large values at frequencies below 1 Hz (Knox, 1969). Integration of the input signal is necessary to explain this relationship, and although no systematic study has yet been made, available evidence identifies a significant integrating step in the mechanics of the eyestalk (Mellon and Lorton, 1977).

The siting of the statocyst on a moveable appendage (the antennule in decapods, the inner ramus of the uropod in mysids) presents the problem of distinguishing between movement of the appendage and movement of the body. The large and regular movements of the antennules of the spiny lobster, *Panulirus argus*, are monitored proprioceptively by a strand organ, and the interaction of proprioceptive signals and those from the statocysts has been studied by measuring the eye reflex induced under different stimulus conditions (Schöne and Schöne, 1967). Tilting the whole animal in the pitch plane activates only the statocyst, which, if intact, induces a compensatory roll of the eyes (Fig. 7, Aa). Tilting the antennule alone produces no eye response, although both statocyst and proprioceptor are activated (Fig. 7, Ba). Tilting the body alone, with the antennule held, stimulates only the proprioceptor, yet eye movements of normal magnitude now occur in the anticompensatory direction (Fig. 7, Ca). An interpretation consistent with

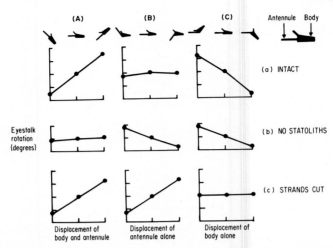

Fig. 7. Interaction between the statocysts and proprioceptors of the antennule in the control of eye movements in the spiny lobster *Panulirus argus*. Eye rotation is plotted against displacement of: A, body and antennules; B, antennules alone; and C, body alone. Experiments were performed with: a, the intact animal with normally filled statocysts; b, animals lacking statoliths; and c, animals with destroyed proprioceptive strands in the antennular basal segments. No eye compensation occurs when neither sensory system is stimulated (Ab, Cc) and when both are stimulated simultaneously (Ba). (After Schöne and Schöne, 1967.)

these and other results is that the inputs from the statocyst and proprioceptor act antagonistically, and they are so calibrated that their effects cancel when the antennule moves alone. Under natural conditions, eye movements will occur only when the body changes its orientation with respect to gravity. The fact that proprioceptive inputs alone cause eye movements suggests that the sensory inputs converge on a central nervous comparator. A cybernetic model of this system has been presented by Schöne (1975a).

Neurophysiological studies at the motor level of eye reflexes induced by stimulation of the statocysts are in accord with behavioral measurements but have so far added little to our knowledge of the underlying control mechanism. Myographic analysis of eye movements in crabs in response to static tilts of the body (Burrows and Horridge, 1968b) reinforces the concept that a given movement is brought about by combined action of the muscles, and that separate pathways of motoneurons for pitch and roll responses do not exist. Optomotor fibers encountered in the optic peduncle of several species during probing with needle electrodes have been classified according to the stimulus that excites them (Wiersma and Oberjat, 1968; York et al., 1972b; Wiersma and Fiore, 1971a; Hisada and Higuchi, 1973; Higuchi and Hisada, 1973). "Head-up" (HU) fibers increase their discharge rates when the animal is pitched head-up and are inhibited when the animal is pitched down, but they show no change from resting levels when the animal is rolled. "Head-down" (HD) fibers have opposite firing characteristics, and distinct categories of fiber responding to "side-up" (SU) and "side-down" (SD) roll tilts also exist. With full-circle turns, firing in these optomotor fibers changes in an approximately sinusoidal manner, and they exhibit properties (such as hysteresis and phasic-tonic decay) that are closely similar to those of the final eye movement itself (Wiersma and Oberjat, 1968).

The needle sampling technique provides an incomplete picture of the total population of fibers and gives no certain indication of the exact destination of a given fiber. A more promising approach is offered by the convenient segregation of motoneurons into different bundles of nerves in the eye of the crayfish (Mellon, 1977a). It has been possible to demonstrate reciprocal discharge in identified antagonistic motoneurons during oscillations of the body (Mellon and Lorton, 1977). Such a preparation should facilitate a more precise analysis of discharge patterns in a number of identified units and a greater understanding of their relationship to firing in sensory and other premotor elements.

B. Rotational Nystagmus

In certain crabs and lobsters, rotation of the legless animal in darkness about a vertical axis induces a nystagmus reaction of the eyes, in which the slow phase compensates for the imposed disturbance (Bethe, 1897; Dijkgraaf, 1956b,c; Sandeman, 1977). These movements persist only during periods of angular acceleration, being initiated by fluid movements within the statocyst sac that deflect sensory hairs projecting inward from the wall (Cohen and Dijkgraaf, 1961; Sandeman, 1975). These fine thread hairs are apparently absent in such species as crayfish and rock lobster, which show no rotational nystagmus (Dijgraaf, 1955a,b; Wiersma and Oberjat, 1968; York et al., 1972a). The region of the statocyst sac bearing thread hairs is modified to form canal-like structures, primitively formed in *Homarus* and *Nephrops* (D. M. Neil, unpublished observations), but highly developed in crabs where two distinct canals are formed (Sandeman and Okajima, 1972; Sandeman, 1975).

Resemblances between the statocyst canal system in crabs and the nonacoustic labyrinth in vertebrates were first noted by Hensen (1863), reemphasized by Dijkgraaf (1956c), and now amply confirmed by physiological studies of Sandeman and co-workers (reviews: Sandeman, 1975, 1976, 1977). An isolated eye-brain preparation has been developed, which allows intracellular recordings to be made from oculomotor neurons during applied rotations (Sandeman and Okajima, 1972), or, more conveniently, by controlled irrigation of the canals (Sandeman and Okajima, 1973a; Silvey and

Sandeman, 1976a). Both stimuli induce normal slow and fast sequences of nystagmus, and it has been established that the response in yaw is driven exclusively by the upper thread hairs lying in the horizontal canal (Sandeman and Okajima, 1972, 1973a; Silvey et al., 1976). Each of the two populations of thread hairs, sensitive to opposite directions of turn, has a specific excitatory connection to the motoneurons of the group of muscles that moves the eye in the compensatory direction, and each has an inhibitory connection that hyperpolarizes the motoneurons of the antagonists (Sandeman and Okajima, 1973b; Silvey and Sandeman, 1976b; Silvey et al., 1976). There is good evidence from sensory and motor firing patterns, neuronal transmission delays, and following frequencies of motor nerves that the excitatory input from thread hairs to eye motoneurons is monosynaptic (Silvey and Sandeman, 1976b,c). Also, profiles of pre- and postsynaptic elements in cobalt-filled preparations have closely matched projections in the brain (Sandeman and Okajima, 1973b).

A model for rotational nystagmus has been proposed in which inputs from thread hairs not only drive slow-phase motoneurons directly, but also synapse onto an interneuronal fast-phase generator which, at threshold, initiates the fast-phase motor output (Sandeman and Okajima, 1973b). The similarity between this model and that for optokinetic nystagmus is striking, and it may be that inputs from visual systems to statocyst systems converge centrally onto common fast-phase generators.

A detailed analysis of the slow-phase component has been made by applying small-amplitude sinusoidal oscillations to the whole crab or to isolated eye-brain preparations (Silvey and Sandeman, 1976a; Janse and Sandeman, 1979a,b). The phase and gain of the eye reflex depend on frequency and amplitude of oscillation, and also on the pitch position of the animal. Phase-lock of eye and body movements and maximum gain of the response occur when the crab is held in its normal pitch position (horizontal canal tipped 20-40° head-up) and oscillated with a peak-to-peak angular displacement of 12.5°, at a frequency of 0.7 Hz (Janse and Sandeman, 1979a). Of major interest is the observation that under these conditions of phase-lock, the thread hairs of the upper canal move in direct antiphase to the body. Since the eye also moves in antiphase to the body, it is not necessary to invoke further significant delays in the nervous system or in the mechanics of the eyecup. The hydrodynamics of the canal and the mechanical properties of the thread hairs together perform the two integrations on the initial input, from acceleration to displacement, which account for the phase relationships of the overall reflex (Janse and Sandeman, 1979a). Apparent contradictions between these findings and earlier results (Silvey and Sandeman, 1976a,b,c; Silvey et al., 1976) have been resolved by the demonstration that the position of thread hairs is affected by linear accelerations.

5. Compensatory Eye Movements

When the statocyst is tilted, the hairs are deflected from a position in which their receptor cells fire tonically (thereby coding displacement) to positions in which they fire phasically (thereby coding velocity) (Janse and Sandeman, 1979b; Fig. 8). This nonlinear behavior of thread hair receptors adequately accounts for the observed changes in the phase and gain of the overall reflex with pitch position of the statocyst. In positions in which receptors are known to be operating in their velocity-sensitive range, the eye movements actually phase-lead a compensatory relationship (Silvey and Sandeman, 1976a; Janse and Sandeman, 1979a). The control of rotational slow-phase nystagmus, therefore, has an elegant simplicity, relying on the polarization and mechanical properties of thread hairs to encode the vector properties of the stimulus in an integrated form, and on a simple hard-wired neuronal circuitry to preserve directionality and timing.

Instructive comparisons can be made between the canal system of crabs and the non-acoustic labyrinth of vertebrates. Although basic similarities exist in the hydrodynamics of these fluid-filled organs of balance, thread hairs of crabs perform two integrations on an impressed angular acceleration of the canal, whereas the cupula-endolymph system of the semicircular canals performs a single integration, angular displacement of the cupula being proportional to velocity at intermediate frequencies (Mayne, 1974). In both groups, the gain and phase of the vestibular eye reflex is frequency-dependent, and studies of the system in crabs emphasize that changes in the hydrodynamic properties of the canal and discharge characteristics of the receptors may separately underlie these effects.

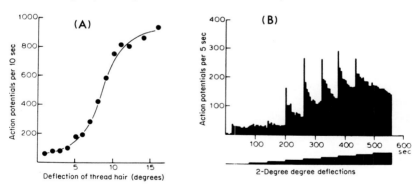

Fig. 8. The effect of gravity on the responses of the thread hairs in the crab, *Scylla serrata*. (A) Response of one receptor, measured at the phasic peak to increasing deflections of the thread hair from its rest position. The response curve has a sigmoid shape with a linear region between 5 and 10° where the receptor codes displacement. Outside this region the receptor is velocity sensitive. (B) Histogram of the receptor response to successive 2° deflections through its working range. Note the large phasic component which saturates beyond 10° deflections. (From Janse and Sandeman, 1979b.)

V. RESPONSES TO SUBSTRATE

In addition to their response to visual and gravitational inputs, the eyes of certain crustaceans are also under the direction of a third avenue of sensory input derived from leg proprioceptors, which provide information about contact with, and orientation to the substrate. Contact with the substrate modifies responses of the eyes to gravity and light (Alverdes, 1926; Kühn, 1914; Stein and Schöne, 1972; Schöne and Neil, 1977), and changes in position of the legs relative to the body produce directed movements of many appendages, including eyes (Schöne et al., 1976). Rotation of a horizontal platform under the legs of a blinded spiny lobster, *Palinurus elephas*, elicits a typical nystagmus, with slow phases in the direction of the imposed rotation (Dijkgraaf, 1956a). Systematic responses also occur about horizontal axes, being well developed in the spiny lobsters, *Palinurus elephas* (Schöne et al., 1976) and *Panulirus interruptus* (York et al., 1972b), and in species of crayfish (Stein and Schöne, 1972; Fay, 1973; Mellon and Lorton, 1977), but less evident in genera such as *Nephrops* and *Homarus* (D. M. Neil, unpublished observations). The effect that orientation to substrate has on movements of a crab's eyes awaits systemic investigation.

A convenient way in which to study responses of eyes to leg proprioceptor stimulation in isolation from other stimuli is to tilt a pivoted platform beneath the legs of a fixed and blinded animal. In response to a tilt of the platform about the long axis of the body, the two eyes of *Palinurus* move conjugately in the direction of tilt, maintaining their positional relationship to the substrate. Movements of single legs produce measurable responses of both eyes, and there is a systematic difference in the contributions of individual legs to the overall response (Schöne et al., 1976; Schöne and Neil, 1977). When leg inputs are combined, their weighted effects summate in an algebraic manner (Schöne et al., 1978), and even complex combinations of stimuli produce eye movements that faithfully follow the resultant of the mechanical inputs (Fig. 9).

As with the statocyst lith system, leg proprioceptors not only code the magnitude of stimulus but also define its plane of action through a synoptic combination of the responses in different legs. Reflex effects are also comparable, since eye movements are directed into the appropriate compensatory plane. Thus, tilt of the platform parallel to the long axis of the body causes the eyes to move vertically, while tilt parallel to the transverse axis of the body induces roll of the eyes about their own long axes (Schöne et al., 1976).

Some insight into the proprioceptive basis of responses to substrate has been obtained from the blocking and controlled movement of specific leg joints, and from the stimulation and ablation of specific receptor strands. The

5. Compensatory Eye Movements

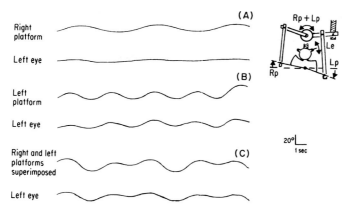

Fig. 9. Eye movement of the spiny lobster *Palinurus elephas* in response to platform tilt. Response of the left eye to tilt of contralateral (A) and ipsilateral (B) halves of a split platform and to simultaneous movement of both halves at different speeds. (C) The complex waveform of the summed stimuli (monitored by the potentiometric arrangement shown in the inset) is faithfully duplicated by the movement of the eye. (H. Schöne and D. M. Neil, unpublished observations.)

receptors at the T-C joint, first described by Alexandrowicz and Whitear (1957), are known to monitor protraction and retraction of the legs (Bush, 1976) and probably underlie the nystagmus responses of the eyes to turntable rotation (Fraser, 1975; Mellon and Lorton, 1977). Of the joints displaced in the vertical plane, C-B exerts the greatest effect on reflex eye movements: flexion and extension at this joint alone (even in an autotomized leg) is sufficient to generate strong eye responses (Schöne et al., 1976; Scapini et al., 1978). The CB chordotonal organ is certainly involved, but its output is apparently gated by stimulation of other receptors at the C-B joint, probably the levator and depressor receptors (Clarac et al., 1976). Changing forces on the leg can, in the absence of joint movement itself, generate reflex eye movements (Neil et al., 1979), but it is not known whether internal muscle tension receptors or cuticular stress detectors are involved.

Receptors at other joints, notably M-C and T-C, modulate the vertical eye reflex such that a given change in C-B angle elicits a larger eye movement when M-C and T-C angles are open than when they are closed (Scapini et al., 1978). Although the neuronal mechanism of this interaction is not yet clear, functionally such effects serve to generate eye movements appropriate to changes in the body-to-substrate orientation, despite changes in the geometrical arrangement of the legs and their standing distance from the body.

The siting of the main detector at the most proximal vertical joint of the leg (C-B) confers upon it a dual sensitivity to the changing stance of the legs on the substrate and to lateral movements of the body against the legs. This latter response may be of biological significance, since long-bodied animals, such as lobsters, are inherently unstable in roll (Alexander, 1971) and must often be subjected to lateral water currents, such as the surge associated with surface waves (Herrnkind and McLean, 1971). In depths occupied by spiny lobsters (5–30 m), such surge effects may reach significant proportions, swell heights of 2–5 m producing surge velocities of 10–60 cm/sec at 20 m (Walton and Herrnkind, 1977).

A frequency analysis of the eye reflex using sinusoidal platform oscillations reveals that at low frequencies the system has high gain, and it phase-leads leg movements by up to 40° (Neil and Schöne, 1979). Interestingly, at the frequency that produces phase-lock (0.25 Hz), the system also displays unity gain, so that eye movements are in perfect compensation for imposed substrate tilt. It is not known if this frequency of tilt simulates some natural stimulus to which the eye reflex is tuned, but it is perhaps significant that the natural frequency of offshore waves is in the range of 0.2–0.3 Hz (Hill, 1962).

Little is known so far about the neuronal mechanisms underlying eye reflexes to substrate tilt. The dynamic characteristics of the system are consistent with the receptor properties of the leg proprioceptors involved (Neil and Schöne, 1979; Bush, 1965), and the rather precise correlation of output with input makes it likely that the pathway between receptors and eye motoneurons is a direct one. The involvement of interneurons, which is suggested by Wiersma's (1958) finding that fibers in the circumoesophageal connectives respond in a "push–pull" fashion to inputs from the C-B joints of the legs, remains to be investigated.

VI. MULTIMODAL INTERACTION

As described in Sections II, III, and IV, the primary orientation of the eyes in stalk-eyed Crustacea is under the control of several sensory systems, and it is maintained with reference to a number of spatial parameters: visual field, gravity, and substrate. These frames of reference generally bear a constant relationship to one another, and the corresponding reflex behaviors produce complementary effects. If an animal standing on a substrate is passively displaced about any axis, the statocyst will report change in body position relative to gravity, the leg proprioceptors will report angular change between body and legs, and the visual receptors will report both the relative displacement of the image over the eye and the changing direction of general illumination. All these reflex effects induce eye movements in the same compensatory direction, and by their synergistic action they will stabilize the spatial orientation of the eye.

5. Compensatory Eye Movements

In some situations the relationship between spatial references is altered, e.g., if an animal stands on an inclined surface or in an enclosed space without overhead light. The visual reference will also change when objects move in the visual field, tending to induce tracking movements of the eye. Final eye position under these conditions is then dependent on the strength of the different reflex inputs and on the nature of their interaction. Such properties have been studied in a number of crustaceans by measuring the eye response to the different sensory inputs delivered separately and in combination.

In mysids (Neil, 1975a,b), decapod shrimps (Schöne, 1961), and crayfish (Hisada et al., 1969), the magnitude of the eye reflex to tilt of the body is increased by congruent visual stimulation (overhead light, stationary stripes), but not always by the simple summation of effects as suggested by Fay (1973). The interplay of gravity and light may be considered in relation to the control systems involved. Control of statocysts is essentially open-loop, with no direct effect of final eye position on the initial sensory input, while visual responses are under the precise control of a closed negative-feedback loop. Although visual responses are thus potentially more effective in dictating eye position, the balance of control must depend on how different inputs are "weighted" in neuronal integration. This "weighting" differs with species, being predominantly toward statocyst input in mysids (Neil, 1975a,b), but being distributed among a large number of sensory inputs in rock lobsters (York et al., 1972b).

More subtle differences exist between the input/output properties of these sensory subsystems. In mysids, the hysteresis in the eye response to roll tilts in opposite directions is significantly different when elicited by stimulation of statocysts (in the dark) than when elicited by visual stimulation (with statocyst removed) (Neil, 1975a). The working range of each sensory system may also be different in terms of both positions and velocity. In the crayfish *Procambarus clarkii,* movements of eyes about the pitch axis are more precisely controlled around the normal position by visual optokinetic input than by input from statocysts, whereas the geotatic system is more effective at larger tilts (Hisada et al., 1969). In decapod shrimps, the effect of a light at a constant angle to the eye is reduced when the body is tilted and the signal from statocysts is increased (Schöne, 1961). One interpretation of this result is that the sensory signals interact not only through their parallel effects on motor outputs, but also by direct inhibitory cross-connections. A similar but more precise effect of the signals from statocysts has been invoked by Wiersma (1966, 1970) to explain the phenomenon of neuronal space constancy in certain visual interneurons of crayfish and rock lobster. This neuronal compensatory mechanism is distinct from, but complementary to the compensatory eye movements described in this chapter (see Chapter 1 of this volume).

Comparison of the gains of the horizontal optokinetic and rotational nystagmus responses of the eyes of crabs reveals that the responses are effectively tuned to different velocity ranges (Silvey and Sandeman, 1976c; Sandeman, 1977). The visual system is extremely sensitive to slow movements, but with increasing velocity its gain decreases, while that of the statocyst response increases (Fig. 10). When operating together under natural conditions, the combined inputs from the two stimulus modalities will effectively extend the range of velocity over which the crab can detect and respond to rotational stimuli. In *Palinurus elephas,* similar relationships exist between the substrate and statocyst responses (Neil and Schöne, 1979; H. Schöne, unpublished observations), and between the substrate and visual responses of *Procambarus* (Olivo and Jazak, 1980); and it is perhaps in terms of such range fractionation that the most relevant interpretation of multimodal interaction can be made.

The neuronal basis of multisensory interaction has so far received little attention. Identified equilibrium interneurons receive input from different sensory systems (Fraser, 1975), but their involvement in the control of eye movement is not known. The clearest evidence has been obtained from recordings of oculomotor neuron activity. In crab optokinetic motor pools, the discharge patterns of the fast and slow phases of nystagmus are closely similar when activated by rotation of a striped drum or by rotation of the body (Wiersma and Fiore, 1971b; Sandeman and Okajima, 1973a). In those species without thread-hair systems (e.g., crayfish, spiny lobster), rotation of

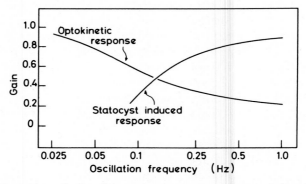

Fig. 10. Comparison of gains of the compensatory eye movements induced by oscillating stripes around a stationary crab (optokinetic response) and oscillating the crab in darkness (statocyst-induced response). The behavioral response to oscillation of the crab in the light will be extended beyond the range of the individual receptor systems. (From Sandeman, 1977; after Silvey and Sandeman, 1976a.)

the body in yaw has never been found to influence the optokinetic fibers, although vigorous effects are produced by inputs from the eyes and from leg proprioceptors (Olivo and Mellon, 1980; York et al., 1972a). Optomotor fibers involved in rolling and lifting the eye also display multimodal inputs. Combinations of body tilt, rotation of a striped drum, overhead illumination, and flexion or extension of the legs can effectively drive individual eye motoneurons in crayfish (Wiersma and Oberjat, 1968; Hisada and Higuchi, 1973; Mellon and Lorton, 1977) and rock lobsters (Wiersma and Yanagisawa, 1971; York et al., 1972b), although the different reflexes are not equal in strength, as judged by the changes in firing level which they produce. Cave crayfish provide a particularly interesting example, for in these troglodytic species the visual apparatus is degenerate and visual input to optomotor fibers absent, yet responses to tilt of the body and rotation of a platform are retained (Mellon, 1977b).

Specific responses of motoneurons to sensory inputs may also be gated by nonspecific inputs from a variety of sources. Firing levels in oculomotor fibers are elevated by vigorous mechanical stimulation (Burrows and Horridge, 1968c; Wiersma and Oberjat, 1968) and also change spontaneously due to a central arousal (Wiersma and Fiore, 1971a). Such effects, which are also observed in visual interneurons (Aréchiga and Wiersma, 1969a,b; Wiersma, 1970), do not significantly influence eye position, since all muscles are affected to an approximately equal extent (Burrows and Horridge, 1968c; Wiersma and Fiore, 1971a). Silvey and Sandeman (1976c) propose a model for this system in which sensory inputs to motoneurons are paralleled by a complex pathway involving a multimodal interneuron, which does not itself produce specific eye movements, but which gates the motor output. Therefore, it appears that motoneurons represent common pathways for compensatory eye reflexes and that multimodal integration is due to the convergence of diverse sensory inputs onto motor or premotor elements. Projection of different sensory inputs onto particular members of a motoneuron pool does not seem to be a general feature, although it applies to the rapid eye withdrawal reflex that is mediated by particular giant axons (Mellon and Lorton, 1977). Determination of the central nervous mechanisms which underlie the various interactive and integrative features of the multimodal system, which have so far been described at the behavioral and motor levels, remains a neurophysiological challenge.

REFERENCES

Alexander, R. McN. (1971). "Size and Shape," Inst. Biol. Stud. Biol., No. 29. Arnold, London.
Alexandrowicz, J. S., and Whitear, M. (1957). Receptor elements in the coxal region of Decapoda Crustacea. *J. Mar. Biol. Assoc. U. K.* **36**, 603–628.

Alverdes, F. (1926). Stato-, Photo- und Tangoreaktionen bei zwei Garneelenarten. *Z. Vergl. Physiol.* **4,** 699–765.
Aréchiga, H., and Wiersma, C. A. G. (1969a). The effect of motor activity on the reactivity of single visual units in the crayfish. *J. Neurobiol.* **1,** 53–69.
Aréchiga, H., and Wiersma, C. A. G. (1969b). Circadian rhythm of responsiveness in crayfish visual units. *J. Neurobiol.* **1,** 71–85.
Barnes, W. J. P., and Horridge, G. A. (1969a). Interaction of the movements of the two eyecups in the crab *Carcinus*. *J. Exp. Biol.* **50,** 651–671.
Barnes, W. J. P., and Horridge, G. A. (1969b). Two-dimensional records of the eyecup movements of the crab *Carcinus*. *J. Exp. Biol.* **50,** 673–682.
Baumann, H. (1921). Des Gefässsystem von *Astacus fluviatilis* (*Potamobius astacus* L.). *Z. Wiss. Zool.* **118,** 246–312.
Berkeley, A. A. (1928). The musculature of *Pandalus danae* (Stimpson). *Trans. R. Can. Inst.* **16,** 181–321.
Bethe, A. (1897). Das Nervensystem von *Carcinus maenas*. *Arch. Mikrosk. Anat.* **50,** 460–546.
Burrows, M., and Horridge, G. A. (1968a). The action of the eyecup muscles of the crab, *Carcinus*, during optokinetic movements. *J. Exp. Biol.* **49,** 223–250.
Burrows, M., and Horridge, G. A. (1968b). Motoneurone discharges to the eyecup muscles of the crab Carcinus. *J. Exp. Biol.* **49,** 251–267.
Burrows, M., and Horridge, G. A. (1968c). Eyecup withdrawal in the crab, *Carcinus,* and its interaction with the optokinetic response. *J. Exp. Biol.* **49,** 285–297.
Bush, B. M. H. (1965). Proprioception by the coxo-basal chordotonal organ, CB, in legs of the crab, *Carcinus maenas*. *J. Exp. Biol.* **42,** 285–197.
Bush, B. M. H. (1976). Non-impulsive thoracic-coxal receptors in crustaceans. In "Structure and Function of Proprioceptors in the Invertebrates" (P. J. Mill, ed.), pp. 115–151. Chapman & Hall, London.
Carpenter, R. M. S. (1977). "Movements of the Eyes." Pion, London.
Clarac, F., Neil, D. M., and Vedel, J.-P. (1976). The control of antennal movements by leg proprioceptors in the rock lobster, *Palinurus vulgaris*. *J. Comp. Physiol.* **107,** 275–292.
Clark, G. P. (1896). On the relation of the otocysts to equilibrium phenomena in *Gelasimus pugilator* and *Platyonichus ocellatus*. *J. Physiol. (London)* **19,** 327–343.
Cochran, D. M. (1935). The skeletal musculature of the blue crab, *Callinectes sapidus*. *Smithson. Misc. Collect.* **92,** 1–76.
Cohen, M. J. (1955). The function of receptors in the statocyst of the lobster *Homarus americanus*. *J. Physiol. (London)* **130,** 9–34.
Cohen, M. J. (1960). The response patterns of single receptors in the crustacean statocyst. *Proc. R. Soc. London, Ser. B* **152,** 30–49.
Cohen, M. J., and Dijkgraaf, S. (1961). Mechanoreception. In "The Physiology of the Crustacea" (T. H. Waterman, ed.), Vol. 2, pp. 65–108. Academic Press, New York.
De Bruin, G. H. P. (1956). Vision in the Crustacea. Ph.D. Thesis, University of Wales.
Dijkgraaf, S. (1955a). Lanterzeugung und Schallwahrnehmung bei der Languste (*Palinurus vulgaris*). *Experientia* **11,** 330–331.
Dijkgraaf, S. (1955b). Rotationssinn nach dem Bogengangsprinzip bei Crustaceen. *Experientia* **11,** 407–409.
Dijkgraaf, S. (1956a). Kompensatorische Augenstieldrehungen und ihre Auslösung bei der Languste (*Palinurus vulgaris*). *Z. Vergl. Physiol.* **38,** 491–520.
Dijkgraaf, S. (1956b). Über die kompensatorischen Augenstielbewegungen bei Brachyuren. *Pubbl. Stn. Zool. Napoli* **28,** 341–350.
Dijkgraaf, S. (1956c). Structure and function of the statocyst in crabs. *Experientia* **12,** 394–396.

5. Compensatory Eye Movements

Elofsson, R. (1964). Some muscle systems in front of the brain in decapod crustaceans. *Crustaceana* **7**, 11-16.

Fay, R. R. (1973). Multisensory interaction in control of eyestalk rotation response in the crayfish (*Procambarus clarkii*). *J. Comp. Physiol. Psychol.* **84**, 527-533.

Fay, R. R. (1975). Dynamic properties of the compensatory eyestalk rotation response of the crayfish (*Procambarus clarkii*). *Comp. Biochem. Physiol. A* **51A**, 101-103.

Fraenkel, G., and Gunn, D. L. (1961). "The Orientation of Animals, Kineses, Taxes and Compass Reactions." Dover, New York.

Fraser, P. J. (1975). Three classes of input to a semicircular canal interneuron in the crab, *Scylla serrata*, and a possible output. *J. Comp. Physiol.* **104**, 261-271.

Hassenstein, B. (1954). Über die Sehschärfe von Superpositionsaugen (Versuche an *Lysmata seticaudata* und *Leander serratus*). *Pubbl. Stn. Zool. Napoli* **25**, 1-8.

Hensen, V. (1863). Studien über das Gehörorgen der Decapoden. *Z. Wiss. Zool.* **13**, 319-412.

Herrnkind, W., and McLean, R. (1971). Field studies of homing, mass emigration and orientation in the spiny lobster, *Panulirus argus*. *Ann. N.Y. Acad. Sci.* **188**, 359-377.

Higuchi, T. (1973). The responses of oculomotor fibres in statocystectomized crayfish, *Procambarus clarki*. *J. Fac. Sci., Hokkaido Univ., Ser. 6* **18**, 507-515.

Higuchi, T., and Hisada, M. (1973). Visual and geotactic contributions to oculomotor responses in the crayfish, *Procambarus clarki*. *J. Fac. Sci., Hokkaido Univ., Ser. 6* **18**, 495-506.

Hill, M. N. (1962). "The Sea," Vol. I. Wiley (Interscience), New York.

Hisada, M., and Higuchi, T. (1973). Basic response pattern and classification of oculomotor nerve in the crayfish, *Procambarus clarki*. *J. Fac. Sci., Hokkaido Univ., Ser. 6* **18**, 481-494.

Hisada, M., Sugawara, K., and Higuchi, T. (1969). Visual and geotactic control of compensatory eyecup movement in the crayfish, *Procambarus clarki*. *J. Fac. Sci., Hokkaido Univ., Ser. 6* **17**, 224-239.

Horridge, G. A. (1966a). Optokinetic memory in the crab, *Carcinus*. *J. Exp. Biol.* **44**, 233-245.

Horridge, G. A. (1966b). Optokinetic responses of the crab, *Carcinus* to a single moving light. *J. Exp. Biol.* **44**, 263-274.

Horridge, G. A. (1966c). Direct response of the crab *Carcinus* to the movement of the sun. *J. Exp. Biol.* **44**, 275-283.

Horridge, G. A. (1966d). Adaptation and other phenomena in the optokinetic response of the crab, *Carcinus*. *J. Exp. Biol.* **44**, 285-295.

Horridge, G. A. (1966e). Study of a system, as illustrated by the optokinetic response. *Symp. Soc. Exp. Biol.* **20**, 179-198.

Horridge, G. A., and Burrows, M. (1968a). Tonic and phasic systems in parallel in the eyecup responses of the crab *Carcinus*. *J. Exp. Biol.* **49**, 269-284.

Horridge, G. A., and Burrows, M. (1968b). The onset of the fast phase in the optokinetic response of the crab, *Carcinus*. *J. Exp. Biol.* **49**, 299-313.

Horridge, G. A., and Sandeman, D. C. (1964). Nervous control of optokinetic responses in the crab *Carcinus*. *Proc. R. Soc. London, Ser. B* **161**, 216-246.

Horridge, G. A., and Shepheard, P. R. B. (1966). Perception of movement by the crab. *Nature (London)* **209**, 267-269.

Janse, C., and Sandeman, D. C. (1979a). The rôle of the fluid-filled balance organs in the induction of phase and gain in the compensatory eye reflex of the crab *Scylla serrata*. *J. Comp. Physiol.* **130**, 95-100.

Janse, C., and Sandeman, D. C. (1979b). The significance of canal-receptor properties for the induction of phase and gain in the fluid-filled balance organs of the crab. *J. Comp. Physiol.* **130**, 101-111.

Kennedy, D., and Davis, W. J. (1977). Organisation of invertebrate motor systems. In "Handbook of Physiology" Sec. 1, Vol. I, pp. 1023-1087. Am. Physiol. Soc., Bethesda, Maryland.

Knox, C. K. (1969). The frequency characteristics of the receptors of the crustacean statocyst organ. In "Systems Analysis in Neurophysiology," pp. 218-224. Laboratory of Neurophysiology, University of Minnesota, Minneapolis.

Kühn, A. (1914). Die reflektorische Erhaltung des Gleichgeurichtes bei Krebsen. Verh. Dtsch. Zool. Ges. **24,** 262-277.

Kunze, P. (1963). Der Einfluss der Grösse bewegter Felder auf den optokinetischen Augenstielnystagmus der Winkelkrabbe. Ergeb. Biol. **26,** 55-62.

Kunze, P. (1964). Eyestalk reactions of the ghost crab, Ocypode. In "Neural Theory and Modeling" (R. F. Reiss, ed.), pp. 293-305. Stanford Univ. Press, Stanford, California.

Lyon, E. P. (1900). A contribution to the comparative physiology of compensatory motions. Am. J. Physiol. **3,** 86-114.

Markl, H. (1974). Invertebrate equilibrium systems. In "Handbook of Sensory Physiology" (H. H. Kornhuber, ed.), Vol. VI/1, pp. 1-67. Springer-Verlag, Berlin and New York.

Mayne, R. (1974). A systems concept of the vestibular organs. In "Handbook of Sensory Physiology" (H. H. Kornhuber, ed.), Vol. VI/2, pp. 494-580. Springer-Verlag, Berlin and New York.

Mayrat, A. (1956). Oeil, centres optiques et Glandes Endocrines de Praunus flexuosus (O. F. Müller) (Crustaces Mysidaces). Arch. Zool. Exp. Gen. **93,** 319-366.

Mellon, DeF. (1977a). The anatomy and motor nerve distribution of the eye muscles in the crayfish. J. Comp. Physiol. **121,** 349-366.

Mellon, DeF. (1977b). Retention of oculomotor reflexes in blind cave-dwelling crayfish. Brain Res. **134,** 191-196.

Mellon, DeF., and Lorton, E. D. (1977). Reflex actions of the functional divisions in the crayfish oculomotor system. J. Comp. Physiol. **121,** 367-380.

Mellon, DeF., Tufty, R. H., and Lorton, E. D. (1976). Analysis of spatial constancy of oculomotor neurons in the crayfish. Brain Res. **109,** 587-594.

Mittlestaedt, H. (1972). Kybernetik der Schwereorientierung. Verh. Dtsch. Zool. Ges. **65,** 185-200.

Mittlestaedt, H. (1975). On the processing of postural information. Fortschr. Zool. **23,** 128-141.

Neil, D. M. (1975a). The control of eyestalk movements in the mysid shrimp Praunus flexuosus. J. Exp. Biol. **62,** 487-504.

Neil, D. M. (1975b). The optokinetic responses of the mysid shrimp Praunus flexuosus. J. Exp. Biol. **62,** 505-518.

Neil, D. M. (1975c). The mechanism of statocyst operation in the mysid shrimp Praunus flexuosus. J. Exp. Biol. **62,** 685-700.

Neil, D. M., and Schöne, H. (1979). Reactions of the spiny lobster, Palinurus vulgaris to substrate tilt. II. Input-output analysis of eyestalk responses. J. Exp. Biol. **79,** 59-67.

Neil, D. M., Schöne, H., and Scapini, F. (1979). Leg resistance reaction as an output and an input. Reactions of the Spiny Lobster, Palinurus vulgaris, to substrate tilt. VI. J. Comp. Physiol. **129,** 217-221.

Olivo, R. F., and Jazak, M. M. (1980). Proprioception provides a major input to the horizontal oculomotor system of crayfish. Vision Res. **20,** 349-353.

Olivo, R. F., and Mellon, DeF. (1980). Oculomotor activity during combined optokinetic and proprioceptive stimulation of the crayfish. Soc. Neurosci. Abstr. **6,** 101.

Ozeki, M., Takahata, M., and Hisada, M. (1978). Afferent response patterns of the crayfish statocyst with ferrite grain statolith to magnetic field stimulation. J. Comp. Physiol. **123,** 1-10.

5. Compensatory Eye Movements

Parker, T. J., and Rich, J. G. (1893). Observations on the myology of *Palinurus edwardsii* Hutton. *Proc. Linn. Soc. N.S. Wales* (Macleay Mem. Vol.) pp. 159-178.

Paterson, N. F. (1968). The anatomy of the cape rock lobster, *Jasus lalandii* (H. Milne Edwards). *Ann. S. Afr. Mus.* **51**, 1-232.

Patton, M. L. (1969). Physiological evidence indicating that decapod statocyst hairs drive three sensory neurones. *Am. Zool.* **9,** 1097.

Precht, W. (1978). Neuronal operations in the vestibular system. *In* "Studies in Brain Function," Vol. II. Springer-Verlag, Berlin and New York.

Prentiss, C. W. (1901). The otocyst of decapod Crustacea. *Bull. Mus. Comp. Zool.* **36,** 167-254.

Robinson, C. A., and Nunnemacher, R. F. (1966). The musculature of the eyestalk of the crayfish *Orconectes virilis*. *Crustaceana* **11,** 77-82.

Sandeman, D. C. (1964). Functional distinction between oculomotor and optic nerves in *Carcinus* (Crustacea). *Nature (London)* **201,** 302-303.

Sandeman, D. C. (1975). Dynamic receptors in the statocysts of crabs. *Fortschr. Zool.* **23,** 185-192.

Sandeman, D. C. (1976). Spatial equilibrium in arthropods. *In* "Structure and Function of Proprioceptors in the Invertebrates" (P. J. Mill, ed.), pp. 485-517. Chapman & Hall, London.

Sandeman, D. C. (1977). Compensatory eye movements in crabs. *In* "Identified Neurons and Behavior in Arthropods" (G. Hoyle, ed.), pp. 131-147. Plenum, New York.

Sandeman, D. C. (1978a). Regionalization in the eye of the crab *Leptograpsus variegatus*: Eye movements evoked by a target moving in different parts of the visual field. *J. Comp. Physiol.* **123,** 299-306.

Sandeman, D. C. (1978b). Eye scanning during walking in the crab *Leptograpsus variegatus*. *J. Comp. Physiol.* **124,** 249-258.

Sandeman, D. C., and Erber, J. (1976). The detection of real and apparent motion by the crab *Leptograpsus variegatus*. I. Behaviour. *J. Comp. Physiol.* **112,** 181-188.

Sandeman, D. C., and Okajima, A. (1972). Statocyst-induced eye movements in the crab *Scylla serrata*. I. The sensory input from the statocyst. *J. Exp. Biol.* **57,** 187-204.

Sandeman, D. C., and Okajima, A. (1973a). Statocyst-induced eye movements in the crab *Scylla serrata*. II. The response of the eye muscles. *J. Exp. Biol.* **58,** 197-212.

Sandeman, D. C., and Okajima, A. (1973b). Statocyst-induced eye movements in the crab *Scylla serrata*. III. The anatomical projections of sensory and motor neurons and the responses of the motor neurons. *J. Exp. Biol.* **59,** 17-38.

Sandeman, D. C., Erber, J., and Kien, J. (1975a). Optokinetic eye movements in the crab, *Carcinus maenas*. I. Eye torque. *J. Comp. Physiol.* **101,** 243-258.

Sandeman, D. C., Kien, J., and Erber, J. (1975b). Optokinetic eye movements in the crab, *Carcinus maenas*. II. Responses of optokinetic interneurons. *J. Comp. Physiol.* **101,** 259-274.

Scapini, F., Neil, D. M., and Schöne, H. (1978). Leg-to-body geometry determines eyestalk reactions to substrate tilt. Substrate orientation in spiny lobsters. IV. *J. Comp. Physiol.* **126,** 287-291.

Schmidt, W. (1915). Die Muskulatur von *Astacus fluviatilis* (*Potamobius astacus* L.). *Z. wiss. Zool.* **113,** 165-251.

Schöne, H. (1951). Die statische Gleichgewichts-orientierung bei dekapoden Crustaceen. *Verh. Dtsch. Zool. Ges.* **16,** 157-162.

Schöne, H. (1954). Statocystenfunktion und statische Lageorientierung bei dekapoden Krebsen. *Z. Vergl. Physiol.* **36,** 241-260.

Schöne, H. (1957). Die Lageorientierung mit Statolithenorganen und Augen. *Ergeb. Biol.* **21,** 161-209.

Schöne, H. (1961). Complex behavior. In "The Physiology of Crustacea" (T. H. Waterman, ed.), Vol. 2, pp. 465-502. Academic Press, New York.

Schöne, H. (1971). Gravity receptors and gravity orientation in Crustacea. In "Gravity and the Organism" (S. A. Gordon and M. J. Cohen, eds.), pp. 223-235. Univ. of Chicago Press, Chicago, Illinois.

Schöne, H. (1975a). Orientation in space: Animals. In "Marine Ecology" (O. Kinne, ed.), Vol. 2, pp. 499-553. Wiley, New York.

Schöne, H. (1975b). On the transformation of the gravity input into reactions by statolith organs of the "fan" type. Fortschr. Zool. **23**, 120-127.

Schöne, H., and Neil, D. M. (1977). The integration of leg position receptors and their interaction with statocyst inputs in spiny lobsters. (Reactions of Palinurus vulgaris to substrate tilt III). Mar. Behav. Physiol. **5**, 45-59.

Schöne, Her, and Schöne, Hed (1967). Integrated function of statocyst and antennular proprioceptive organ in the spiny lobster. Naturwissenschaften **54**, 289-290.

Schöne, H., and Steinbrecht, R. A. (1968). Fine structure of statocyst receptor of Astacus fluviatilis. Nature (London) **220**, 184-186.

Schöne, H., Neil, D. M., Stein, A., and Carlstead, M. K. (1976). Reactions of the spiny lobster, Palinurus vulgaris, to substrate tilt (I.). J. Comp. Physiol. **107**, 113-128.

Schöne, H., Neil, D. M., and Scapini, F. (1978). The influence of substrate contact on gravity orientation. Substrate orientation in spiny lobsters. V. J. Comp. Physiol. **126**, 293-295.

Silvey, G. E., and Sandeman, D. C. (1976a). Integration between statocyst sensory neurons and oculomotor neurons in the crab Scylla serrata. I. Horizontal compensatory eye movements. J. Comp. Physiol. **108**, 35-43.

Silvey, G. E., and Sandeman, D. C. (1976b). Integration between statocyst sensory neurons and oculomotor neurons in the crab Scylla serrata. III. The sensory to motor synapse. J. Comp. Physiol. **108**, 53-65.

Silvey, G. E., and Sandeman, D. C. (1976c). Integration between statocyst sensory neurons and oculomotor neurons in the crab Scylla serrata. IV. Integration, phase lags and conjugate eye movements. J. Comp. Physiol. **108**, 67-73.

Silvey, G. E., Dunn, P. A., and Sandeman, D. C. (1976). Integration between statocyst sensory neurons and oculomotor neurons in the crab Scylla serrata. II. The thread hair sensory receptors. J. Comp. Physiol. **108**, 45-52.

Stein, A. (1975). Attainment of positional information in the crayfish statocyst. Fortschr. Zool. **23**, 109-119.

Stein, A., and Schöne, H. (1972). Über das Zusammenspiel von Schwereorientierung und Orientierung zur Unterlage beim Flusskrebs. Verh. Dtsch. Zool. Ges. **65**, 225-229.

Steinacker, A. (1975). Proprioceptive feedback in the oculomotor system of the crab. Brain Res. **89**, 353-357.

von Buddenbrock, W., and Friedrich, H. (1933). Neue Beobachtung über die Kompensatorischen Augenbewegung und der Farbensinn der Taschenkrabben (Carcinus maenas). Z. Vergl. Physiol. **19**, 747-761.

Walton, A., and Herrnkind, W. (1977). Hydrodynamic orientation of spiny lobster, Panulirus argus. Wave surge and monodirectional currents. Mem. Univ. Newfoundland Mar. Sci. Res. Lab. Tech. Rep. **20**, 184-211.

Waterman, T. H. (1961). Light sensitivity and vision. In "The Physiology of Crustacea" (T. H. Waterman, ed.) Vol. 2, pp. 1-64. Academic Press, New York.

Waterman, T. H., and Horch, K. W. (1966). Mechanism of polarized light perception. Science **154**, 467-475.

Wiersma, C. A. G. (1958). On the functional connections of single units in the C.N.S. of the crayfish Procambarus clarkii Girard. J. Comp. Neurol. **110**, 421-471.

Wiersma, C. A. G. (1966). Integration in the visual pathway of Crustacea. *Symp. Soc. Exp. Biol.* **20,** 151–177.

Wiersma, C. A. G. (1970). Reactivity changes in crustacean neural systems. *In* "Short-Term Changes in Neural Activity and Behavior" (G. Horn and R. A. Hinde, eds.), pp. 211–236. Cambridge Univ. Press, London and New York.

Wiersma, C. A. G., and Bush, A. (1963). On the movements of the eyestalks of crabs: Particularly of *Calappa hepatica* (L.). *Proc. K. Ned. Akad. Wet.* **66,** 13–17.

Wiersma, C. A. G., and Fiore, L. (1971a). Factors regulating the discharge frequency in optomotor fibres of *Carcinus maenas*. *J. Exp. Biol.* **54,** 497–505.

Wiersma, C. A. G., and Fiore, L. (1971b). Unidirectional rotation neurones in the optomotor system of the crab, *Carcinus*. *J. Exp. Biol.* **54,** 507–513.

Wiersma, C. A. G., and Hirsh, R. (1974). Memory evoked optomotor responses in crustaceans. *J. Neurobiol.* **5,** 213–230.

Wiersma, C. A. G., and Oberjat, T. (1968). The selective responsiveness of various crayfish oculomotor fibers to sensory stimuli. *Comp. Biochem. Physiol.* **26,** 1–16.

Wiersma, C. A. G., and Yanagisawa, K. (1971). On types of interneurons responding to visual stimulation present in the optic nerve of the rock lobster, *Panulirus interruptus*. *J. Neurobiol.* **2,** 291–309.

York, B., Wiersma, C. A. G., and Yanagisawa, K. (1972a). Properties of the optokinetic motor fibres in the rock lobster: Build-up, flipback, after discharge and memory, shown by their firing patterns. *J. Exp. Biol.* **57,** 217–227.

York, B., Yanagisawa, K., and Wiersma, C. A. G. (1972b). Input sources and properties of position-sensitive oculomotor fibres in the rock lobster, *Panulirus interruptus* (Randall). *J. Exp. Biol.* **57,** 229–238.

6

Control of Mouthparts and Gut

W. WALES

<table>
<tr><td>I.</td><td colspan="2">Control of the Mouthparts</td><td>166</td></tr>
<tr><td></td><td>A.</td><td>Mandibles</td><td>167</td></tr>
<tr><td></td><td>B.</td><td>Labrum</td><td>168</td></tr>
<tr><td></td><td>C.</td><td>Paragnatha</td><td>169</td></tr>
<tr><td></td><td>D.</td><td>Receptors of the Mouthparts</td><td>169</td></tr>
<tr><td>II.</td><td colspan="2">Control of the Foregut</td><td>170</td></tr>
<tr><td></td><td>A.</td><td>Esophagus</td><td>172</td></tr>
<tr><td></td><td>B.</td><td>Cardiac Sac</td><td>175</td></tr>
<tr><td></td><td>C.</td><td>Pyloric Press and Filter</td><td>176</td></tr>
<tr><td></td><td>D.</td><td>Gastric Mill</td><td>179</td></tr>
<tr><td>III.</td><td colspan="2">Control of the Midgut</td><td>186</td></tr>
<tr><td>IV.</td><td colspan="2">Movements of the Hindgut and Anus</td><td>186</td></tr>
<tr><td>V.</td><td colspan="2">Perspectives and Conclusion</td><td>188</td></tr>
<tr><td></td><td colspan="2">References</td><td>189</td></tr>
</table>

In Crustacea, investigations into the control of gut* and mouthparts have, for reasons of size, accessibility, and availability, been performed almost exclusively on the larger and more common decapods. This has strictly limited the comparative data available. Although crustaceans have adopted

---

*In this chapter, and in the following one, terminology of foregut structures is that currently in use by neurobiologists. The most recent revision of the terminology can be found in a later volume of this series.

many feeding methods, including filter-feeding, parasitism, active predation, and scavenging, this review is necessarily restricted to a group of animals with a fairly uniform type of gut.

This chapter concentrates on description of the components of the digestive system and their activities. Analysis of the cellular mechanisms involved in generation of rhythmic activities is pursued in Chapter 7.

I. CONTROL OF THE MOUTHPARTS

The decapods studied so far are scavenger-predators with an ability to handle and masticate large pieces of food, and their anterior gut is specialized to process particulate matter. The mouth, or anterior opening of the gut, lies between the large mandibles (Fig. 1). The anterior wall of the mouth cavity is formed by the labrum, a muscular mobile structure suspended between the epistoma, the anterior medial edge of the mandibles, and the esophagus. To the rear of the mouth lie the paragnatha, which may be considered part of the mouth, as they are not among the segmental series of appendages. The mouth is connected to the gut by a short esophagus. Although different mouthparts have separate musculature, the movements of each must be affected by the others to some degree. It is, therefore, not surprising to find that there are proprioceptors responding to movement of the group as a whole (see below). Wales (1976) offers the most recent survey of proprioceptors in the mouthparts of arthropods.

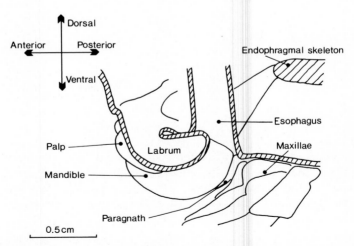

Fig. 1. A sagittal section through anterior cephalothorax of *Homarus gammarus* to show relative positions of labrum, mandibles, esophagus, and paragnatha.

A. Mandibles

The mandibles act to reduce food particles to a size suitable for ingestion. In *Homarus gammarus,* these mouthparts possess both cutting and crushing processes, and in normal eating movements, they act in coordination with accessory feeding appendages, especially the second and third maxillipeds, which hold and pull the food being acted upon by the mandibles (Wales, 1972). Movements of the mandibles and their control are described by Wales et al. (1976a,b) and Macmillan et al. (1976).

Although the mandibles are capable of independent movement, they normally move in unison. Observations of the electromyographic pattern from the major mandibular muscles show that although the time course and magnitude of activity in the bilateral counterparts are similar, there is no evidence of coupling between corresponding motoneurons. Exceptions to this rule are the common inhibitor neurons, which innervate some muscle fibers in all of the major mandibular muscles (Ferrero and Wales, 1976; Wales and Ferrero, 1976).

When the food to be processed is large and of uniform structure, the mandibles exhibit cyclical activity. Because of the regular nature of this rhythmic activity, Wales et al. (1976b) were able to gather large samples of electromyograms for statistical analysis and thus provide a quantitative description of mandibular movement. They observed that, although the degree of opening and the time at which closing commences may not be identical for the two mandibles in any given biting cycle, engagement of substrate appears to be bilaterally balanced. When necessary, the substrate is rapidly centered by the bite. The mechanism for this can be demonstrated by artificially loading one mandible: the electromyographic activity for the loaded mandible is much greater. There is, therefore, a high degree of coordination between mandibles, particularly during biting.

Presentation of different substrates to the mandibles demonstrates that muscular activity can change dramatically to cope with different materials (Macmillan et al., 1976). Incompressible substrates considerably shorten the length of the burst, whereas elastic substrates increase it (see Fig. 2). The greatest change occurs not in the time taken for the mandibles to close, but in the time spent compressing the substrate. These results led us to conclude that mandibular rhythm is set by a central "program" or "score," which is subject to sensory input regarding position and stress in the mandibles. Positional input has the predominant role prior to engagement of substrate, but tension afference subsequently plays an increasing role.

The mandibles are well equipped with proprioceptors capable of providing sensory information necessary to such a control system (Wales, 1972; Wales et al., 1976a). The most sophisticated of these is the mandibular

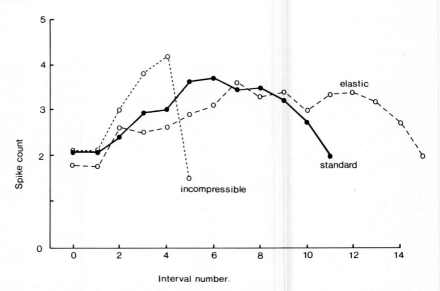

Fig. 2. Effect of "food" texture on burst structure of mandibular muscle. Note shortening of burst with incompressible substrate and lengthening with elastic substrate. (After MacMillan et al., 1976.)

muscle receptor organ (Wales and Laverack, 1972a,b). The source of tension afference has not yet been demonstrated, but it may be significant that the biting edge receives a massive innervation, which may serve a role similar to that of the intradental sensory endings in mammalian teeth (Anderson et al., 1970). The mandibular muscle receptor organ may also provide an indirect measure of tension.

B. Labrum

The labrum is an extremely mobile structure partially suspended from the mandibles and connected to the esophagus. Earlier workers have suggested that the labrum may be analogous to the vertebrate tongue (Nicol, 1932; Marshall and Orr, 1960), though its ability to perform similar tasks may be limited by its small size relative to the large food particles manipulated by the mandibles (Fryer, 1977).

Robertson (1978) has made a study of the control of labral and esophageal movements in *Homarus gammarus*. The labrum (Fig. 1) tends to be retracted between bites, a situation opposite of that to be expected if the labrum were aiding the passage of food particles into the esophagus. This function is apparently accomplished by the more posterior accessory feeding appendages. During biting, the labrum lies close to its resting position. Following the chewing sequence, the labrum rapidly moves posteriorly to occlude the opening into the esophagus, then slowly relaxes to its resting position. During swallowing, the labrum can be seen to undergo rhythmical movements of small amplitude which are thought to be related to esophageal peristalsis. These rhythmical contractions decay in frequency with time, but presentation of more food toward the end of the swallowing sequence causes an increase in the frequency of these movements. Moreover, the duration of these swallowing sequences decreases with satiation. The above data suggest that control of the labrum is similar to that of the mandible, in that it is driven by a central program. Control is also modified by sensory input from mandibular and esophageal receptors, in addition to labral receptors.

The labrum of *Homarus gammarus*, unlike that of *Austropotamobius pallipes* (Thomas, 1970), has no external sensillae, but vital staining techniques have revealed small uniterminal cells believed to be chemosensory (Dando and Laverack, 1969) and three bilateral groups of larger bipolar or tripolar cells that may be mechanosensory (Dando, 1969; Robertson, 1978). These authors have shown that mechanical deformation of the labrum alters the activity in the inner labral nerve, but it remains to be shown which cells are responding to this stimulus.

C. Paragnatha

Movements of the paragnatha are undescribed. Yet it is clear that the paragnatha are strongly mechanically coupled to the mandibles, and one would expect to find considerable neural coordination between the generators of mandibular and paragnathal rhythm.

D. Receptors of the Mouthparts

In addition to the proprioceptors monitoring discrete parts of the mouth, a complex system of proprioceptors monitoring movements of the mouth as a whole has been described for a number of decapods (Laverack and Dando, 1968; Moulins et al., 1976; Dando and Maynard, 1974) and for the stomatopod, *Squilla mantis* (Wales and Ferrero, 1982). The receptor system of the mouthparts consists of three bilaterally paired groups of multiterminal neurons designated mouthpart receptors (MPR) 1, 2, and 3, associated with a pair of elastic strands lying in an anterior-posterior orientation around the ventral side of the esophagus and running from the labrum to the paragnatha. The number of cells in each receptor group is small (five to ten), except in *Squilla mantis,* which has a few large cells plus a large number of smaller bipolar cells. In methylene blue stained preparations, these small cells appear to be uniterminal.

The receptors of the mouthparts respond to movement of the labrum, mandibles, esophagus/buccal cavity, and paragnatha. The response of the various receptors within the group differs with the source and type of movement, and groups of receptors may respond asynchronously. Afferent activity from MPR 1 enters the CNS via the inferior esophageal nerve to the commissural ganglion and that of MPR 2/3 enters the CNS via the paragnathal nerve to the subesophageal ganglion. As shown below, the commissural ganglion plays an important role in the coordination of foregut activity, whereas motoneurons of the mandible and paragnatha appear to lie in the subesophageal ganglion. Thus, the two pathways followed by MPR inputs may indicate a difference in function between MPR 1 and MPR 2/3. The role of this input in controlling movement of gut or mouthparts is unknown.

II. CONTROL OF THE FOREGUT

The foregut of decapods, which is of ectodermal origin, consists of a short esophagus opening into a large cardiac sac through the esophageal-cardiac sac valve. The stomach can be divided into three sections, the cardiac sac, the gastric mill, and the pyloric filter. The foregut possesses a chitinized cuticle, and the gastric mill has large calcified cuticular ossicles, which function as teeth in aiding the breakdown of food particles. The foregut is separated from the midgut by a pyloric press and filter, which separate the particles according to size.

Unlike the mouthparts, which are innervated by the large subesophageal ganglion, the foregut is largely innervated by a discrete extension of the central nervous system, the stomatogastric nervous system. This consists of bilaterally paired commissural ganglia, the medial esophageal ganglion and the single stomatogastric ganglion plus their associated nerve trunks (Fig. 3) (Maynard and Dando, 1974). The most detailed anatomical studies of this system are those of Orlov (1926a,b) and more recently of Maynard and Dando (1974). The stomatogastric ganglion has received the greatest attention (see Chapter 7) because it consists of a small number of neurons that interact to produce patterned activity even in isolated ganglia.

Movements of the foregut are produced by a complex pattern of extrinsic and intrinsic striated muscles innervated by a small number of motoneurons located in the ganglia of the stomatogastric system. Most muscles of the foregut are innervated by a single motor axon (Govind et al., 1975), although some may receive as many as five axons (see Chapter 2 of Volume

6. Control of Mouthparts and Gut

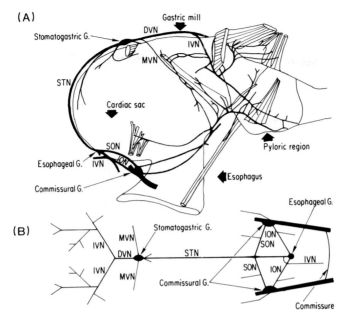

Fig. 3. Stomatogastric nervous system of palinurids showing (A) relationship to foregut (after Maynard and Dando, 1974); and (B) topography and connections between ganglia (after Vedel and Moulins, 1977). DVN, dorsoventricular nerve; ION, inferior esophageal nerve; IVN, inferior ventricular nerve; LVN, lateral ventricular nerve; MVN, median ventricular nerve; SON, superior esophageal nerve; STN, stomatogastric nerve; G, ganglion.

3). Each motor neuron may innervate up to five discrete muscles, indicating that those muscles function as an anatomical unit (Govind et al., 1975). The stomatogastric and esophageal ganglia do not contain paired motoneurons; therefore, single motoneurons normally innervate bilaterally paired muscles (Maynard, 1972; Moulins and Vedel, 1977). Innervation of muscles of the foregut is excitatory, inhibitory motoneurons being unknown (Maynard, 1972).

The four main regions of the foregut (esophagus, cardiac sac, gastric mill, and pyloric filter), exhibit neurogenic rhythmic movements (Fig. 4), each controlled by a separate group of neurons which may generate the pattern in isolation. The neurons involved are primarily motoneurones, which generate the pattern by virtue of their intrinsic properties and synaptic connections (Russell and Hartline, 1978; also see Chapter 7). Interneurons are involved, either as part of the pattern-generating circuit, or as pathways for external control. Sensory input has been shown to play an important role, and there is a degree of central interaction between the regional pattern generators.

Fig. 4. Simultaneous recording of output from three rhythmical centers of the foregut: (A and B) cardiac sac; (C) pylorus; (D) gastric mill. (From Vedel and Moulins, 1977.)

A. Esophagus

The control of esophageal movement has been studied in the spiny lobsters, *Palinurus elephas* (Moulins and Vedel, 1977) and *Panulirus argus* (Selverston et al., 1976), the crayfish *Procambarus clarkii* (Spirito, 1975), and the European lobster *Homarus gammarus* (Robertson, 1978). Esophageal movements result from the action of extrinsic dilators and intrinsic constrictors, which produce a peristaltic wave. The neuronal network controlling these muscles lies in the commissural and esophageal ganglia. In *Procambarus* (Spirito, 1975), the source of patterned output appears to be the commissural ganglia, which continue to produce a synchronous rhythmic output following section of the superior and inferior esophageal nerves connecting them to the esophageal ganglion. Synchrony of the bilateral output is only lost following section of the esophageal commissure. The patterned motor output (Fig. 5A) travels in both inferior and superior nerves and passes through the esophageal ganglion to the stomatogastric ganglion via the stomatogastric nerve. In *Palinurus* (Moulins and Vedel, 1977), the somata of several motoneurons that generate esophageal movements have been shown to lie in the esophageal ganglia. This appears to contradict the findings of Spirito (1975), but Moulins and Vedel have demonstrated that although the neuron somata lie in one ganglion, they may have spike-initiating zones in other ganglia, e.g., OD 1 has three zones, one in the esophageal ganglion and one in each of the commissural ganglia. Moulins and Vedel (1977) have identified three esophageal dilators (OD 1-3) and two constrictors (C,C') (Table I). In isolated preparations of the stomatogastric system of *Panulirus* (Selverston et al., 1976), coupling between outputs of commissural ganglia occurs in the absence of the esophageal commissure. In these preparations, coupling is subserved by units in the superior esophageal nerve.

In *Homarus*, esophageal peristalsis is controlled by esophageal chemoreceptive organs (Robertson, 1978; Robertson and Laverack, 1978). Presumptive chemoreceptors have been described for *Palinurus, Astacus,* and *Homarus* (see Dando and Maynard, 1974, for review). There is some variation among species in position and structure of these organs, and a

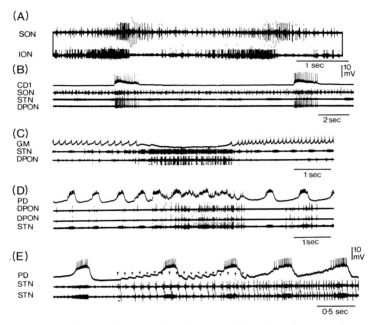

Fig. 5. Activity of the esophagus and cardiac sac. (A) Esophageal rhythm, demonstrating coordinated output in superior and inferior esophageal nerves (after Selverston et al., 1976). (B) Rhythmical activity of the cardiac sac, recorded intracellularly from CD 1 and extracellularly at three locations in the stomatogastric nervous system. (C) Coordination among motoneurons of the gastric mill (GM) and cardiac sac, the GM neurons being inhibited during a burst of the cardiac sac. (D) Input from the cardiac sac onto pyloric dilator neuron (PD), disrupting pyloric cycling. (E) Common input from IVN through-fibers onto pyloric dilator neuron, with concomitant increased CD 2 activity in stomatogastric nerve. Dots indicate individual stimuli to IVN. DPON, dorsal posterior esophageal nerve; other abbreviations as per Fig. 3 (B–E after Moulins and Vedel, 1977).

degree of caution is warranted when drawing homologies. The function of these receptors is best described for *Homarus,* which has two bilaterally arranged pairs of chemosensory organs named the anterior and posterior esophageal sensors (AOS and POS) in a ring around the esophagus close to the esophageal-cardiac sac valve (OCSV). The AOS are separated from the main lumen of the esophagus by folds of tissue, except when the cardiac sac of the stomach is fully distended. Chemical stimulation of the POS increases the peristaltic rate, whereas chemical stimulation of AOS produces a decrease.

Robertson (1978) has proposed the following mechanism for control of esophageal movements. When the stomach is empty, food will rapidly pass through the esophagus and OCSV, stimulating the POS but not the AOS.

TABLE I

Identified Foregut Neurons[a,b]

Abbreviation	Name	Function	Location of cell soma
Esophagus			
OD 1	Esophageal dilator 1	Dilator	OG
OD 2	Esophageal dilator 2	Dilator	?
OD 3	Esophageal dilator 3	Dilator	?
C	Constrictor	Constrictor	OG
C'	Constrictor	Constrictor	OG
Cardiac sac			
CD 1	Cardiac sac dilator 1	Dilator	OG
CD 2	Cardiac sac dilator 2	Dilator	STGG
VD	Ventral dilator	Dilator	STGG
AM	Anterior median	Constrictor	STGG
IC	Inferior cardiac	Constrictor	STGG
Gastric mill (a) lateral teeth subset			
LG	Lateral gastric	Closer	STGG
MG	Median gastric	Closer	STGG
LPG(2)	Lateral posterior gastric	Opener	STGG
Gastric mill (b) medial tooth subset			
GM(4)	Gastric mill	Power stroke	STGG
DG	Dorsal gastric	Return stroke	STGG
Pyloric filter			
PD(2)	Pyloric dilator	Dilator	STGG
VD	Ventral dilator	Dilator	STGG
LP	Lateral pyloric	Constrictor	STGG
PY(8)	Pyloric	Constrictor	STGG
Interneurons			
INT 1	Interneuron 1		STGG
INT 2	Interneuron 2		STGG
EX(6)	Excitatory neuron		STGG
AB	Anterior Burster		STGG
P (?)	Pyloric neuron		CG
E (2)	Excitatory neuron		CG
L (2)	Large neuron		CG

[a] The information in this table was extracted from the work of Moulins and Vedel (1977) and Selverston et al. (1976), and it refers only to the palinurids.

[b] A summary of the stomatogastric neurons identified to date. Note that they are distributed among three ganglia: the stomatogastric ganglion (STGG), the esophageal ganglion, (OG), and the commissural ganglion (CG). Where more than one of a category of neuron exists, the number is given in brackets.

6. Control of Mouthparts and Gut

Thus, the introduction of food into the esophagus may be required for the onset of peristalsis. This supposition is supported by Robertson's observation that peristalsis appears to start after the commencement of chewing. However, as we have seen above, strong proprioceptive inputs during mandibular movement may also be involved. As feeding proceeds, the partially filled cardiac sac will slow the passage of food through the OCSV; this leads to prolonged stimulation of the POS and possibly to adaptation and slowing of the rate of peristalsis. On filling of the cardiac sac, the AOS receptors will become exposed to food in the esophagus, and peristaltic movements will be inhibited. This clearly is a partial picture, since it takes no account of mechanosensory input (see Dando and Maynard, 1974), and since esophageal control must be coordinated with the total feeding mechanism.

The above mechanism may not apply to the palinurids, as Russell (in Selverston et al., 1976) has shown that in *Panulirus interruptus,* electrical stimulation of the nerve of a chemoreceptor organ on the anterior esophageal wall produces a two- to threefold increase in the rate of gastric and possibly esophageal rhythms. If this receptor is homologous to that in *Homarus,* its effect appears to be opposite. However, this nerve also contains motor axons of the esophageal and possibly cardiac-sac systems, and the observed response may have been partly or entirely due to antidromic stimulation of the esophageal oscillators. It is possible that the esophageal organs have a wider function in controlling the activity of the foregut, but the pathways and central connections are as yet unknown.

B. Cardiac Sac

The cardiac sac is the main storage region of the foregut, yet it is capable of independent movement. Its musculature is relatively simple. Moulins and Vedel (1977) have demonstrated the presence of three dilator motoneurons in *Palinurus elephas,* the cardiac dilators (CD 1 and CD 2) and ventral dilator (VD); and two constrictor motoneurons, the anterior median (AN) and inferior cardiac (IC) (see Table I). The VD and IC neurons are also involved in the pyloric rhythm (see below). Cardiac dilator 1 is located in the esophageal ganglion, whereas all other motoneurons of the cardiac sac are in the stomatogastric ganglion. Cardiac dilator motor activity (Fig. 5B) appears to be partly organized in the esophageal ganglion where the central branches of CD 1 and CD 2 converge (Vedel and Moulins, 1977), and the antidromic input to the stomatogastric ganglion via CD 2 may be a pathway by which activity of the cardiac sac affects other cycles of the foregut. CD 2, like OD 1, receives input in more than one ganglion and has two centers of spontaneous activity, one in the stomatogastric and one in the esophageal ganglion. In *Panulirus interruptus,* CD 1 makes no connection in the stomatogastric ganglion (see Vedel and Moulins, 1977), and this may also apply to *Palinurus elephas*. Moulins and Vedel (1977) have shown that cardiac motor activity has a strong modulating affect on the esophageal and pyloric cycles, and its activity is to some degree coordinated with other rhythms of the foregut (Fig. 5C,D). Motor activity of the cardiac sac is also sensitive to input, as shown in Fig. 5E.

C. Pyloric Press and Filter

Although, in a functional sense, the gastric mill follows the cardiac sac, the circuitry of the pyloric filter was first analyzed and is better understood. Control of movements of the pyloric filter has been extensively studied in the spiny lobsters *Panulirus argus* and *P. interruptus* (Hartline and Maynard, 1975; Maynard, 1972; Maynard and Selverston, 1975). Some comparative data are available for the pyloric system of other decapods (Dando et al., 1974; Hermann and Dando, 1977; Maynard and Burke, 1966; Nagy, 1977; Nagy and Dando, 1973), but the description presented here is restricted to spiny lobsters.

There are thirteen motoneurons and one interneuron in the pyloric group (see Table I). The motoneurons can be divided into two antagonistic groups, though it must be stated that the function of some neurons and muscles is not yet clear. The dilator group consists of two pyloric dilators (PD), one ventral dilator (VD), and the anterior burster (AB), which is possibly an interneuron, since it has not yet been found to innervate a muscle. The constrictor group consists of the lateral pyloric (LP), the inferior cardiac (IC), and eight pyloric neurons (PY). The function of VD and IC appears to be related to movements of both pylorus and cardiac sac (Maynard, 1972; Moulins and Vedel, 1977), although these neurons participate in the pyloric cycle. The sequence of pyloric neuronal activity is shown in Fig. 6.

Neurons of the pyloric filter usually exhibit rhythmic activity in the deafferented stomatogastric ganglion, and the activity pattern (Fig. 7a) is similar to that observed in the intact nervous system (Morris and Maynard, 1970). The duration of the cycle varies from 0.65/sec to 1.2/sec. By using intracellular techniques, Maynard and colleagues have systematically studied the connections between individual neurons to provide the data in Table II. Basically, two types of connections exist, chemical inhibitory and nonrectifying electrical excitatory synapses. If we divide the constrictor neurons into two sets, as shown in Table II, we can see that, as a general rule, neurons that fire together are electrically coupled and each group inhibits the neurons that fire in a different phase. Main exceptions to this rule are described below.

The pyloric filter transports food from foregut to midgut and filters the food

6. Control of Mouthparts and Gut

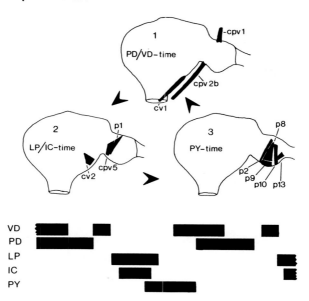

Fig. 6. Sequence of activity for pyloric motoneurons and muscles during the pyloric cycle. Upper diagrams show muscles active in three phases, and lower diagram the relative phasing of neurons. See Table I for abbreviations of neurons and Maynard and Dando (1974) for nomenclature of muscles. (After Hartline and Maynard, 1975.)

particles. During transport, activity is led by the PD–AB group (Fig. 6), which causes the pyloric filter to open its entrance from the gastric region, thus allowing material to enter. This is followed by activity in LP, which innervates a muscle antagonistic to PD action, thereby causing the entrance to close. Finally, the PY neurons fire and give rise to a peristaltic-like wave of contraction that drives the contents into the midgut or hepatopancreas. The sequence of motoneuron activity observed during filtering is similar. Because of its filtering role, the duration of pyloric activity is greater than for the other sections of the foregut.

Mechanoreceptors have been described on the pyloric stomach and its musculature (Dando and Maynard, 1974), but the effect of their input on the pyloric rhythm is unknown. The pyloric rhythm is, however, modulated by other inputs to the stomatogastric ganglion. Studies on the feeding of intact lobsters by use of chronically inplanted electrodes (Morris and Maynard, 1970) have shown that pyloric activity is intensified when the animal feeds. The changes in activity, which include increased rate of cycling and intensified, long, high-frequency bursts from the individual units, can be mimicked in acute preparations by stimulation of the posterior stomach nerve (Fig. 7E); this nerve carries proprioceptive information during the

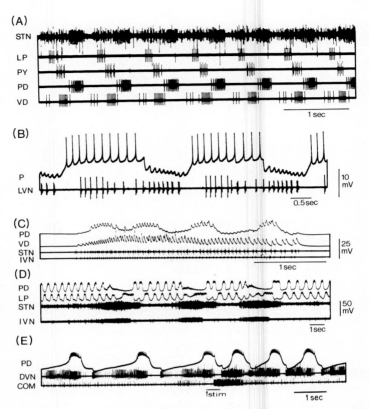

Fig. 7. Pyloric activity. (A) Simultaneous extracellular recordings from several nerves to show phase relationships of the different neurons. (B) P neuron in commissural ganglion firing in bursts synchronized to pyloric rhythm, as seen in lateral ventricular nerve. (C and D) IVN through-fiber modulation of pyloric rhythm. In C, the input produces large EPSP's, which drive the cycling. In D, more intense activity inhibits pyloric activity. (E) Stimulation of posterior nerve of the stomach results in increased rate of pyloric cycling. See Table I and Fig. 3 for abbreviations. (A-D after Selverston et al., 1976; E after Nagy, 1977.)

power stroke of the medial tooth in the gastric mill (Dando et al., 1974; Nagy, 1977). The pathway for this reflex is described by Hermann and Dando (1977) and Nagy (1977). Direct synaptic input can be detected in the PD's but not in LP or PY neurons, and the observed changes in output pattern are mediated via the PD group.

Dando and Selverston (1972) have described a further input to the pyloric neurons via the stomatogastric nerve and inferior ventricular nerve from the brain. This input is derived from neurons referred to as "IVN through-fibers," because the axons pass uninterrupted through the esophageal ganglion; but

TABLE II

Synaptic Connections among Pyloric Neurons in the Stomatogastric Ganglion[a,b]

Presynaptic	Postsynaptic						
	PD(2)	AB	VD	LP	IC	PY(4)	PY(4)
PD(2)	ec	ec	i,ec	i	i	i	i
AB	ec	—	i,ec	i	i	i	i
VD	ec	ec	—	—	i	i	—
LP	i	i	i	—	—	i	—
IC	—	—	i	—	—	—	—
PY(4)	—	—	—	i	i	ec	ec
PY(4)	—	—	—	—	—	ec	ec

[a] This is a simplified summary of information presented by Maynard (1972) and Maynard and Selverston (1975). It does not take into account the different types of inhibitory input.
[b] i = inhibitory; ec = electronic coupling. See Table I and text for other abbreviations

Selverston et al. (1976) have since shown that IVN through-fibers also send branches into the superior and inferior esophageal nerves. Thus, their distribution and correspondingly their effect is more widespread than first thought. Moreover, these fibers have integrative regions in the esophageal ganglion, where they generate intense bursts. Activity in the IVN through-fibers has a powerful excitatory effect on the pyloric neurons, particularly to PD and VD. The effect of IVN input is complex and may excite or inhibit pyloric activity (Fig. 7C,D), depending on the nature of IVN activity. The functional significance of this input in modulating pyloric activity is unknown, and the pyloric system remains active in the absence of these bursts.

One pyloric neuron, AB, has a process traveling centrally in the stomatogastric nerve to the commissural ganglion, and pyloric activity can be detected along this pathway. Selverston et al. (1976) have described the presence of neurons in the commissural ganglia (P cell, Fig. 7B) that participate in pyloric activity, with bursts synchronized 1:1 with the pyloric cycle. The P cells are an integral part of the pyloric network, as hyperpolarization of P cells stops pyloric cycling. The bursts in these cells occur before LP and cease on commencement of the PD-AB burst. Again, the function of this input is not known, but the more vigorous activity observed in preparations containing the commissural ganglia may be due in part to P neuron input.

D. Gastric Mill

Although the gastric mill forms a single functional unit (see Fig. 8), its control system can be divided into two subsystems, which have been ob-

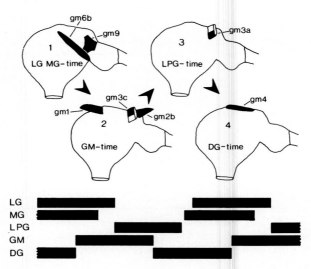

Fig. 8. Sequence of activity for gastric motoneurons and muscles during cycling of the gastric mill. Upper diagrams show muscles active in four phases, and lower diagram shows relative phasing of neurons. See Table I for abbreviations of neurons and Maynard and Dando (1974) for nomenclature of muscles. (After Hartline and Maynard, 1975.)

served to function independently (Hartline and Maynard, 1975). These systems are referred to as the "lateral teeth" and "medial tooth" subsets (Table I). Gastric movements in *Panulirus interruptus* are produced by a group of 12 neurons in the stomatogastric ganglion (Mulloney and Selverston 1974a; Selverston and Mulloney, 1974). The gastric cycle (Figs. 8 and 9A,B) is largely independent of the pyloric cycle and much more labile, occurring only infrequently in the deafferented stomatogastric ganglion. The period of the gastric rhythm is longer than that of the pyloric, ranging from 15 to 25 sec in *Homarus americanus,* in which the pyloric period was 1.33 ± 0.18 sec in the intact nervous system (Morris and Maynard, 1970). In *P. interruptus,* the gastric rhythm period of the deafferented ganglion ranges from 2 to 40 sec (Selverston et al., 1976).

1. LATERAL TEETH

The lateral teeth of *P. interruptus* (Mulloney and Selverston, 1974a) are driven by the following groups of motoneurons: the lateral gastric (LG), which causes the teeth to part; the median gastric (MG), the function of which is not clear; and the two lateral posterior gastric neurons (LPG), which cause the teeth to rotate together (Hartline and Maynard, 1970). The LG and LPG produce alternate bursts of activity, which cause the respective muscles to contract (Fig. 8). As with pyloric neurons, the antagonists reciprocally

6. Control of Mouthparts and Gut

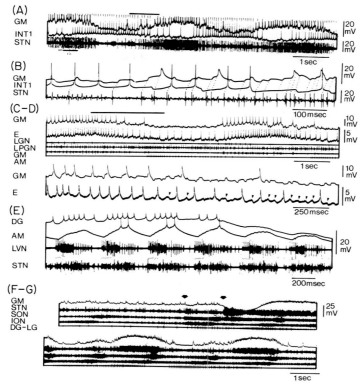

Fig. 9. Gastric mill activity. (A) Gastric cycling as seen extracellularly in stomatogastric nerve and intracellularly in GM motoneurons and INT 1 interneuron. (B) Expanded form of section marked by bar in A; demonstrates interaction between GM and INT 1, the former receiving IPSP's from the latter. (C) E neuron in commissural ganglion firing in phase with GM neuron. (D) Expansion of section marked by bar in C, showing inhibitory input to E neuron. (E) Gastric neurons (DG and AM) in phase with pyloric rhythm. (F-G) Activation of gastric and esophageal cycling by crushing nerve from esophageal chemoreceptors. See Table I and Fig. 3 for abbreviations. (After Selverston et al., 1976).

inhibit one another and synergists are electrically coupled (Table III). In addition, MG and LG reciprocally inhibit each other and are electrically coupled, and the LPG's do not inhibit LG. This circuit has the essential elements required for generating an alternate bursting pattern in response to an unpatterned excitatory input.

The two antagonistic groups interact with two interneurons in the stomatogastric ganglion. These are referred to as interneurons 1 and 2 (INT 1, INT 2). INT 1 and the LG-MG group reciprocally inhibit each other, and it is suggested by Mulloney and Selverston (1974a) that the reciprocal inhibition acts as a switching mechanism. INT 2 receives no input from any lateral teeth motoneurons but excites INT 1 and inhibits LPG's. INT 2 activity directly inhibits the LPG's and indirectly inhibits the LG-MG group, thus switching the system on. INT 2 is, therefore, the highest order neuron in the subset controlling the others.

Russell and Hartline (1978) have demonstrated that LG and MG exhibit "plateau potentials" suggestive of pacemaker activity; this may explain how the system is switched on and how burst period is regulated. Activity is occasionally seen in the isolated deafferented stomatogastric ganglion, although plateau potentials are not observed in such preparations.

2. MEDIAL TOOTH

Movements of the medial tooth are similarly produced by two sets of neurons in the stomatogastric ganglion driving antagonistic muscles (Selverston and Mulloney, 1974). The four motoneurons of the gastric mill (GM motoneurons) produce an anterior-ventral movement of the medial tooth; a single dorsal gastric (DG) motoneuron resets the tooth to its starting position. The GM neurons are electrically coupled but have no input to neurons of the antagonistic group. DG is electrically coupled to the anterior median (AM) neuron, which produces movement of the cardiac sac, the activity of which is synchronized to the gastric mill cycle. AM activity constricts the cardiac sac, moving its contents forward toward the gastric mill. Both DG and AM inhibit the GM group.

The medial tooth subset also interacts with INT 1 and INT 2. INT 1 unilaterally inhibits the GM group (Fig. 9A,B) and appears to directly excite DG. INT 2 also inhibits GM but excites INT 1. Thus, INT 2 is again the highest order neuron, receiving no input from the other neurons in the subset; but, in this case, INT 1 also directly innervates the two antagonistic groups.

The GM neurons are inhibited by all other units of the subset and exhibit post-inhibitory rebound excitation. When the GM's are active, in the absence of bursting activity from the subset, they fire continuously. As with the "lateral teeth" subset, two types of motoneuron (DG and AM) exhibit plateau potentials (Russell and Hartline, 1978), but in this case, the reciprocal inhibitory network between antagonists is absent. Both interneurons are capable of switching on DG activity, with INT 2 producing the greater inhibition of the GM group.

3. COORDINATION BETWEEN SUBSETS OF THE GASTRIC MILL

Although the origin of bursting activity within the subsets has yet to be ascertained, both subsets have been observed to exhibit rhythmical independent activity; this suggests that each subset has the ability to generate

6. Control of Mouthparts and Gut

alternating activity. However, the gastric cycle is much more labile. It is seldom active in the isolated ganglion, whereas it proves reliable when its connections to the esophageal and commissural ganglia are intact (Russell, 1976). As can be seen in Table III, there is a high degree of synaptic coupling between the two subsets, particularly between the GM's and the LG-MG group, which are electrically coupled, and between the GM's and the LPG's, which are electrically coupled by a rectifying synapse. There are also inhibitory inputs from LG to the GM's and possibly direct inhibitory pathways from LG to DG, AM to MG and LPG, and DG to LG, MG, and LPG. The interactions are, therefore, complex and difficult to interpret in a functional sense. Moreover, this may not be the complete list of connections (Mulloney and Selverston, 1974b).

In normal activity, the two subsets are coordinated, the lateral teeth holding the food while the medial tooth grinds it. As shown in Fig. 8, the cycles of lateral and medial teeth are phased so that the medial tooth (GM activity) lags behind closure of the lateral tooth (LG activity) (Hartline and Maynard (1975). However, this coordinated pattern is not always observed (Powers, 1973; Dando et al., 1974).

Mulloney and Selverston (1974b) have described three main types of burst patterns during coordinated activity between subsets; this emphasizes the flexibility of the system. The reciprocal inhibitory connections between neurons do not produce a sharp changeover in activity, so that a degree of overlap may be observed in the activity of antagonistic neurons. Moreover, the responsiveness of the neurons to inhibition exhibits temporal changes.

TABLE III

Synaptic Connections among Gastric Neurons[a,b]

Presynaptic	Post synaptic							
	LG	MG	LPG(2)	AM	DG	GM(4)	INT 1	INT 2
LG	—	i,ec	i	—	i	i,ec	i	
MG	i,ec	—	i	—	—	i(?),ec	i	
LPG(2)	—	i	ec	—	—	rec	—	
AM	—	i	i	—	ec	i	—	
DG	i	i	i	ec	—	i	—	
GM(4)	ec	ec	—	—	—	ec	—	
INT 1	i	i	—	—	e	i	—	
INT 2	—	—	i	—	—	i	e	

[a] From information extracted from Mulloney Selverston (1974a,b) and Selverston and Mulloney (1974).

[b] e = excitatory, i = inhibitory, ec = electronic coupling, and rec = rectifying electronic coupling. See Table I and text for other abbreviations.

Connections between gastric neurons are consistent with their observed activity: a mechanism exists to produce the basic pattern. Details relating to the period of the cycle and the on or off switching of the gastric rhythm remain to be worked out. Russell and Hartline (1978) have demonstrated that some neurons in both medial and lateral subsets exhibit endogenous activity, but this is seen only in the presence of central connections and not in the isolated stomatogastric ganglion. Further studies of the gastric system in the intact nervous system are required before we can understand how the gastric rhythm is regulated.

The gastric mill is monitored by both mechanoreceptors and chemoreceptors (Dando and Maynard, 1974). The function of proposed chemoreceptors is yet to be demonstrated. There is no proprioceptive input directly to the motoneurons; all input passes to the commissural ganglion by one of three routes. The largest and possibly the most important input from the posterior stomach receptor (Dando and Laverack, 1969; Dando and Maynard, 1974) travels via the posterior stomach nerve (PSN). Somata of these proprioceptive cells lie in the PSN and number 30 to 100. The sensory endings respond to movement of the gastric mill, but due to the branching of the cell processes, some neurons may also innervate the mandible (Wales et al., 1976a; Wales and Ferrero, 1976). Electrical stimulation of the PSN evokes changes in the gastric rhythm, which is initially inhibited and then exhibits rebound excitation. Therefore, the gastric rhythm, like the pyloric, is modulated by a reflex pathway that involves the commissural and stomatogastric ganglia. Additional sense organs known to modulate gastric mill activity are chemoreceptors located in the esophageal wall. Electrical stimulation of these receptors may increase rate of gastric cycling two- to threefold and intensify neuronal output.

The activity of the gastric mill is also modulated by IVN through-fibers. Bursts in these fibers disturb and reset the gastric cycle. Dando and Selverston (1972) found the effect of whole IVN stimulation to be frequency-dependent, the gastric cycle being more strongly inhibited with stimuli of higher frequency. IVN input to gastric neurons is extensive (Selverston et al., 1976).

As with the pyloric system, the system controlling gastric rhythm has participating neurons in the commissural ganglia (E neurons, Fig. 9C,D). The function of these neurons has not yet been described, but they most probably account for the higher reliability of the gastric rhythm in preparations that include the commissural ganglia (Selverston et al., 1976).

4. COORDINATION OF RHYTHMS OF THE FOREGUT

Although the functional divisions of the foregut possess independent control centers, there is growing evidence that their rhythmic activity is coordi-

6. Control of Mouthparts and Gut

nated. This is achieved by a number of pathways involving (1) functional synaptic contact between motoneurons (direct contact has not yet been demonstrated), (2) interneuronal pathways, and (3) reflex modulation based on sensory input from other foregut sections (Selverston et al., 1976).

Within the stomatogastric ganglion, there is interaction between pyloric and gastric cycles (Fig. 9E) in the form of weak electrotomic synaptic connections involving a number of motoneurons. Of course, both the pyloric and gastric systems have participating neurons outside of the stomatogastric ganglion (P cells and E cells), and these offer another pathway for interaction. Indeed, the pyloric cycle modulates the gastric cycle via the P neurons.

The esophageal rhythm generated in the commissural and esophageal ganglia passes to the stomatogastric ganglion (Spirito, 1975) and can be detected in pyloric neurons (Selverston et al., 1976). The gastric rhythm is likewise modulated by esophageal input, and the two rhythms are loosely coupled in dissected preparations of the stomatogastric nervous system. The E neurons in particular have been shown to receive trains of EPSP's synchronized to the esophageal cycle (Selverston et al., 1976).

Selverston et al. (1976) have suggested that the esophageal rhythm may act as a pacemaker for pyloric and gastric rhythms. Moulins and Vedel (1977) reiterate the concept of a functional hierarchy between the rhythmical centers and offer evidence that the rhythm of the cardiac sac modulates the activity of other foregut centers. These authors also suggest that the IVN through-fibers may act as a master clock for the four rhythmical centers of the foregut. A further synchronizing factor is the common sensory input shared by certain centers, e.g., the esophageal chemoreceptors (Fig. 9F,G) and the posterior stomach receptors (Fig. 7E).

The observation of Bethe (1897) and Morris and Maynard (1970) suggests that there may be inputs to the stomatogastric system from the higher central nervous system. This is an area in which little research has been done, and it is essential to our overall understanding of foregut function that we define the role played by the higher nervous system. Moreover, it is highly probable that functional linkages will be found between the rhythms of the foregut and those of other rhythmical centers employed in food processing, e.g., the mandibles, labrum, paragnatha, midgut, and hindgut.

5. GASTROPYLORIC MOVEMENTS IN INTACT ANIMALS

The normal patterns of gastropyloric movements in intact animals have been investigated in the lobster, *Homarus americanus* (Morris and Maynard, 1970), and in the crabs, *Cancer magister, Cancer productus,* and *Callinectes sapidus* (Powers, 1973). In *Homarus,* the pyloric rhythm is more frequently active than is the gastric rhythm. The regular rhythm of the pylorus increases dramatically when the animal is fed. The gastric rhythm is frequently absent but is recruited on feeding, and its period is longer than that of the pyloric rhythm. Morris and Maynard (1970) quote a mean pyloric period of 1.33 sec and a mean gastric period of 15 sec, measured simultaneously in a lobster. Feeding, or the presentation of chemical stimuli to the legs, antennae, or mouthparts, increases the frequency and intensity of the pyloric cycle and recruits gastric units. The gastric rhythm was examined in greater detail by Powers (1973), who noted that the sensitivity of gastropyloric activity to chemical stimulation exhibited habituation. The observed changes occurred less rapidly than in isolated preparations.

III. CONTROL OF THE MIDGUT

To date, the midgut has received very little attention, and the nature of movements of the midgut or their control is not known. The presence of sensory innervation of the midgut has been noted but not described in detail (Dando and Maynard, 1974).

IV. MOVEMENTS OF THE HINDGUT AND ANUS

Control of movements of the hindgut has been investigated in the crayfish, *Procambarus clarkii* (Wolfe and Larimer, 1971; Muramoto, 1977), and the lobster, *Homarus gammarus* (Winlow and Laverack, 1972a,b,c). In lobsters, the hindgut and anus are innervated by the posterior intestinal nerves (PIN's) and anal nerves, the latter being sensory only. The anatomy of innervation of the hindgut has been described by Alexandrowicz (1909) and Orlov (1926c) as motor and sensory fibers that form a plexus on the gut. Alexandrowicz suggests that these fibers make contact with one another, but this is not confirmed by the observations of Orlov.

Movements of the hindgut and anus are coordinated to produce defecation. The sequence is started by contraction of the longitudinal rectal muscles, followed by a powerful contraction of the circular muscles. As this wave of contraction passes posteriorly, the anus opens to expel the fecal pellet. Section of the gut does not destroy the coordination of this sequence, but section of the PIN's abolishes all coordinated movement. This indicates that the peristaltic wave is neurogenic. Rhythmic movements of the hindgut and anus have been shown to result from activity in the PIN's, since bursts of activity are coincident with the start of the defecatory sequence, and since the coordinated pattern can be elicited by stimulation of the ventral nerve cord. Very short bursts of stimuli produce anal movements of small amplitude, which increase in size with duration of the stimulus. Longer trains of stimuli produce sequential series of rhythmical movements that appear to be largely independent of the duration of the stimulus. The outputs of the hindgut and anal motoneurons are synchronized, both bilaterally and between ipsilateral branches of the PIN's.

Winlock and Laverack (1972b,c) have shown that neurons involved in hindgut movements can be divided into three groups; (1) motoneurons, which can be subdivided into bursting and non-bursting units (though this division is not a distinct one); (2) drivers, which, when stimulated intracellularly, cause the excitation of a number of motoneurons; and (3) interneurons, of which there are at least two pairs. Interneurons are differentiated from drivers in that the latter have their soma in the sixth abdominal ganglion, whereas the former probably do not. Furthermore, the interneurons probably run the length of the central nerve cord, since stimulation of the esophageal connectives initiates coordinated movements of the hindgut and anus. Among the motoneurons, the bursters initiate powerful peristaltic movements. The non-bursters fire irregularly at low frequencies and are considered to prime the longitudinal muscles to facilitate defecation, but they may also be involved in the production of rhythmical movements. The motoneurons laterally excite their bilateral counterparts to produce bilateral coupled output, and some are multibranched, sending axons into all branches of the PIN's. Winlow and Laverack (1972b) suggest that bursting and non-bursting motoneurons may be driven by different interneurons, and they consider at length the interaction between units. Stimulation of the motoneurons in the ventral nerve cord produces a series of rhythmical movements, which may exceed the duration of the stimulus, and there is a maximum duration for which the system can be driven, after which the amplitude of movement is greatly reduced.

In the crayfish species (Muramoto, 1977), anal movements have similarly been shown to be produced by activity of the PIN's. Motoneurons of the anus have been identified anatomically and physiologically: there is one large (L) and one small (S) neuron in each hemiganglion. Burst activity in large motoneurons always precedes anal movement. L motoneurons make contact with a number of other neurons, including interneurons that may correspond to the driver neurons above, and several interganglionic efferent interneurons. Muramoto describes the action of only one interganglionic interneuron that drives the anal rhythm at a frequency independent of the stimulus frequency, though the latency to onset of anal movement is frequency-dependent. Wolfe and Larimer (1971) have distinguished several interganglionic efferent interneurons controlling hindgut activity that are diverse in their effect on motoneuron activity. These efferent interneurons excite different numbers of motoneurons and may be excitatory or inhibitory. Wolfe and Larimer have also identified as least three neurons in each

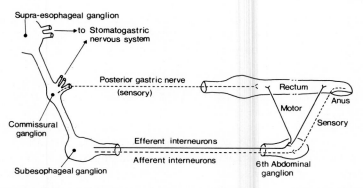

Fig. 10. Innervation of hindgut in relation to central nervous system of *Homarus gammarus*. (After Winlow and Laverack, 1972c.)

connective, which run directly to the intestine without synapsing in the sixth abdominal ganglion. Muramoto (1977) has identified two interneurons in the sixth abdominal ganglion that make contact with motoneuron L. These may correspond to the driver neurons identified physiologically by Winlow and Laverack (1972b), especially since they have points of contact with fibers that pass into other branches of the intestinal nerve.

Proprioceptors that monitor anal movements have been identified (Winlow and Laverack, 1970, 1972a; Muramoto, 1971, 1977), but these appear to play no reflex function in controlling anal or intestinal movements. The receptors involved are simple bipolar cells with multiterminal endings; these cells respond to distortion of the soft cuticle bordering the anus.

It is generally considered that there is a specific relationship between activity of the hindgut and that of the foregut (Fig. 10) (Winlow and Laverack, 1972a; Wolfe and Larimer, 1971), though this relationship is yet to be investigated.

V. PERSPECTIVES AND CONCLUSION

The pioneering work of D. M. Maynard identified the stomatogastric ganglion as a suitable preparation for the study of neural integration (see Chapter 7), and the information on gut control has, to some extent, risen indirectly from this work. The investigation of gut control has therefore evolved centrifugally from the stomatogastric ganglion giving an exaggerated emphasis to the importance and autonomy of this neural unit. As our knowledge of interganglionic pathways has increased, the stomatogastric ganglion has been found to be less autonomous than the literature at first suggested. It is important that the relationship of the stomatogastric system with the remainder of the nervous system be elucidated if we are to understand the mechanisms of gut control.

The role of sensory input (mechanoreceptive, proprioceptive, and chemoreceptive) is becoming more clearly defined, but much more remains to be done in this area, as gut control is likely to be affected by input from a wide range of sense organs both without (antennae and mouthparts) and within the stomatogastric system (see Dando and Maynard, 1974).

The more posterior regions of the gut, particularly the midgut, have received sparse attention, and these regions are clearly worthy of further investigation.

Finally, those neural control systems investigated to date have been shown to modulate mechanical movements and successful food processing requires actions other than transport of food substrates. The degree of neural control of enzyme or mucus release, water absorption, etc., remains to be elucidated.

REFERENCES

Alexandrowicz, J. S. (1909). Zur kenntnis des sympathischen nervensystem der Crustaceen. *Jena. Z. Naturwiss.* **45**, 395–444.

Anderson, D. J., Hannam, A. G., and Matthews, B. (1970). Sensory mechanisms in mammalian teeth and their supporting structures. *Physiol. Rev.* **50**, 171–195.

Bethe, A. (1897). Vergleichende Untersuchungen über die Functionen des Centralnervensystems der Arthropoden. *Arch. Gesamte. Physiol. Menschen Tiere* **68**, 449–545.

Dando, M. R. (1969). Studies an the structure and function of mechanoreceptors in the stomatogastric nervous system of some Decapoda Crustacea. Ph.D. Thesis, University of St. Andrews, Scotland.

Dando, M. R., and Laverack, M. S. (1969). The anatomy and physiology of the posterior stomach nerve (p.s.n.) in some decapod crustacea. *Proc. R. Soc. London, Ser. B* **171**, 465–482.

Dando, M. R., and Maynard, D. M. (1974). The sensory innervation of the foregut of *Panulirus argus* (Decapoda Crustacea). *Mar. Behav. Physiol.* **2**, 283–305.

Dando, M. R., and Selverston, A. I. (1972). Command fibres from the supraoesophageal to the stomatogastric ganglion in *Panulirus argus*. *J. Comp. Physiol.* **78**, 138–175.

Dando, M. R., Chanussot, B., and Nagy, F. (1974). Activation of command fibres to the stomatogastric ganglion by input from a gastric mill proprioceptor in the crab, *Cancer pagurus*. *Mar. Behav. Physiol.* **2**, 197–228.

Ferrero, E., and Wales, W. (1976). The mandibular common inhibitor system. I. Axon topography and the nature of coupling. *J. Comp. Physiol.* **109**, 123–134.

Fryer, G. (1977). Studies of the functional morphology and ecology of the Atyid prawns of Dominica. *Philos. Trans. R. Soc. London, Ser. B* **277**, 57–128.

Govind, C. K., Atwood, H. L., and Maynard, D. M. (1975). Innervation and neuromuscular physiology of intrinsic foregut muscles in the blue crab and spiny lobster. *J. Comp. Physiol.* **96**, 185–204.

Hartline, D. K., and Maynard, D. M. (1975). Motor patterns in the stomatogastric ganglion of the lobster *Panulirus argus*. *J. Exp. Biol.* **62,** 405–420.

Hermann, A., and Dando, M. R. (1977). Mechanism of command fibre operation onto bursting pacemaker neurones in the stomatogastric ganglion of the crab, *Cancer pagurus*. *J. Comp. Physiol.* **114,** 15–33.

Laverack, M. S., and Dando, M. R. (1968). The anatomy and physiology of mouthpart receptors in the lobster, *Homarus vulgaris*. *Z. Vergl. Physiol.* **61,** 176–195.

Macmillan, D. L., Wales, W., and Laverack, M. S. (1976). Mandibular movements and their control in *Homarus gammarus*. III. Effects of load changes. *J. Comp. Physiol.* **106,** 207–221.

Marshall, S. M., and Orr, A. P. (1960). Feeding and nutrition. *In* "The Physiology of Crustacea" (T. H. Waterman, ed.), Vol. 1, pp. 227–258. Academic Press, New York.

Maynard, D. M. (1972). Simpler networks. *Ann. N.Y. Acad. Sci.* **193,** 59–72.

Maynard, D. M., and Burke, W. (1966). Electronic junction and negative feedback in the stomatogastric ganglion of the mud crab, *Scylla serrata*. *Am. Zool.* **6,** 526.

Maynard, D. M., and Dando, M. R. (1974). The structure of the stomatogastric neuromuscular system in *Callinectes sapidus, Homarus americanus* and *Panulirus argus* (Decapoda Crustacea). *Philos. Trans. R. Soc. London, Ser. B* **268,** 161–220.

Maynard, D. M., and Selverston, A. I. (1975). Organization of the stomatogastric ganglion of the spiny lobster. IV. The pyloric system. *J. Comp. Physiol.* **100,** 161–182.

Morris, J., and Maynard, D. M. (1970). Recordings from the stomatogastric nervous system in intact lobsters. *Comp. Biochem. Physiol.* **33,** 969–974.

Moulins, M., and Vedel, J.-P. (1977). Programmation centrale de l'activité motrice rythmique du tube digestif antérieur chez les Crustacés décapodes. *J. Physiol. (Paris)* **73,** 471–510.

Moulins, M., Dando, M. R., and Laverack, M. S. (1976). Further studies on mouthpart receptors in decapod crustacea. *Z. Vergl. Physiol.* **69,** 225–248.

Mulloney, B., and Selverston, A. I. (1974a). Organization of the stomatogastric ganglion of the spiny lobster. I. Neurones driving the lateral teeth. *J. Comp. Physiol.* **91,** 1–32.

Mulloney, B., and Selverston, A. I. (1974b). Organization of the stomatogastric ganglion of the spiny lobster. III. Co-ordination of the two subsets of the gastric system. *J. Comp. Physiol.* **91,** 53–78.

Muramoto, A. (1971). The afferent nerve response of the movement receptor around the anal region of the crayfish. *J. Fac. Sci., Hokkaido Univ., Ser. 6* **17,** 545–563.

Muramoto, A. (1977). Neural control of rhythmic anal contraction in the crayfish. *Comp. Biochem. Physiol. A* **56A,** 551–557.

Nagy, F. (1977). Modulation sensorielle d'une activité motrice programmée par le système nerveux stomatogastrique de la Langouste. *C.R. Hebd. Seances Acad. Sci.* **285,** 921–924.

Nagy, F., and Dando, M. R. (1973). Modification du programme moteur du ganglion stomatogastrique par stimulation d'un nerf sensoriel stomacal chez le Crabe, *Cancer pagurus*. *C.R. Hebd. Seances Acad. Sci.* **276,** 599–601.

Nicol, E. A. T. (1932). The feeding habits of the Galatheidea. *J. Mar. Biol. Assoc. U.K.* **18,** 87–106.

Orlov, J. (1926a). Die Innervation des Darmes des Flusskrebses. *Z. Mikrosk.-Anat. Forsch.* **4,** 101–149.

Orlov, J. (1926b). Systeme nerveux intestinal de l'écrevisse. *Bull. Inst. Rech. Biol. Perm.* **5,** 29–47.

Orlov, J. (1926c). Système nerveux intestinal de l'ecrevisse. *Izv. Biol. Nauchno-Issled. Inst. Biol. Stn. Permsk. Gos. Univ.* **5,** 29–32.

Powers, L. W. (1973). Gastric mill rhythms in intact crabs. *Comp. Biochem. Physiol. A* **46A,** 767–783.

Robertson, R. M. (1978). The anatomy and physiology of organs involved in food ingestion in the lobster *Homarus gammarus*. Ph.D. Thesis, University of St. Andrews, Scotland.

Robertson, R. M., and Laverack, M. S. (1978). Inhibition of oesophageal peristalsis in the lobster after chemical stimulation. *Nature (London)* **271**, 239-240.

Russell, D. F., and Hartline, D. K. (1978). Bursting neural networks: a reexamination. *Science* **200**, 453-456.

Selverston, A. I., and Mulloney, B. (1974). Organization of the stomatogastric ganglion of the spiny lobster. II. Neurons driving the medial tooth. *J. Comp. Physiol.* **91**, 33-51.

Selverston, A. I., Russell, D. F., Miller, J. P., and King, D. G. (1976). The stomatogastric nervous system: Structure and function of a small neural network. *Prog. Neurobiol.* **7**, 215-290.

Spirito, C. P. (1975). The organization of the crayfish oesophageal nervous system. *J. Comp. Physiol.* **102**, 237-249.

Thomas, W. J. (1970). The setae of *Austropotamobius pallipes* (Crustacea : Astacidae). *J. Zool.* **160**, 91-142.

Vedel, J.-P., and Moulins, M. (1977). Functional properties of interganglionic motorneurones in the stomatogastric nervous system of the rock lobster. *J. Comp. Physiol.* **118**, 307-325.

Wales, W. (1972). A comparative study of proprioception in the appendages of decapod crustaceans. Ph.D. Thesis, University of St. Andrews, Scotland.

Wales, W. (1976). Receptors of the mouthparts and gut of arthropods. *In* "Structure and Function of Proprioceptors in the Invertebrates" (P. J. Mill, ed.). Chapman & Hall, London.

Wales, W., and Ferrero, E. (1976). The mandibular common inhibitor system. II. Input sensitivity. *J. Comp. Physiol.* **109**, 135-146.

Wales, W., and Ferrero, E. (1981). In preparation.

Wales, W., and Laverack, M. S. (1972a). The mandibular muscle receptor organ of *Homarus gammarus* (L) (Crustacea, Decapoda). *Z. Morphol. Tiere* **73**, 145-162.

Wales, W., and Laverack, M. S. (1972b). Sensory activity of the mandibular muscle receptor organ of *Homarus gammarus* (L). 1. Response to receptor muscle stretch. *Mar. Behav. Physiol.* **1**, 239-255.

Wales, W., Macmillan, D. L., and Laverack, M. S. (1976a). Mandibular movements and their control in *Homarus gammarus*. 1. Mandible morphology. *J. Comp. Physiol.* **106**, 177-191.

Wales, W., Macmillan, D. L., and Laverack, M. S. (1976b). Mandibular movements and their control in *Homarus gammarus*. II. The normal cycle. *J. Comp. Physiol.* **106**, 193-206.

Winlow, W., and Laverack, M. S. (1970). The occurrence of an anal proprioceptor in the decapod crustacea *Homarus gammarus* (L). (Syn. *H. vulgaris* M. ed) and *Nephrops norvegicus* (Leach). *Life Sci.* **9**, 93-97.

Winlow, W., and Laverack, M. S. (1972a). The control of hindgut motility in the Lobster, *Homarus gammarus* (L). 1. Analysis of hindgut movements and receptor activity. *Mar. Behav. Physiol.* **1**, 1-27.

Winlow, W., and Laverack, M. S. (1972b). The control of hindgut motility in the lobster, *Homarus gammarus* (L). 2. Motor output. *Mar. Behav. Physiol.* **1**, 29-47.

Winlow, W., and Laverack, M. S. (1972c). The control of hindgut motility in the lobster *Homarus gammarus* (L). 3. Structure of the sixth abdominal ganglion (6A.G.) and associated ablation and microelectrode studies. *Mar. Behav. Physiol.* **1**, 93-121.

Wolfe, G. E., and Larimer, J. L. (1971). The intestinal control system in the crayfish, *Procambarus clarkii*. *Am. Zool.* **11**, 666.

7

Small Systems of Neurons: Control of Rhythmic and Reflex Activities

T. J. WIENS

I.	Introduction	193
II.	Several Small Systems: Motor Output and Its Mechanisms	194
	A. Rhythmic Systems	194
	B. Non-Rhythmic Systems	214
III.	Principles of Operation	220
	A. Central Role of Motoneurons in Output Formulation	220
	B. Mechanisms of Rhythmicity	225
IV.	Trends and Prospects	230
	References	232

I. INTRODUCTION

The preceding chapters have outlined the nervous, muscular, and sensory equipment of crustaceans and have raised the question: how does this equipment generate behavior? In attempting an answer, one can argue that crustacean behavior is to some degree "modular" in its mechanism and its central origin: a small subset of the nervous system can, more or less independently, generate a basic unit of behavior through its particular motor output pattern. To be sure, central mechanisms must exist to coordinate the activities of many such subsystems. Still, learning how these component subsystems work will substantially advance our understanding of the nervous system as a whole. An analogy might be drawn with a complex computer program in which simple subroutines are used upon appropriate command to execute simple well-defined tasks. In this chapter, several small subsystems will be introduced. In each case, known neuronal properties and organization will be shown to account for some aspect of the animal's behavior.

The behaviors considered here fall into two categories. Rhythmic activities, featuring automatic repetition of a motor sequence, are important in locomotion, grooming, feeding, and visceral function. An important question here concerns the source of the rhythm—does it arise from an individual cell's membrane or does it emerge from the network interactions of intrinsically non-rhythmic cells? Two small rhythmic neural assemblies—the cardiac ganglion controlling the heartbeat and the stomatogastric ganglion controlling movements of the upper digestive tract—will be examined in the following section in an attempt to answer this question.

In non-rhythmic or episodic (usually reflex) activity, sensory or central drive evokes a single motor sequence (though it may do so repeatedly). Chapters 2 and 8 describe how the crayfish abdomen, for example, may be held tonically in an appropriate posture and how it may be rapidly flexed in an escape response. The present chapter will examine the reflex movements of the crayfish claw and compare these with the defensive withdrawal of the crab's eyecup.

With these descriptions in hand, the third section of this chapter will make comparisons among these systems and seek common rules in their organization.

II. SEVERAL SMALL SYSTEMS: MOTOR OUTPUT AND ITS MECHANISMS

A. Rhythmic Systems

1. STOMATOGASTRIC NERVOUS SYSTEM (*PANULIRUS*)

   a. *Activity Patterns.* In *Panulirus*, as in most malacostracans, food from the cardiac sac of the stomach is ground in a gastric mill before being filtered through the pylorus (see Chapter 6 of this volume). These two independent motor rhythms are imposed on the walls and calcareous ossicles of the stomach by a complex striated musculature. Control is effected by about 23 motoneurons, identifiable (though not in all cases uniquely) by the muscles that they innervate. These 23, together with about 9 interneurons and

7. Small Neural Systems and Control of Activity

perhaps a few sensory neurons, make up the stomatogastric ganglion. The structure (Maynard and Dando, 1974) and physiology (Selverston et al., 1976) of the stomatogastric system in *Panulirus* have recently been described in detail. While the many ossicles and muscles interact in a complex way, a simple interpretation of stomach mechanics and underlying motoneuronal activities can be given, largely following Hartline and Maynard (1975). The neuronal nomenclature employed is that used by Selverston et al. (1976) (Chapter 6 gives a more detailed description of the details of motor output; the present summary is given as a basis for further analysis of neuronal mechanisms).

The gastric mill functions through the contraction of the cardiac sac and the coordinated grinding movements of three "teeth," each comprising several ossicles. In a typical observed sequence:

(1) the two lateral teeth are pulled together (clamping coarse food particles between them) by intrinsic muscles activated by a burst of spikes in the lateral gastric (LG) motoneuron. Another neuron, the median gastric (MG), is activated at the same time but the effect of its target muscle is not clear.

(2) extrinsic muscles (i.e., having their origin on the carapace rather than within the stomach) protract the single medial tooth over the clamped food in a grinding action driven by four gastric mill (GM) motoneurons. Two E interneurons with somata in the commissural ganglia burst with the GM's.

(3) LG and MG cease firing and their antagonists, the two lateral posterior gastric (LPG) motoneurons, commence excitation of extrinsic muscles that pull the lateral teeth apart. An interneuron, INT 1, fires along with LPG.

(4) the GM's cease firing and their antagonist, the dorsal gastric (DG) neuron, drives an intrinsic muscle to retract the medial tooth. At the same time, a single anterior median (AM) motoneuron fires a burst constricting the cardiac sac, thus presumably circulating more coarse food particles between the lateral teeth and propelling fine particles on into the pylorus.

This sequence of neuronal and mechanical activities, summarized in Fig. 1, repeats typically every 7 sec. The existence of a second interneuron, INT 2, has been inferred only from its synaptic effects.

In the pyloric rhythm, about ten times faster, regions of the pylorus and the ventral cardiac sac are driven to constrict, somewhat asynchronously, by eight pyloric (PY), one lateral pyloric (LP), and one inferior cardiac (IC) motoneuron. A period of general dilation is then produced by the firing of two pyloric dilator (PD) neurons and one ventricular dilator (VD) neuron. The effect is to filter coarsely ground food to the midgut and fine particles to the hepatopancreas. The movement is no doubt more complex than a simple posteriad peristalsis, since, for example, the most posterior pyloric constrictor muscles seem to be activated earlier in the cycle than some more anterior

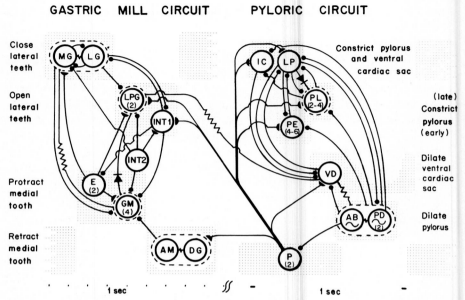

Fig. 1. Activity cycles and synaptic connections of the stomatogastric neurons and the P and E neurons of the commissural ganglia in *Panulirus argus*. The horizontal axis represents time, and the stippled domain surrounding each group of neurons represents their period of spike activity (following Hartline and Maynard, 1975). Note two time scales. Broken lines enclose some electrically-coupled groups of neurons; input or output synapses terminating or originating on the broken line are common to all enclosed cells. Other symbols: —●, inhibitory chemical synapse; —◀, excitatory chemical synapse; ─w─, ─▶─, non-rectifying and rectifying electrical connections; ~, endogenous oscillator neurons. The circuit is revised from that of Selverston *et al.* (1976) according to data of Maynard (1972), Mulloney (1977), Graubard *et al.* (1977), Mulloney and Sigvardt (1978), Hartline *et al.* (1979), and A. I. Selverston and J. P. Miller (unpublished). Known "functional" connections (see text) include an excitatory pathway from INT 1 to DG, and inhibitory pathways from LG to DG, DG to LG, LPG to MG, LG to LP, PD to INT 1, and VD to IC.

constrictors. Hartline *et al.* (1979) have accordingly subdivided the PY class into "late" (PL) and "early" (PE) categories. Several interneurons figure in the pyloric rhythm. The anterior burster (AB) interneuron fires with the PD's and reports the pyloric rhythm as "efference copy" to higher centers (Selverston, 1977; Russell, 1978). Under its inhibitory influence, two P neurons in the commissural ganglia fire in bursts alternating with those of PD/AB.

The neural and mechanical activities of the pyloric rhythm are summarized in Fig. 1.

  b. Neural Organization. Contained in the stomatogastric ganglion are the somata of the 26 neurons already enumerated (excluding P and E) and those of six non-motor EX neurons. Anatomical counts of cell bodies (King, 1976a) reveal some variation in numbers: from 27 to 32 large monopolar cells (probably motoneurons, AB, and EX cells), 1 to 5 smaller monopolar cells (probably other interneurons), and a few bi- or tripolar cells have been observed. These last may be sensory (Larimer and Kennedy, 1966; Dando and Maynard, 1974).

The stomatogastric ganglion is connected to the CNS via the stomatogastric nerve and the esophageal and commissural ganglia. Ascending the stomatogastric nerve are centripetal axons from some interneurons (AB, INT 1, INT 2, EX) and a few sensory cells. Hartline et al. (1979), for example, report an ascending sensory (HD) unit from the duct of the midgut gland (hepatopancreas), which excites PD, AB, and PE neurons in passage. About 120 sheathed centrifugal axons descend the nerve. Several of these (e.g., P and E neurons, IVN through-fibers) participate in, and have profound activating effects on, the stomatogastric rhythms. Indeed, normal pyloric and gastric rhythms are observed in vitro only if the esophageal and commissural ganglia remain attached to the stomatogastric ganglion (in the so-called combined preparation of Russell, 1976a; see also Dando and Selverston, 1972). These ganglia regulate stomatogastric output, acting in part on feedback information in the form of efference copy and sensory input (Selverston et al., 1976). Descending control is exercised in at least two ways: through phasic synaptic effects (see Fig. 1: P and E neurons fire bursts in the pyloric and gastric rhythms respectively, and synaptically excite many stomatogastric neurons) and through tonic chemical effects that somehow unmask bursting properties in stomatogastric neurons (see below).

Maynard and his colleagues first discovered that the motoneurons of the ganglion communicate with each other synaptically (Maynard and Burke, 1966; Maynard, 1967; Maynard and Atwood, 1969). Maynard's scientific heirs (Mulloney and Selverston, 1974a,b; Selverston and Mulloney, 1974; Maynard and Selverston, 1975; Mulloney, 1977; Hartline et al., 1979; Selverston and Miller, 1981) pursued these studies, largely through paired microelectrode penetrations of pre- and postsynaptic cell somata. Unitary postsynaptic potentials (PSP's) following presynaptic spikes were commonly observed, indicating a probable monosynaptic linkage (e.g., Fig. 2). The neural circuit formed by all such known connections is shown in Fig. 1. The circuit is dominated by inhibitory connections and electrical junctions between motoneurons; the few interneurons in the system participate in excitatory, inhibitory, and electrical junctions with each other and with motoneurons.

The complexity of the circuit does not permit an intuitive prediction of firing sequences (Maynard, 1972). However, the connections seem generally consistent with the relative firing phase of each neuron in either the pyloric or gastric cycle (see Sections III,B,4 and 5). Thus, for example, LG

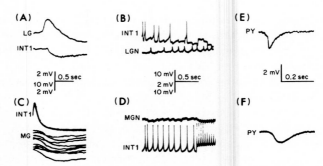

Fig. 2. Evidence for synaptic connections from paired penetrations, showing variation in postsynaptic effects of presynaptic spikes. (A and B) IPSP's in INT 1 following LG spikes, from two different studies. (C and D) IPSP's in MG following INT 1 spikes, from the same two studies. (E and F) IPSP's of different time courses in a PY neuron due to spikes in PD and AB, respectively. (A and C, adapted from Mulloney and Selverston, 1974a; B and D, adapted from Selverston et al., 1976; E and F, adapted from Hartline and Gassie, 1979.)

inhibits and is inhibited by INT 1 as they fire alternating bursts. Several other pairs of neurons also display this reciprocal inhibition, which, as we shall see in Section III, is conducive to rhythmicity. Longer closed chains of inhibitory connections, producing recurrent cyclic inhibition, can also be detected in the circuit, increasing the potential for oscillatory bursting. It is generally agreed (see below) that the gastric mill rhythm is largely determined by such network interactions.

Several connections exist between the pyloric and gastric subcircuits: the pyloric VD neuron is electrically coupled to the gastric LPG's, while the P neurons excite several cells in both circuits. Thus, while the two subcircuits cycle at very different frequencies, their two rhythms do interact weakly (Mulloney, 1977; Russell and Hartline, 1978; Powers, 1973). This interaction may help to synchronize interfacing movements of pyloric and gastric muscles: the bridging LPG neurons, for instance, apparently innervate a pyloric muscle as well as their primary target in the gastric mill (Hartline et al., 1979).

With some cell pairs, a gradual non-unitary potential change occurs in one cell when the other is depolarized by a train of spikes or injected current (e.g., Mulloney and Selverston, 1974a). These so-called "functional" synaptic effects are now believed to result from polysynaptic pathways involving unspecified neurons, and they are listed in the legend of Fig. 1 (see Selverston et al., 1976). They may play crucial roles in the circuit: AM/DG, for instance receive only functional input from other gastric neurons (INT 1, LG).

7. Small Neural Systems and Control of Activity

c. *Synaptic and Cellular Properties.* How these circuits actually generate rhythm is determined by the detailed properties of the neurons and their synapses; these, therefore, merit a closer examination.

i. TRANSMITTER CHEMISTRY. Biochemical and pharmacological studies led Marder (1974, 1976) to conclude that acetylcholine is the neuromuscular transmitter of the PD neurons and probably of the VD, LPG, and GM neurons as well. Glutamate, though, seems the best candidate for the LP, PY, IC, and AM neurons, and now (Lingle, 1980) for LG, MG, and DG also. [Strangely, the target muscles of the last three possess extra-junctional acetylcholine sensitivity in addition to their junctional glutamate receptors (Lingle, 1980); this led Marder to suggest originally that these neurons were cholinergic.]

From Dale's principle, we would expect these neurons to release the same transmitter at their central synapses as at their neuromuscular terminals. Indeed, when either glutamate or cholinergic agonists are iontophoretically applied to stomatogastric neuropil, increases in potassium conductance (as well as other inhibitory and excitatory effects) are observed in local impaled neurons (Marder and Paupardin-Tritsch, 1978a,b).

ii. IPSP TIME COURSE. Maynard (1972) noted that, while most IPSP's in the stomatogastric ganglion are of "short" rise time and duration (Fig. 2E) and are blocked by picrotoxin, some pyloric neurons produce "long" IPSP's (Fig. 2F) that are insensitive to picrotoxin and slightly different in reversal potential. The most recent evidence (Bidaut et al., 1978; Hartline and Gassie, 1979; Selverston and Miller, 1981) suggests that only the cholinergic PD and VD neurons exert "long" inhibition. (The remaining cholinergic cells— GM's, LPG's—have no known inhibitory outputs.) Conflicting observations ("long" inhibition by AB: Maynard, 1972; "short" inhibition by PD: Selverston et al., 1976) may result from the strong electrotonic connections and hence spike synchrony between PD and AB. Secondary variation in PSP time courses arises from differences in membrane time constants (Mulloney et al., 1979).

iii. SYNAPTIC MORPHOLOGY. King's (1976a) ultrastructural study of identified stomatogastric motoneurons revealed two different types of synaptic terminals: type A terminals (found on PD, VD, and LPG neurons) differed from type B terminals (on LP, PY, AM, and DG neurons) in having somewhat smaller and less regular synaptic vesicles. As far as they go, these groupings correspond precisely to the cholinergic and glutamatergic neurons of Marder and Lingle. These three admittedly incomplete lines of evidence thus suggest the generalization that glutamate-releasing cells have type B terminals and produce "short" IPSP's, while type A terminals and long inhibition are characteristic of cholinergic neurons.

In further significant observations, King (1976b) found that synaptic contacts are made exclusively on the distal, unsheathed portions of neuropilar processes, and that the same dendritic twig often contains both pre- and postsynaptic specializations. Unexpectedly (in view of the prevalence of electrical coupling in the ganglion), no gap junctions were observed morphologically.

iv. NON-SPIKE-MEDIATED INHIBITION (NSI). Several stomatogastric neurons can release inhibitory transmitter from their central terminals without spiking; the release is graded by the degree of terminal depolarization above a certain release threshold. This "non-spike inhibition" (Maynard, 1972), or "chemotonic" inhibition (Hartline and Gassie, 1979), can be exerted by spiking cells (PD, LP) during subthreshold depolarization (Mulloney and Selverston, 1974a; Maynard and Walton, 1975) or when treated with TTX (Graubard et al., 1977; Selverston et al., 1976), as well as by the non-spiking EX 1 cells (Maynard and Walton, 1975; Graubard, 1978).

When EX 1 neurons are step-depolarized, hyperpolarization of the postsynaptic GM neurons sets in after a latency of 12-25 msec, builds to a peak over about 100 msec, and then decays to a plateau level (sometimes almost indistinguishable from resting voltage) with an estimated time constant of

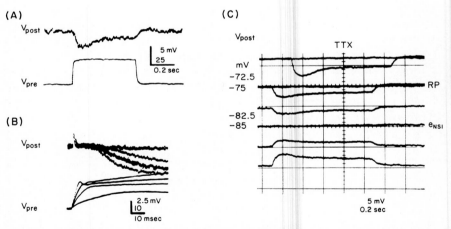

Fig. 3. Characteristics of non-spike-mediated inhibition produced in GM by EX 1. (A and B) Pre- and postsynaptic voltage during presynaptic current injection, on slow and fast time scales. (C) Reversal of postsynaptic effect with postsynaptic hyperpolarization, in the presence of TTX. See text. (Adapted from Graubard, 1978.)

150-200 msec (Fig. 3; Graubard, 1978). Both peak and plateau have the same reversal potential (−76 mV, 17 mV negative to rest). When the presynaptic depolarization ends, the postsynaptic cell shows a transient depolarizing rebound (see also Maynard and Walton, 1975). In a related temporal effect, Graubard found that repeated depolarizing steps elicited declining peaks of postsynaptic hyperpolarization, although plateaus did not diminish.

Finally, Graubard et al. (1977) observed that, in TTX-treated PD neurons, the release threshold was more negative than the resting voltage. Thus, the PD neurons, like vertebrate photoreceptors, could exert a tonic inhibition on their target cells, so that presynaptic hyperpolarization would result in postsynaptic depolarization through disinhibition.

Raper (1978) has shown that non-spike-mediated transmission can produce nearly normal pyloric cycling in TTX-poisoned ganglia (treated with 1 mM dopamine). This, with reports of non-spiking oscillator networks in insects (Pearson and Fourtner, 1975; Burrows, 1979b), hints at the potential significance of this mode of transmission.

v. SYNAPTIC STRENGTH AND FACILITATION PROPERTIES. Clearly relevant to a neural circuit's function are the relative strengths of its synapses. Figure 4 summarizes PSP amplitudes gathered or estimated from the literature for a number of stomatogastric synapses. This measure of synaptic strength is

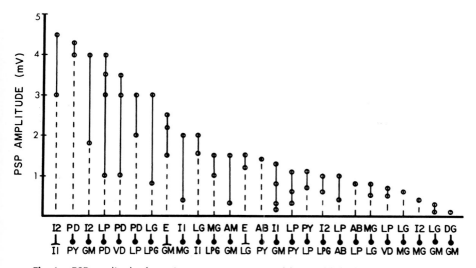

Fig. 4. PSP amplitudes for various synapses, estimated from published records of Mulloney and Selverston (1974a,b), Selverston and Mulloney (1974), Maynard and Selverston (1975), and works listed under Fig. 1. Synaptic symbols as in Fig. 1.

imperfect (ignoring as it does the duration and the electrotonic decrement of the PSP, as well as reversal effects) and variable (Figs. 2, 4). (An alternative measure of "synaptic efficacy" has been utilized by Hartline and Gassie, 1979.) Still, data of this sort have been useful in constructing and improving simulations of these circuits (e.g., Hartline, 1979b; Warshaw and Hartline, 1976; Hartline and Cooke, 1969). Modeling in turn extends our capacity to predict the output of a circuit, given a set of experimentally known or assumed properties. Discrepancies between predicted and observed output can suggest new experimental approaches and lead to improvements in the model. Hartline and colleagues (Hartline, 1979b; Hartline et al., 1979), for example, were led by such discrepancies to discover a "delaying conductance" in some cells (see below).

Facilitation and antifacilitation are evident at only a few of these synapses. The inhibition of LG by MG (Mulloney and Selverston, 1974a, Fig. 12) and that of LP by VD (Selverston and Miller, 1980) weakens with repetition. Facilitation occurs at the synapses of INT 1 and AM onto GM (Selverston and Mulloney, 1974, Figs. 2, 5) and of LG onto LPG (Mulloney and Selverston, 1974a, Fig. 8). These properties have generally been ignored by modellers.

vi. ADAPTATION AND POST-INHIBITORY REBOUND. When held depolarized, most neurons in the isolated stomatogastric ganglion will fire a train of spikes (Fig. 5A) that declines or adapts in frequency (to about 30% of the peak rate for pyloric neurons: Hartline and Gassie, 1979). Similarly, when a hyperpolarization (produced by either synaptic inhibition or injected current) ends, these neurons show post-inhibitory rebound: their spike frequency increases transiently from its inhibited value to a level greater than that at rest (Selverston et al., 1976; Hartline and Gassie, 1979). Figure 5B shows that the strength of the rebound increases with the level of hyperpolarization and further that both a lowering of the spike threshold and a positive shift in "resting" voltage can contribute to the rebound. Through post-inhibitory rebound, a transient hyperpolarization can actually have a net (delayed) excitatory effect: normally silent cells can be made to fire after a bout of inhibition (Hartline and Gassie, 1979) or even to fire alternating bursts indefinitely when linked in pairs by reciprocal inhibition (Perkel and Mulloney, 1974). These properties are thus of great import for network rhythm generation (see Section III,B,3).

vii. DELAYING CONDUCTANCE. In some cells (PY, PD, GM), the recovery of the membrane voltage following imposed hyperpolarization occurs by a slow ramp rather than by the expected abrupt step (Hartline et al., 1979; Hartline, 1979b). Apparently, hyperpolarization *enables* and subsequent

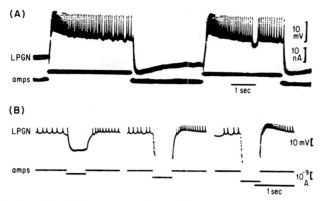

Fig. 5. Transient electrical properties of a typical gastric neuron, LPG. (A) Spike adaptation during injection of constant depolarizing current. (B) Post-inhibitory rebound after passage of increasing hyperpolarizing currents. (Adapted from Selverston et al., 1976.)

depolarization *activates* a "delaying conductance" increase, which slows depolarization through a transient increase in outward current flow. This slowly decaying current thus behaves like the molluscan "early" K current (see Section III,B,1). It could help explain why the PY neurons recover from PD/AB inhibition more slowly than do LP and IC (Fig. 1; Maynard, 1972); it may also contribute to PD's endogenous rhythm (see Section II,A,1,c,ix; Selverston, 1977).

viii. PLATEAU POTENTIALS. If the ganglion is not totally isolated, but left with intact connections to the commissural ganglia, the stomatogastric rhythms are expressed much more powerfully and consistently (Russell, 1976a,b). Russell and Hartline (1978) discovered one underlying mechanism: transient depolarization of most motoneuron somata in the combined preparation triggers plateau potentials, defined as "prolonged regenerative depolarizations resulting from intrinsic membrane properties and contributing to the production of bursts." The 5–20 mV depolarization of the plateau state is stable for several hundred milliseconds before it terminates either spontaneously or through imposed hyperpolarization (Fig. 6A).

Plateaus occur only when "unmasked" by specific descending inputs (IVN through-fibers, P cells) or by input from the hepatopancreas duct unit (Hartline and Russell, 1978). As the unmasking effect persists even when the PSP's of the input fibers are blocked by curare or picrotoxin, one may conclude that unmasking occurs by a direct chemical effect on the membrane, while subsequent triggering is electrical. In keeping with this notion, Raper (1978) suggested that the induction by bath-applied dopamine of

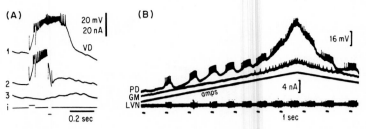

Fig. 6. Burst-producing properties in two pyloric neurons. (A) Plateau potentials in VD in a combined preparation. The plateau is triggered by a depolarizing current pulse (see trace i); it terminates spontaneously in 1 and through a hyperpolarizing pulse (see trace i) in 2. 3 shows a similarly phased but "untriggered" control recording (adapted from Russell and Hartline, 1978, copyright 1978 by the American Association for the Advancement of Science). (B) Endogenous bursting in PD (top and bottom traces) is accelerated when PD is increasingly depolarized (top trace) through "triangular" current injection (third trace). The gastric GM cell (second trace) fires tonically when so depolarized (from Selverston, 1974).

graded, TTX-resistant, pyloric cycling was partly due to the unmasking of plateaus, but this has not yet been confirmed (see also Raper, 1979; Anderson and Barker, 1977; Barker et al., 1979; Selverston and Miller, 1980).

Observed in all pyloric motoneurons as well as in AM, DG, LG, and MG, these plateau potentials will clearly enhance burst production in driven cells. They would not, however, seem capable of producing endogenously rhythmic bursting, as suggested by Russell and Hartline (1978), without an auxiliary interburst depolarizing mechanism.

ix. ENDOGENOUS RHYTHMICITY. Many neurons, known as pacemakers, tend intrinsically to fire tonic, uninterrupted trains of spikes (e.g., INT 1, LPG, GM: Mulloney and Selverston, 1974a; Selverston and Mulloney, 1974). Other neurons, repetitive bursters, generate a more complex rhythm: bursts or trains of spikes interspersed by periods of silence. Such intrinsically bursting cells could, in principle, drive the motor rhythms of the pyloric and gastric systems. In fact, the circuit of the gastric mill appears to lack endogenous bursters. However, the PD/AB neurons of the pyloric circuit are generally conceded to be endogenously rhythmic bursters. PD soma recordings reveal a slow interburst depolarizing ramp that regularly triggers a large and prolonged "burst (or driver) potential," which in turn triggers spikes (Selverston, 1977). This rhythm continues in the isolated ganglion even when rhythmic activity in other pyloric neurons is suppressed by hyperpolarization; on the other hand, "hyperpolarizing out" the PD/AB bursts disrupts the entire pyloric rhythm (Selverston, 1974, 1977; Maynard, 1972). Similarly, the uncoupling of chemical synapses by a low Ca–high Mg bath destroys the bursting rhythm in all but the PD/AB cells (Selverston et al.,

1976). Finally, depolarizing current injected into the PD/AB neurons increases their burst frequency (Fig. 6B), whereas in other cells in the isolated ganglion (Fig. 5A), it simply increases their tonic firing rate. (How unmasked plateau-producing cells respond to sustained depolarizing current is presently not clear.)

While the PD/AB neurons thus seem to be unique in the stomatogastric ganglion as self-contained oscillators, they are not, contrary to erstwhile belief, absolutely essential for the generation of pyloric rhythm. This possibility, suggested by computer simulations of Warshaw and Hartline (1976), was dramatically confirmed by the recent microablation studies of Selverston and Miller (1980). These authors were able to kill individual cells selectively by intracellular injection of Lucifer Yellow and subsequent exposure to ultraviolet radiation (Miller and Selverston, 1979). They found that when both PD's and AB were in this way totally eliminated from the pyloric circuit (in the combined preparation), rhythmic bursting of the remaining pyloric neurons continued, though more weakly. The surviving rhythm could be terminated either by blockage of descending inputs in the stomatogastric nerve (especially P cells) or by the added inactivation of VD (see Section III,B,4).

Nonetheless, the endogenous PD/AB rhythm remains an important determinant of pyloric rhythm. Selverston (1977) reports preliminary attempts at defining its mechanism. The oscillating membrane seems not to be located in or near these neurons' somata (as in molluscan bursters), but in their neuropilar processes; thus, it is inaccessible to true voltage clamping through the soma. Still, the obtainable data suggest a mechanism similar to that in molluscan bursters (Section III,B,1). The interburst depolarization seems to be produced by a decreasing outward K flow (membrane conductance decreases throughout this period, and TEA prolongs the cycle) coupled with inward Na and Ca currents (zero Na reduces the slow wave amplitude as does TTX; Co, a Ca channel blocker, stops the slow wave; Ba enhances it). Hermann's (1979a,b) data from *Cancer pagurus* are consistent with this picture. It will be interesting to see how closely the delaying conductance and plateau potentials that are observed in many non-rhythmic cells (e.g., PY cells have both) correspond to the interburst and burst mechanisms of the rhythmic PD/AB neurons.

2. CARDIAC GANGLION

The crustacean heart, a simple sac of striated muscle, inflates passively and then contracts about once per second through the synchronous discharge of neurons forming the cardiac ganglion in its dorsal wall. Tantalizingly simple in its construction and activity, this ganglion has attracted electrophysiological investigation since the early 1950s (Welsh and Maynard,

1951; Maynard, 1960; Hagiwara, 1961, has reviewed the important early findings). Although its study has been valuable in the development of neurophysiology, the cardiac ganglion is still not completely understood. This description will focus on the lobster genera, *Panulirus* and *Homarus*, and comparisons will be made with data from various crabs and the mantis shrimp *Squilla*.

a. *Anatomy.* The nine neurons that make up the cardiac ganglion of lobsters fall into two groups: an anterior group of five large (c. 54μm × 84μm) motoneurons and a posterior group of four small (c. 22μm × 43μm) neurons apparently lacking motor function (Fig. 7A; Alexandrowicz, 1932; Maynard, 1953, 1955). Their somata are well separated along the ganglion, spanning about a centimeter, and are numbered sequentially from the front. In *Panulirus*, the ganglionic trunk is linear; but in *Homarus*, it bifurcates anteriorly: cells 1 and 2 lie, respectively, in the right and left branches of the resulting "Y" (Hartline, 1967). The ganglion of crabs resembles that of *Homarus*, except that the four small cells and two of the five large cells are clustered at the posterior end of the ganglion.

The neurons themselves may be unipolar, bipolar, or multipolar (Alexan-

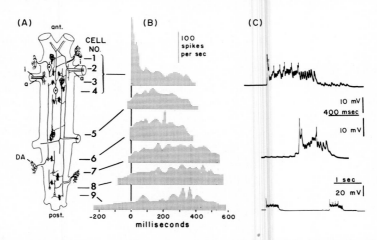

Fig. 7. Schematic structure of the cardiac ganglion and of its activity burst in *Panulirus*. (A) Placement and branching of large (1-5) and small (6-9) cells. Also shown are the entry of the extrinsic accelerator (a) and inhibitor (i) axons, and the latter's distribution. Note dendritic arborizations (DA) inside and outside of the ganglionic trunk (modified from Maynard, 1961). (B) Activation sequence and spike frequency progression of the intrinsic neurons during their burst. Assembled from Friesen's (1975a) data for several bursts, with the beginning of the burst in the large cells taken as the time origin (vertical line). (C) Intracellular recordings from large cell 1 or 2 at different time scales. Note the abrupt onset and the smaller EPSP's preceding and following some bursts (adapted from Bullock and Terzuolo, 1957).

drowicz, 1932), possessing one to three independent axons (Hartline, 1967; Friesen, 1975a), small axon collaterals, neuropilar dendrites, and dendritic arborizations (possibly stretch-sensitive) embedded in the heart muscle (Fig. 7). Large cell axons exit the ganglion through antero- and posterolateral roots. Small cell axons are confined to the ganglionic trunk, travelling forward into, but not beyond, the region of the large cell bodies.

Besides these nine "intrinsic" neurons, three pairs of central cardioregulatory neurons (one inhibitory, two acceleratory) send "extrinsic" axons into the ganglion. These ramify over large cell bodies and over proximal axons of large and small cells, even following their processes into the neuropil and the myocardium (Maynard, 1961; Fig. 7).

   b. *Activity Patterns.* Through extracellular recording, Maynard (1955) demonstrated that each normal heartbeat is triggered by a 300 msec burst of spike activity in the cardiac ganglion. A small-amplitude spike initiates a burst of larger spikes, each of which is followed after 14 msec by a heart muscle potential. Carefully adjusted stimulation of the extrinsic inhibitor axon can silence all ganglion cells except one small unit, which continues to burst in functional isolation, apparently endogenously. The physiological mapping of each neuron's axon(s) later permitted the identification of each neuron's extracellular spike during the burst (Hartline, 1967). In *Panulirus* (Friesen, 1975a; Fig. 7), normal burst activity begins in small cell 9 and spreads anteriad through cells 8 and 7 to cells 5 and 6; cell 6's first spike is followed immediately by a synchronous burst in large cells 1–4, each decelerating rapidly from a peak spike frequency of over 200 impulses per second. In *Homarus* (Hartline, 1979a), the sequence is similar but somewhat less orderly, generally beginning in cell 7.

Intracellular recordings showed that a prolonged and simultaneous depolarization of all the inexcitable large cell somata accompanies the burst (Fig. 7; Hagiwara and Bullock, 1957). This "burst depolarization" normally comprises an abrupt and large onset (20 mV) declining to a lower plateau phase, upon which are superimposed decremented spikes and other brief positive potentials. These last, as well as the onset deflection, are correlated one for one with impinging spikes in small cell axons and have been interpreted as the driving synaptic input through which small cells trigger the large cell burst (see below).

Large cells can remain silent between bursts ("simple follower" activity) or develop pacemaker activity in the late interburst period ("follower with spontaneity"), particularly when tonically excited by stretch of the ganglion. Bullock and Terzuolo (1957) also occasionally observed "follower activity without sustained depolarization," and they noted that a given cell can shift between these modes of activity. Finally, several authors (Bullock and

Terzuolo, 1957; Watanabe, 1958; Connor, 1969; Tazaki, 1971a,c) have observed that slow periodic depolarizations, sometimes triggering bursts of spikes, can occur in large cells in the apparent absence of synaptic input. These observations and their implications will be examined in the following section.

c. *Neural Organization and the Basis of Rhythmicity.* What are the neural properties and connections that produce these activity patterns? A picture currently acceptable for the marine decapods is that the small cells (at least one) are endogenous bursters: a pacemaker depolarization recurrently triggers a slow, electrically excitable "driver potential" that generates a burst of spikes. This burst in turn synaptically drives all large cells in concert, triggering in them slow burst depolarizations, of which a similar active driver potential is probably a component. The resulting output bursts are further synchronized by electrical connections among large cells and between large and small cells. The following subsections will elaborate the various elements of this model and their supporting evidence.

i. SYNAPTIC DRIVING OF LARGE BY SMALL CELLS. The early intracellular recordings suggested strongly that small cells trigger large cell bursts. By recording simultaneously inside large cell somata and outside identified small cell axons of *Panulirus interruptus,* Friesen (1975b) indeed found that all small cells (6-9) synaptically excite all large cells (1-5) (cf. Tazaki, 1971c; Tameyasu, 1976).

Cell 6 elicits a particularly powerful but rapidly antifacilitating EPSP in cells 1 through 4, accounting for the abrupt onset of their burst depolarization. The EPSP's due to cells 7-9 are smaller, but they antifacilitate less and often commence prior to the motor burst onset (Fig. 7; Friesen, 1975b). Hagiwara et al. (1959) measured the reversal potential of these EPSP's to be about -10 mV.

Electrotonic connections also occur between large and small cells (see Section II,A,2,c,iv). In the crabs (Tazaki and Cooke, 1979a), these are important in triggering the motor burst.

ii. SYNAPTIC INTERACTION AMONG SMALL CELLS. The technical difficulty of recording intracellularly from the small cells has been the major obstacle to the study of this system. Still, Friesen (1975b) managed to record EPSP's in cell 6 of *Panulirus* correlated with spikes in cells 7-9. Spike train cross-correlations supported these synaptic influences and, further, suggested excitation of cell 7 by cell 8 and of cell 8 by cell 9. All of these are "forward directed" synapses, consistent with the forward burst progression (Fig. 7). Tameyasu (1976), on the other hand, detected "backward" excitation of cell

7. Small Neural Systems and Control of Activity

7 by cell 6 and of cell 9 by unidentified cells (but his cell identification was less rigorous and thus open to question).

iii. SYNAPTIC INTERACTIONS AMONG LARGE CELLS. An important early result of the intracellular studies was the demonstration of "low pass" electrotonic connections among the large cells (Watanabe, 1958; Hagiwara et al., 1959). Long polarizations due to injected current pass between neighboring somata with attenuation factors as low as 2; shorter events are severely attenuated. This latter characteristic, coupled with the fact (Tazaki, 1972b) that the excitable axons seem to be linked only via their inexcitable somata, prevents action potentials from propagating between cells in *Panulirus* (compare with *Squilla*, below). Hartline (1979a), though, found that spike trains in cells 1 and 2 of *Homarus* induce EPSP's and eventually spikes in cells 3, 4, and 5, but he could not determine whether the transmission was chemical or electrical. In any event, the spread of depolarization among large cells during the burst clearly further synchronizes their activity, which is already simultaneously initiated by common presynaptic drive.

iv. SYNAPTIC ACTIONS OF LARGE CELLS ON SMALL CELLS. When large cell somata are depolarized by injected current, the succeeding small cell burst (evidenced by the EPSP's it produces in large cells) is accelerated; hyperpolarization has the opposite effect (Watanabe, 1958; Watanabe and Bullock, 1960; Tazaki, 1971c). This implies an electrotonic feedback (or other non-spike-mediated excitation) of large cells upon small cells. Accordingly, when a large cell is held depolarized, the small cell axons' threshold for spike production in response to extracellular stimulation is lowered (Watanabe and Bullock, 1960). Similarly, antidromic spike trains in large cells can elicit spikes and hasten bursts in small cells (Mayeri, 1973a). Finally, in crabs, Tazaki and Cooke (1979a) have directly measured electrotonic spread from large to small cells. Thus, there is little doubt that the motor output feeds back upon and influences the rhythm generator in this system.

v. BURST-INITIATING FUNCTION OF SMALL CELLS. From evidence already cited (the isolated bursting of a single small cell during extrinsic inhibition; the commencement of normal burst activity in small cells; the synaptic excitation of large by small cells during the burst), it seems clear that the small cells can burst endogenously and that they normally initiate the motor burst. Two further lines of evidence support this inference. First, if the small cell bodies are surgically isolated by ganglion section (Hagiwara and Bullock, 1957; Connor, 1969; Mayeri, 1973a) or by a ligature (Cooke and Tazaki, 1979), they continue to burst normally, but the large cell rhythm is disrupted (at least temporarily—see below). Chemically silencing the small cells [by topical procaine (Watanabe, 1958) or by DNP (Mayeri, 1973a)] has a similar effect on large cells. Secondly, fragmentary intracellular recordings (Tameyasu, 1976; Tazaki, 1973) from small cell somata in *Panulirus* have revealed slow burst depolarizations, devoid of spikes or EPSP's in some cases, and ramp-like pacemaker depolarizations have triggered spikes in other recordings. Much stronger intracellular data have been obtained from small cells of the crab, *Portunus sanguinolentus*, by Tazaki and Cooke (1979a; Fig. 8). They observed a gradual pacemaker depolarization, which triggered a smooth burst depolarization ("driver potential"), which in turn produced spikes. These spikes produced EPSP's in a simultaneously recorded large cell, triggering its own burst depolarization. Although the authors report that PSP's are not seen in small cells, Fig. 8 seems to show two very small depolarizing potentials (arrows) correlated with spikes in a second active small cell (as would be expected if small cells are synaptically interconnected as in *Panulirus*). This single recording thus encapsulates much of the physiology of the cardiac ganglion of decapods.

vi. ENDOGENOUS BURSTING PROPERTIES OF LARGE CELLS.   What is the origin of the prolonged depolarization underlying the large cells' motor burst? Is it purely a compound synaptic potential due to small-cell excitation, or does an endogenous, active membrane response ("driver potential") contribute to it? Further, if partially endogenous, can the large cell burst be initiated and repeated physiologically without any small-cell input?

Three major lines of evidence support an active component in the large

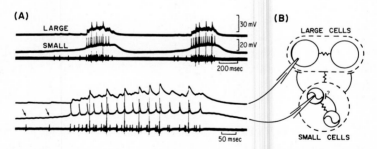

Fig. 8.   Relation of small to large cells in *Portunus sanguinolentus*. (A) Simultaneous recordings of burst activity as seen inside a large and a small cell (top traces), and outside the ganglion (bottom trace—several large and small axons), on two time scales. Note: the interburst depolarization in the small cell but not in the large; EPSP's in the large cell following small cell spikes seen in the lower two traces; possible EPSP's (arrows) in the impaled small cell following spikes in a second, extracellularly-recorded small cell (adapted from Tazaki and Cooke, 1979a). (B) Possible synaptic organization of the cardiac ganglion in lobsters and crabs. Symbols as in Fig. 1. The nature of the synaptic interactions among the small cells is not well established.

cells' burst depolarization and, in abnormal circumstances, endogenous rhythmicity. First, following disruption of the large cell rhythm by the chemical or surgical ablation of small cell somata (see above), large cells may sometimes resume their rhythmic bursting (Maynard, 1966; Watanabe, 1958; Connor, 1969; Mayeri, 1973a). Secondly, intracellular recordings from large cells in the intact or sectioned ganglion sometimes reveal burst depolarizations in which PSP's are absent or not prominent (Tazaki, 1971a,c, 1973); interburst depolarizing ramps appear to trigger these bursts. Thirdly, depolarizing current injected into large cell somata tends to trigger or hasten burst depolarizations sometimes without apparent synaptic input, while hyperpolarizing current delays the burst (Tazaki, 1971a; Matsui et al., 1977). Cooke and Tazaki (1979) reported similar results in large cells of lobster ganglia ligatured to block communication between large and small cells; TTX fails to block this slow burst depolarization.

All of these observations argue for a prolonged, electrically excitable large cell response, triggered by a preceding depolarization that may itself be either synaptic or (abnormally) endogenous in origin. This interpretation is gaining acceptance (Friesen and Stent, 1978; Hartline, 1979a) and has the merit of simplicity.

But one cannot at present exclude the possibility that the burst depolarization is generated entirely by small-cell synaptic input, since large and small cells form an electrical syncytium. Stumps of small cell axons are always present among the large cell somata and can resume rhythmic driving of large cells even in severed or ligatured ganglia (Maynard, 1955; Mayeri, 1973a; Matsui et al., 1977); hence these stumps may well produce slow driver potentials endogenous to themselves. Through the known electrotonic coupling, current injected into large cells may excite the small cell stumps, hastening their bursts. Injury to large or small cells would have a similar depolarizing effect (Connor, 1969). Large cells would then in turn be depolarized through the known excitatory synapses that all small cells make on all large cells (Friesen, 1975b). This synaptic excitation may not always be visible as unitary EPSP's in large cell somata because it may be partially non-spike-mediated (electrotonic or chemotonic) or, if spike-mediated, it may be electrotonically smoothed. [Thus, Tazaki (1973, Fig. 8) saw no EPSP's in large cells of Panulirus japonicus even when small cells were spiking.]

If the burst depolarization is purely synaptically generated in large cells, then its amplitude and time course should be predictable from the known presynaptic spike patterns and input synapse properties. Hartline and Cooke (1969), using real synaptic data in a computer simulation of the system, found that in some cases, the large cell burst depolarization thus predicted agreed almost exactly with that experimentally measured. Thus, to rule out this purely synaptic mechanism of large cell burst generation, it may be necessary to show that large cells can still undergo burst depolarizations in a synaptically decoupled ganglion. Interestingly, in crab ganglia, a bath low in Ca and high in Mg does prevent the burst depolarization, but possibly for other reasons (see below).

The above data are from the lobsters. Recent work on crab ganglia has greatly strengthened the case for an endogenous driver potential in large cells. The crab preparation has the great advantage that small cells can be penetrated more successfully. It has been found that the synaptic (chemical and electrical) organization of the crab ganglion is similar to that in lobsters (Tazaki and Cooke, 1979a), although electrical synchronization of large cell activity seems even more powerful here.

Seemingly spontaneous bursting in large cells of crabs was first noted by Bullock and Terzuolo (1957, "driver without synaptic activity" in *Cancer anthonyi*). Tazaki (1970, 1971b, 1972a,c) later recorded, in the large cells of *Eriocheir japonicus*, both burst depolarizations and interburst pacemaker depolarizations; these were avowedly (but not always convincingly) uncontaminated by synaptic input and, being resistant to TTX, were presumed endogenous. A recent important study (Tazaki and Cooke, 1979a,b,c) in *Portunus sanguinolentus* paints a somewhat different and clearer picture. When large cells are depolarized to a certain threshold by synaptic (chemical or electrical) input or by injected current, "driver potentials" of 20 mV amplitude and several hundred milliseconds duration are triggered; these normally give rise to bursts of spikes (Fig. 8). Usually, however, no endogenous interburst pacemaker depolarization is observed in large cells. When the ganglion is bathed in TTX, spikes are blocked but the driver potentials persist. Their amplitude is graded with the strength and frequency of stimulation. The driver depolarization appears to be dependent on Ca entry, as baths containing low Ca and high Mg, or low concentrations (4 mM) of Mn, block the driver potential (while 4 mM Mn has no apparent effect on EPSP's). Repolarization and a subsequent two-stage after-hyperpolarization involve two probable K conductances, triggered, respectively, by depolarization and by internal Ca buildup.

Small cells produce similar driver potentials, though they are normally triggered differently: synaptic potentials are usually not prominent in small cell somata, but rather an interburst pacemaker depolarization unique to the small cells repeatedly brings their membrane voltage to the threshold for initiation of driver potentials (Fig. 8). These observations clearly define the role of the small cells as generators of burst rhythm and the role of the large cells as followers that possess intrinsic burstiness but no endogenous mechanism for rhythmically initiating bursts. Formally, the small cells behave as relaxation oscillators, in which one state (the interburst) develops in such a way (through slow depolarization) as to trigger a second state (the burst), which in turn eventually reinitiates the first state (upon repolarization). Mayeri (1973b) and J. A. Benson (unpublished) have explored the implications of this property.

These data round out the evidence for the model outlined above and in Fig. 8; it accommodates most of the data from crabs and lobsters, if suitable allowances are made for experimental error. While other interpretations have not been totally excluded, this theory of cardiac ganglionic function is simple and can be tested by further experimentation.

It is interesting to compare these findings in marine decapods with those from the stomatopod *Squilla oratoria* (the mantis shrimp). Its cardiac ganglion is composed of about 14 motoneurons, one per segment (Alexandrowicz, 1934). They are electrically coupled, both through somato-dendritic junctions (in the rostral part of the ganglion) that pass slow voltages, and through axo-axonal "side junctions" (Irisawa and Hama, 1965) that pass and synchronize spike activity (Watanabe and Takeda, 1963; Watanabe et al., 1967a). Unlike the decapod ganglion, chemical synapses have not been demonstrated among these cells, and the cell bodies are electrically excitable. The rostral cells normally trigger the ganglionic burst, which then sweeps posteriad (Watanabe et al., 1967b), but a secondary "pacemaker" at the caudal end of the ganglion can also trigger bursts. The rostral "pacemaker" cells, much like the crab's small cells, undergo a gradual interburst depolarizing ramp which triggers a slow burst depolarization crowned with spikes ("mammalian heart activity"); the rostral burst is delayed or prevented by artificial hyperpolarization (suggesting it is electrically excited) or by extrinsic inhibitor activity (Watanabe et al., 1968). Extrinsic accelerator activity increases frequency and duration of bursts and, at high frequency, "locks" the rostral cells into the burst depolarization plateau state (Watanabe et al., 1969).

Does this simple system, in which each segment is served by one neuron, often clearly combining "pacemaker" and motor function, exemplify a primitive state from which the marine decapod scheme has condensed? If so, it seems that the decapod large and small cells have evolved divergently, one class retaining motor output and the other specializing in rhythm generation.

d. Modulation of Rhythmicity

i. INTRINSIC REFLEX. When the cardiac ganglion is stretched through inflation of the isolated heart, the heartbeat is accelerated and strengthened (Bullock et al., 1954; Maynard, 1961; Cooke, 1966), and individual motoneurons become tonically active between bursts (Bullock and Terzuolo, 1957). Alexandrowicz (1932) noted that some dendrites of motoneurons are embedded in the heart muscle. It seems quite likely that these serve a sensory function, monitoring distension of the heart and modulating the ganglion's motor output through an intrinsic, "asynaptic" stretch reflex.

ii. CARDIOREGULATORY INNERVATION. The cardiac rhythm is also modulated by the inhibitory and acceleratory axons already mentioned. A full discussion of their interesting effects is unfortunately beyond the scope of this chapter (see Maynard, 1953, 1961; Terzuolo and Bullock, 1958; Otani and Bullock, 1959; Watanabe et al., 1968, 1969), but the influences mentioned for *Squilla* are representative (see previous section). A less direct influence of the CNS is exerted by neurosecretion from the pericardial organ. Its secretion, which probably contains both monoamines and peptides (Maynard and Welsh, 1959; Berlind and Cooke, 1970; Cooke and Goldstone, 1970; Kravitz et al., 1976) generally increases burst frequency when applied topically to the proximal axonal regions of small cells, and it has varying effects on burst duration (Cooke and Hartline, 1975).

iii. ELECTROGENIC SODIUM PUMP. Livengood and Kusano (1972) presented strong evidence for a hyperpolarizing Na-K pump in large cardiac ganglion cells of *Homarus*. When activated by intracellular Na injection or by restoration of external K, this pump produces hyperpolarizations of 5-20 mV and stops burst activity. As in other neurons, only two of three extruded Na ions seem to be coupled to inward-transported K ions; a net steady-state hyperpolarization of about 5 mV results, which disappears when ouabain or DNP blocks the pump.

Interestingly, when the normal bath is exchanged for one lacking K, the large cells again hyperpolarize (as expected from the increase in the potassium equilibrium potential), but the ganglion's burst frequency *rises*. This increase presumably results from a net excitation of small cells, through blockage of the electrogenic pump, when external K is removed. Whether the pump plays a significant role in the physiological dynamics of the heart is not known.

B. Non-Rhythmic Systems

Let us now consider two very simple neural subsystems in which motor sequences are not automatically repetitive but are elicited individually by environmental stimuli. More complex behaviors of this sort are treated in the following chapters.

1. EYECUP WITHDRAWAL IN CRABS

The crab's eye can be rapidly withdrawn into its protective socket during a strong tactile stimulus to the eye or to the surrounding carapace (Burrows, 1967; Sandeman, 1967; Burrows and Horridge, 1968). Two motoneurons

7. Small Neural Systems and Control of Activity

activate several muscles (also involved in eyecup rotation; see Chapter 2 of Volume 3 and Chapter 5 of this volume) to effect the withdrawal of each eye. Motor output is elicited by deformation-sensitive cuticular receptors when activated by touch or by electrical stimulation of their axons, but the course of the withdrawal is unaffected by proprioceptive feedback.

Sandeman (1969a,b,c, 1971) has studied the integrative physiology of these motoneurons in *Carcinus maenas* and *Scylla serrata*. Effectively combining intracellular and extracellular recording (Sandeman, 1969b) and dye injection (Sandeman, 1969c), he demonstrated elegantly the existence of a large, non-excitable "integrating segment" in the central motor axon or neurite. Stimulation of the sensory nerve initially produces a graded, short-latency (less than 1 msec) depolarization, upon which are superimposed positive deflections of various sizes: the latter disappear when the motoneuron is artificially hyperpolarized (Sandeman, 1969a,c). Sandeman inferred that the sensory neurons make excitatory synaptic contacts on side branches arising from the motoneuron's integrating segment. Local spikes initiated in the side branches fail to invade the main axon but account for the unitary deflections observed; through summation, the decremented branch spikes can trigger efferent spikes.

This initial reflex depolarization is followed by two further phases in the postsynaptic potential: a repolarization (sometimes into the hyperpolarizing range) and finally, given an adequate initial stimulus, a second depolarization that is generally prolonged and gives rise to a reiterative burst of spikes. The mechanisms of these events are not understood, but Sandeman speculates that a recurrent excitation through an interneuron loop might underly the reiterative burst (which seems to be essential for a complete eye withdrawal).

In *Scylla*, Sandeman (1971) found that the somata of the eye-withdrawal motoneurons are the largest in the brain and yield useful intracellular recordings. Paired penetrations revealed that the two ipsilateral motoneurons are electrically coupled and further that they receive strikingly parallel synaptic inputs. Injection of Procion Yellow and serial sectioning showed that the main axons and branches of the two cells lie close together, sometimes even wrapping around each other; however, actual membrane contact was observed only once, between integrating segments. (Each cell's structure is entirely ipsilateral to its axon, and accordingly no electrical coupling is observed between contralateral homologues.) What appear to be presynaptic terminals sometimes penetrate between the adjacent membranes of the ipsilateral pairs, giving a probable anatomical substrate for their shared synaptic input.

Two central effects on these motoneurons should be noted. First, "spontaneous" bursts of EPSP's and consequent spikes occur irregularly in intact crabs, but they occur rhythmically (at 3–50 sec intervals) if the ipsilateral statocyst is removed (Sandeman, 1967). This input sums with sensory excitation to trigger spikes at an apparently common spike-initiating point (Sandeman, 1969a). Secondly, stimulation of the esophageal connectives also exerts a biphasic, excitatory-then-inhibitory, synaptic effect on the motoneuron (Sandeman, 1969c). The inhibitory phase may summate with suprathreshold sensory input to prevent spike production; it is characterized by a measurable conductance increase and sometimes by hyperpolarization. These last facts, together with the lack of significant conductance increase during synaptic excitation (and other properties of excitatory input—see above) suggested to Sandeman that, in contrast to excitatory synapses, inhibitory contacts occur on or near the integrating segment. The decrement, and hence the efficacy, of any excitatory input would then be dependent not only on the position of the receiving side branch with respect to the spike-initiating zone, but also on concurrent inhibitory short-circuiting of the integrating segment's membrane.

This very simple reflex system, even when taken out of context, thus contains a number of lessons concerning neural integration. These will be brought up again in Section III.

2. CLAW CONTROL IN CRAYFISH

The simply equipped decapod claw has proven valuable in several pioneering lines of research, notably those on crustacean neuromuscular organization (see Chapter 2 of Volume 3), on peripheral post- and presynaptic inhibition (Biedermann, 1887; Hoffmann, 1914; Dudel and Kuffler, 1961) and on the relation between peripheral synaptic structure and integration (Chapter 3 of Volume 3). This description will focus on another aspect of the claw's function: the central interactions of the controlling sensory and motor neurons, particularly in reflex behavior. References will be to the crayfish (*Procambarus clarkii*) unless otherwise specified.

Two muscles, an opener and a closer, rotate the dactyl of the claw about a hinge on the propodite. An excitatory (OE) and an inhibitory (OI) motoneuron control the opener muscle, while a fast and a slow excitor (FCE, SCE) and an inhibitor (CI) innervate the closer muscle (Wiersma and Ripley, 1952; Hill and Lang, 1979). These efferents, whose somata lie in the first thoracic ganglion, can be activated by central interneurons or by sensory inputs of tactile and proprioceptive modalities.

a. Proprioceptive Reflexes. An important resistance reflex derives from the chordotonal P-D organ spanning the propodite-dactyl joint: passive (imposed) opening of the claw activates one population of P-D afferents, which reflexly excites the synergistic "closing" motoneurons FCE, SCE, and OI

(Eckert, 1959; Bush 1963; Chapter 9 of Volume 3). By cross-correlating simultaneously-recorded sensory and motor spike trains, and by measuring delays in axonal conduction, Wiens and Gerstein (1976) showed that a strong narrow peak in probability of motoneuron firing follows each P-D spike, at the latency expected for a monosynaptic reflex pathway. Lindsey and Gerstein (1977, 1979b) recorded unitary EPSP's in both OI and SCE at identical and fixed latencies after spikes in P-D afferents, thus supporting a picture of divergent and direct sensory-to-motor synapses (Fig. 9). They found further that OI and SCE respond to small opening or closing steps at different dactyl positions as though driven by a linear summation of inputs from many P-D afferents, each activated by a unique combination of position and movement (Wiersma and Boettiger, 1959; Mill and Lowe, 1972; Lindsey and Gerstein, 1979a). Those P-D units sensitive to passive opening uniformly excite OI and SCE. Very interestingly, however, their intracellular recordings also revealed inhibition (fixed latency IPSP's in neuropilar processes—see Fig. 9) of OI and SCE by those P-D afferents activated by closing movements; this is logically equivalent to a resistance reflex against closing. That these IPSP's might be mediated by interposed interneurons (e.g., Burrows, 1979a) was suggested by their occasional failure and their fairly long (2.8 msec) latency following the afferent terminal field potential.

b. Tactile Reflexes. Touching the biting surfaces of the claw evokes reflex closing, again through activation of the three closing synergists (Bush,

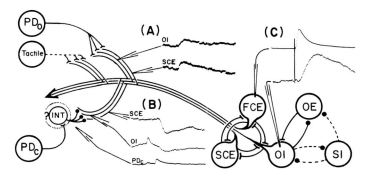

Fig. 9. Synaptic relationships among crayfish claw neurons. Recordings A and B show divergent proprioceptive reflex excitation and inhibition of OI and SCE by opening-sensitive (PD$_o$) and closing-sensitive (PD$_c$) PD afferents. PD spikes trigger the traces; note equal latencies for PSP's in OI and SCE (adapted from Lindsey and Gerstein, 1979b). The triplotomically branching dendrites of the closing synergists (Wiens, 1976) are suggested as the site of this divergent input. Recording C shows averaged excitation of OI by antidromic FCE spikes (from Wiens and Atwood, 1978). Tactile reflex input and other motoneuronal interactions deduced from cross-correlation alone are indicated by broken lines. Possible interposed interneurons (INT) are only inferred. See text. Horizontal calibration: 40 msec duration for all traces.

1963) and again with a sufficiently short sensory-to-motor spike latency to suggest monosynaptic excitation (Wiens and Gerstein, 1976). Touching almost any other part of the carapace causes reflex opening, but concurrent touch inside the claw will reduce this response through central as well as peripheral integrative effects (Bush, 1962: *Carcinus;* Wiens and Gerstein, 1976).

c. *Central Inputs.* Central interneurons activate the claw motoneurons in various combinations and may produce opening or closing of the claw (Atwood and Wiersma, 1967; Smith, 1972, 1974). Often the behavioral effect on the claws is only part of a larger behavior "commanded" by that interneuron: for example, the "defense reflex," obtainable by the stimulation of a single interneuron, involves the raising, extending, and opening of the claws (Wiersma, 1952; Atwood and Wiersma, 1967).

Smith (1972) found that one central interneuron ("B"), whose behavioral effect was to close the claw, produced paired spikes in OI and OE, with OI's generally leading by a few milliseconds. Such timing, necessary for presynaptic inhibition of the opener muscle (Dudel and Kuffler, 1961), had been sought in physiological spike trains in the crayfish but without success (Wilson and Davis, 1965; Smith, 1972). In the fiddler crab, however, Spirito (1970) did detect such phasing between the excitor and inhibitor spikes to the stretcher muscle during reflex activity.

d. *Motoneuron Interactions.* A peculiarity of motor output to the cheliped is that antagonistic efferents (e.g., OE and OI) are often activated in concurrent bursts (Eckert, 1959; Bush, 1963; Wilson and Davis, 1965; Field, 1974); the net behavioral effect depends on their frequency ratio (Ripley and Wiersma, 1953). The occurrence of antagonistic coactivation is related to the following facts: (1) several motoneurons serve multiple roles (e.g., the opener excitor is also the stretcher excitor), (2) peripheral inhibition is used to prevent excessive mechanical interference between antagonistic muscles, and (3) many claw motoneurons are cross-connected synaptically (Fig. 9). This last fact emerged from cross-correlation of concurrent motor spike trains (Wiens and Gerstein, 1975) and was later confirmed by intracellular recording during antidromic stimulation (Wiens, 1976; Wiens and Atwood, 1978). It was found that all of the synergistic closing effectors, OI, SCE, and FCE, mutally excite each other. This cross-excitation, together with common sensory input (see above), tends to synchronize spikes in OI and SCE. Further, OI inhibits OE centrally and OE excites OI. This last finding, unexpected in an antagonistic pair, helps to explain these neurons' coactivation (Bush, 1962).

Two other efferents to the claw participate in such correlations. One of

7. Small Neural Systems and Control of Activity

these, the stretcher inhibitor (SI), until recently was considered also to be the (ineffectual) closer inhibitor and was so labelled by Wiens and Gerstein (1975) on the basis of its anatomy (van Harreveld and Wiersma, 1937; Hoffmann, 1914). However, Hill and Lang (1979) have shown that this designation is incorrect and that SI does not inhibit the closer muscle. This is consistent with Hoffmann's (1914) anatomical finding that SI, while it appeared to join the closer bundle, usually disappeared before reaching the closer muscle; only in 2 out of 50 preparations did it actually branch into the closer muscle with the closer bundle. Moreover, a third, small closer bundle efferent was often noted anatomically by Hoffmann, and physiologically by Wiens and Gerstein (1975), who denoted it as CX. Hill and Lang (1979) have now shown convincingly (in lobsters and crayfish) that a small closer bundle efferent is the true closer inhibitor and is shared with the bender and extensor muscles. It will therefore be denoted herein as CI and can very likely be identified with CX.

This identification lends significance to Wiens and Gerstein's (1975) finding that, especially during tactile opening reflexes, CI (their CX) spikes were closely coupled to those of OE, preceding them by a few milliseconds. This shows that concurrent closer inhibition is used physiologically to enhance claw opening. Whether the synergistic coactivation of CI and OE occurs through cross-connection or shared input is not clear.

The same study revealed a deep central trough in the OI–SI (their CI) cross-correlogram, suggesting reciprocal inhibition, or at least inhibition of OI by SI (Fig. 9). This would tend to prevent coactivation of the two inhibitors, which seems appropriate since the opener and stretcher muscles share a single excitor axon; the inhibitors OI and SI can disable either muscle to allow specific contraction of the other, but it would normally seem undesirable to inhibit both targets simultaneously. A few similar observations also suggested that SI inhibited its antagonist OE (SE).

The functional relationships of the claw motoneurons are reflected closely in their structures, as revealed by cobalt injection (Wiens 1976; see Chapter 1 of Volume 3, Fig. 14). The neurites of the three closing synergists converge and send their major dendrites out in register, branching triplotomically. The most prominent dendrites form a triple hoop encircling the large root, through which the tactile and proprioceptive afferents enter the ganglion. It is then hardly surprising that these afferents can provide divergent input with identical latency to OI and SCE (Fig. 9). The opener excitor's antagonism to this trio is reflected in a totally different structure. The structures of CI and SI have not been studied sufficiently, but their somata lie near the midline of the ganglion (T. J. Wiens, unpublished results). With their possible exception, none of the claw motoneurons appear to cross the midline.

Thus, the claw motoneurons interact in simple ways with each other and a few classes of sensory neurons; the resulting neural circuit seems designed to ensure efficient claw movements in response to a variety of sensory and central activating influences.

III. PRINCIPLES OF OPERATION

This chapter's focus on small motor systems has heavily biased the sample of neurons considered toward motoneurons. This bias is reflected below and should be recognized. Other chapters will redress this imbalance.

A. Central Role of Motoneurons in Output Formulation

A striking generality that emerges from an overview of the systems described above is the large role that many crustacean motoneurons play in determining their own output. The early discovery of command interneurons and their important effects led naturally to an emphasis on pre-motor integration (e.g., Davis and Kennedy, 1972b,c; Kennedy, 1971; Wilson and Davis, 1965). But it is now apparent that motoneurons do not merely relay to muscle activity patterns that have been determined more centrally; they also integrate and transmit centrally large amounts of information from a variety of sources, including their own variously endowed membranes.

1. INTRINSIC PROPERTIES OF MOTONEURONS

   a. *Structure and Integrative Strategy.* The functional anatomy of crustacean motoneurons is dealt with in Chapters 1 and 3 of Volume 3. A few structural features particularly relevant to their integrative role bear reiteration. (1) The large extent of many motoneurons' dendritic fields attests to their assimilation of a great deal of diverse synaptic input. Related to this is the dendritic spike mechanism (e.g., Section II,B,1; Kennedy and Mellon, 1964; Takeda and Kennedy, 1964; Sandeman, 1969c); it may be that purely electrotonic integration via very long dendrites entails too much attenuation of some important inputs, so that branch spikes become necessary. (2) The dual pre- and postsynaptic specializations of single motor dendrites (King, 1976b) fit the motoneuron for a central transmitting role. (3) The parallel dendritic structures in some synergistic neurons (e.g., Section II,B) form an anatomical substrate for divergent sensory input. Furthermore, it has been suggested by Sandeman (1971), citing calculations of Bennett and Auerbach (1969), that the close and extensive apposition of these structures could produce electrical coupling, even without specialized electrical junctions [which have been conspicuously indetectible in several anatomical studies

(King, 1976a,b; Sandeman, 1971)]. (4) Most motoneurons have structures totally ipsilateral to their target muscle, but some (e.g., certain abdominal flexor and extensor motoneurons) are contralaterally deployed. This necessitates decussation of bilateral homologues and provides a possible coupling site (e.g., Wine et al., 1974; Tatton and Sokolove, 1975). It seems that this latter arrangement may be reserved for systems where left and right effectors are rigidly coupled rather than independently maneuverable.

b. *Size Principle.* Davis (1971) found that motoneurons with smaller axons and somata were more tonically active and more easily recruited by excitatory drive than their larger synergists in the swimmeret system of *Homarus.* He inferred that Henneman's (Henneman et al., 1965) "size principle" could be extended to this invertebrate system. Similar tendencies exist among the abdominal slow flexor excitors (Evoy et al., 1967; Wine et al., 1974) and the claw closer excitors (Wiens and Gerstein, 1976). One should bear in mind, however, that crustacean motoneurons, unlike those of vertebrates, tend to be unique and few in number; the slow flexor excitors, for instance, differ in their connections to command fibers (Evoy and Kennedy, 1967) and to the muscle receptor organ (Chapters 2 and 8 of this volume). Factors other than size will, therefore, influence their recruitment during different behaviors.

c. *Oscillatory and Bursting Properties.* The evident importance of motoneuron membrane properties in the generation of rhythms and bursts will be discussed in Section III,B.

d. *Sensory Functions of Motoneurons.* The large cells of the cardiac ganglion probably monitor inflation of the heart through stretch-sensitive dendrites, stretch increasing their excitability. A similar "asynaptic" stretch reflex seems to be mediated by the salivary burster motoneuron of the mollusk *Limax,* which both monitors and excites contractions of the salivary duct (Beltz and Gelperin, 1978). Also, in *Caenorhabditis elegans,* a sensory neuron forms a neuromuscular synapse (Ware et al., 1975). While something of a novelty, this combined sensory-motor capability may be important in very simple control situations.

2. THE VARIETY OF MOTONEURON SYNAPTIC INTERACTIONS

a. *Sensory Input.* Motoneurons appear to receive monosynaptic sensory excitation in a number of systems and with a variety of effects. The simplest examples may be the tactile reflexes, such as the crab's eyecup withdrawal and (possibly) the tactile claw-closing reflex. Closely related is the eyestalk rotation monosynaptically induced by statocyst afferents (Chapter 5 of this volume). The proprioceptive resistance reflexes opposing passive opening and closing of the crayfish claw probably operate, respectively, through monosynaptic excitation and disynaptic inhibition of the closing effector motoneurons. The stretch reflex opposing remotion of the crab coxa (Chapter 9 of Volume 3) is particularly significant because it has provided a long-standing demonstration of the potential of non-spike-mediated transmission, as well as a convenient model of such synapses (Blight and Llinas, 1978).

In locust flight (Burrows, 1975; Wilson and Gettrup, 1963), forewing stretch receptors that sense wing elevation monosynaptically excite the depressor motoneurons of the ipsilateral fore- and hindwings; they inhibit the levators with a somewhat longer latency, again suggesting an interposed "signal-inverting" neuron, perhaps non-spiking (see Burrows, 1979a). Thus, proprioceptive feedback here monosynaptically modulates and coordinates rhythms of the wing. In the crayfish abdomen, the extension-sensing muscle receptor organs (Chapter 2 of this volume) play similar roles: each excites a slow extensor motoneuron (as well as fast extensors and the fast flexor inhibitor—see Chapter 8) in its own segment, effecting a resistance reflex. Additionally, each excites inhibitory ("accessory") efferents to the MRO's of adjacent segments, so that during voluntary postural flexions, which begin posteriorly and progress forward, the now inappropriate resistance reflex will be weakened or eliminated in each succeeding segment. Finally, a presumably sensory unit in the duct of the midgut gland (hepatopancreas) excites monosynaptically the AB/PD neurons, accelerating the pyloric rhythm (Hartline et al., 1979).

These examples, chosen for their monosynaptic character, illustrate many of the general roles played by sensory inputs in motor systems. Other roles may be mentioned. Through variations in strength of their output synapses, sensory neurons may confer plasticity on reflex behavior (Chapter 8 of this volume). Another sensory function may be load compensation during centrally commanded output (Chapters 2 and 3 of this volume). Additionally, some behavioral rhythms rely on sensory input as a generator of rhythm rather than simply a modulator. Thus, swimming in scallops (Mellon, 1969) seems to comprise a series of stretch reflexes, triggered by elastic opening of the shell and "gated" by noxious chemical stimulation. Less extremely, when sand crabs "tread water" (Paul, 1976), the power stroke seems to be a resistance reflex movement elicited by the preceding, centrally-driven return stroke. Finally, many complex behaviors (e.g., the defense reflex) are triggered by sensory input filtered through an interneuron network often culminating in a command neuron.

7. Small Neural Systems and Control of Activity 223

b. Interactions of Motoneurons with Interneurons. Motoneurons receive direct synaptic input from interneurons in many known cases, and it seems likely that most motor drive comes via interneurons. For instance, in the crayfish escape response (a behavior requiring a high degree of speed, sensory convergence, and motor divergence), the lateral giant neuron supplies excitation to all the fast flexor motoneurons (though not exclusively monosynaptically, as once thought—see Chapter 8). In such cases, interneurons seem to usurp much of the motoneurons' integrative role.

The command neurons driving motor activity of swimmerets may also do so monosynaptically (Paton, 1978; Davis and Kennedy, 1972a), and the extrinsic modulation of cardiac motoneurons is partly monosynaptic. Oscillatory interneurons, spiking and non-spiking, inhibit or excite motoneurons monosynaptically in the cardiac and stomatogastric ganglia, the swimmeret system (Paul, 1981), (probably) the scaphognathite system (Simmers, 1979; Mendelson, 1971), and walking systems of insects (Burrows and Siegler, 1978; Pearson and Fourtner, 1975). Interneurons can be used for "signal-inversion": whenever a certain non-spiking interneuron in the locust receives an EPSP, it transmits an IPSP to a postsynaptic motoneuron (Burrows, 1979a). Perhaps delayed inhibitory effects of sensory spikes on motoneurons to the claw (above) and to locust wings (Burrows, 1975) are relayed in this way.

In turn, motoneurons transmit synaptic output to interneurons. This occurs in the oscillating circuits controlling the heart, stomach, and swimmerets, and allows the actual motor output to modulate the circuit's oscillatory activity; the motoneuron may thus feed back to the oscillator information from other sources (e.g., sensory input, endogenously generated potentials), and, at the same time, supply efference copy to higher centers. Even the abdominal fast flexor motoneurons, whose activities are largely centrally determined, seem to feed back their output to interneurons, which in turn inhibit other motoneurons (Chapter 8).

While many of the general roles of interneurons are illustrated in these examples (roles in oscillation, sensory integration, and motor coordination of complex behaviors—see Chapter 9), a few others that involve interaction with sensory neurons or other interneurons should be mentioned. Central interneurons act as sensory "feature extractors" (e.g., Chapter 1 of this volume; Calabrese, 1976). They can inhibit sensory terminals, suppressing anticipated input, as during the swimming of crayfish (Chapter 8). They can act as "delay lines" (Maynard, 1966; Wine, 1977). Finally, they are involved in behavioral plasticity; depression of sensory input to interneurons underlies habituation of the crayfish escape response (Chapter 8) and, in *Aplysia*, an interneuron can resensitize a depressed sensory-motor synapse (Klein and

Kandel, 1978). Thus, unique functions performed by interneurons can increase the subtlety and adaptiveness of simple motor system responses many-fold.

c. *Motor–Motor Interactions.* Finally, motoneurons synaptically influence and are influenced by other motoneurons. The stomatogastric ganglion alone will amply convince the skeptic, and many more examples are found in the cardiac ganglion, the swimmeret system (Heitler, 1978), the abdominal postural system (Chapter 2), the claw control system, the eyecup withdrawal system, and others. These cross-connections serve many purposes. In both reflex and rhythmic systems, they promote coactivation of synergists and sometimes of bilateral homologues, and the decoupling of antagonists. In some circuits that are rhythmic through connectivity, motoneuron cross-connections are essential to rhythm generation. In other cases they may transmit, coordinate, and even modulate rhythm generated elsewhere. The very existence of motor–motor connections attests to the motoneuron's deep involvement in the integrative process.

d. *Roles of Inhibition in Small Neural Systems.* Synaptic inhibition is put to a fascinating variety of uses in crustacean systems. Peripherally, motoneurons exert post- and presynaptic inhibition on muscle and on terminals of other motoneurons. Central inhibition can occur even between motoneurons excitatory to muscle (in the stomatogastric system) and may be reciprocal; it helps assure advantageous sequences and combinations of activity in rhythmic and non-rhythmic systems. Through post-inhibitory rebound, inhibition may have a delayed excitatory effect important for rhythm generation (Section II,A,1,c,vi). Inhibition of the dendrites or the synaptic terminals of sensory cells (if carefully timed—Wine, 1977) prevents both inappropriate reflex responses (Chapters 2 and 8) and synaptic depletion during anticipated sensory input. More generally, inhibition can "gate out" whole behavior patterns, either rhythms (in the cardiac ganglion or molluscan bursters—see below) or reflexes (e.g., the eyecup withdrawal response). In some reflexes (see above), sensory input exerts inhibition on certain motoneurons (perhaps indirectly) during excitation of their antagonists. In sensory processing, lateral inhibitory arrays sharpen contrast detection, as the classic studies in *Limulus* demonstrate (Hartline et al., 1961).

It is appropriate to stress here our growing awareness of non-spike-mediated inhibition (and excitation) and its enormous potential importance. Utilized by interneurons, sensory cells, and motor cells, non-spiking transmission makes possible a subtle, continuously graded control over post-synaptic voltage: when release is tonic (Graubard, 1978; Burrows and

Siegler, 1978), both positive and negative influences relative to background are possible through the same synapse. Non-spiking inhibitory cells can act as sensitive signal inverters, or, when linked in circuits, as rhythm generators (Burrows, 1979a,b): they are involved in the rhythms of the swimmerets (Paul, 1981), the scaphognathites (Simmers, 1979), and insect limbs. Non-spike-mediated chemical excitation underlies the crab's coxal stretch reflex (Chapter 9 of Volume 3) and may occur more generally. Clearly, the full implications of non-spike-mediated transmission remain to be discovered, but they will be great.

B. Mechanisms of Rhythmicity

Rhythmic bursting occurs in some circuits by virtue of the endogenous "driver" properties of one or more participating cells; other circuits, lacking such special cells, burst only through their cyclical synaptic arrangement. Let us examine these mechanisms more closely.

1. RHYTHMICITY INTRINSIC TO SINGLE NEURONS

While the mechanism of endogenous polarization rhythms has been little studied in crustaceans, a useful working model has emerged from studies on mollusks. In molluscan bursting cells, the "bursting pacemaker potential" originates in or near the soma and can, therefore, be studied by a voltage clamp on the soma membrane (see Kandel, 1976). Such studies have revealed ionic conductance properties, enumerated below, that vary with voltage, time, and intracellular Ca levels; their periodic variation produces a gradual interburst depolarization, which triggers the sustained bursting pacemaker potential (with accompanying spikes), which in turn yields to a repolarization and a new cycle (Fig. 10A).

To begin, spikes in these cells are due to the familiar voltage-dependent sodium and potassium conductances: a "rapid G_{Na} (V)" that, together with a similar Ca conductance, produces the regenerative upswing of the spike, and a "delayed G_K (V)" that repolarizes the membrane (Connor and Stevens, 1971a). A second component of delayed K conductance, "delayed G_K (Ca)," is triggered by the elevated intracellular Ca levels arising from the inward spike current (Thompson, 1977; Meech and Standen, 1975). Strongly activated by a burst of spikes, this would clearly help terminate the burst and also act as a burst-spacing mechanism. A third potassium current, "early G_K," also helps shape the interburst depolarization. It is "enabled" by the post-burst hyperpolarization and "activated" by a small subsequent depolarization; it thus acts as a brake on the interburst depolarizing drift, until it inactivates spontaneously (Connor and Stevens, 1971b; Neher, 1971; Thompson, 1977). A rather high "resting G_{Na}" (Smith et al., 1975; Carpenter and Gunn, 1970), which is persistent and voltage-independent, then

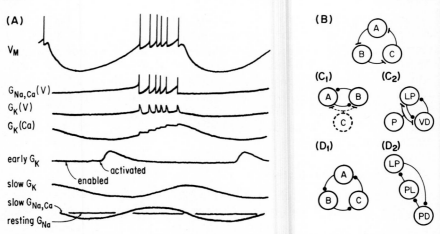

Fig. 10. Mechanisms for endogenous and network rhythmicity. (A) A model for the endogenous burst cycle in a molluscan neuron and the underlying conductance changes (based on Kandel, 1976, and references cited in text). (B, C, and D) Cyclical synaptic arrangements for generating network rhythm. (B) Recurrent cyclic excitation; (C_1) reciprocal inhibition with possible shared excitatory drive; (C_2) a pyloric example; (D_1) recurrent cyclic inhibition; (D_2) a pyloric example.

gradually depolarizes the membrane as all the K currents die away. This interburst depolarization, in turn, activates a persistent, voltage-dependent "slow G_{Na}" (and sometimes a parallel Ca current: Eckert and Lux, 1976) that initiates the burst, producing a sustained depolarization to about −30 mV (Smith et al., 1975). Finally, a voltage-dependent "slow G_K" is slowly turned on by the burst depolarization and, together with the delayed G_K's, repolarizes the membrane, terminating the burst (Smith et al., 1975) before decaying with a time constant of 13 msec.

What we have here, then, is a rapid (spike) oscillation superimposed on an analogous, but much slower, oscillation that results from the slow Na and K conductances and the high resting G_{Na}; the early G_K and the delayed G_K (Ca) help determine the burst frequency. The special character of this slow oscillation is clear from Strumwasser's (1971) observation that burst potentials persist in TTX-poisoned *Aplysia* cells though spikes are suppressed.

Since most of the underlying conductances are voltage-dependent, the slow oscillation can be predictably influenced by imposed polarization: hyperpolarization slows and depolarization accelerates the rhythm. (This constitutes one criterion for endogenous rhythmicity.) An implication is that the cell's bursting may be modulated by electrical or chemical transmission through their polarization effects (e.g., Ayers and Selverston, 1979). Chemical transmission further seems capable of affecting the membrane's ionic conductances chemically, through metabolic effects: "long" IPSP's in *Aplysia* bursters can suppress bursting either by eliminating the slow inward current of Na and Ca (Wilson and Wachtel, 1978) or by inducing a prolonged K conductance increase (Parnas and Strumwasser, 1974).

How much of this molluscan physiology applies to crustaceans is presently uncertain, but Selverston's (1977) data suggest a closely parallel burst mechanism in the pyloric bursters, and cardiac cells of crabs may have similar properties (Section II,A).

2. NON-RHYTHMIC ENDOGENOUS BURSTING

A number of cells in the stomatogastric and cardiac ganglia, although they do not generate rhythmic bursts endogenously, do produce slow "driver" or "plateau" potentials and bursts of spikes when synaptically excited. The properties that Tazaki and Cooke (1979a,b,c) found to underlie the cardiac large-cell burst in crabs are reminiscent of some of the molluscan burster characteristics: the driver potential is induced by depolarization, is regenerative, TTX-resistant, dependent on Ca entry, and is probably terminated by K conductances activated by voltage and elevated calcium. It seems possible that such cells may differ from true endogenous bursters mainly by the absence of an interburst depolarizing ("pacemaker") mechanism, such as the high resting G_{Na} of the molluscan cells. Small cells of crabs, on the other hand, evidently do possess some such mechanism. In the stomatogastric ganglion, chemical modulation may be required to unmask some of the burst-producing conductances. Much still needs to be learned about these plateau mechanisms.

3. RHYTHMICITY EMERGENT FROM NETWORK CONNECTIVITY

Appropriate cyclic synaptic configurations can produce repetitive bursting even in endogenously non-rhythmic cells (Wilson, 1966). The reader is referred to Friesen and Stent's (1978) excellent review of the relevant principles and experimental examples. Burst activity may be transmitted around a circuit either by synaptic excitation or inhibition. Recurrent excitation (Fig. 10B) produces, through positive feedback, an accelerating simultaneous burst in all the linked neurons; it requires a burst-terminating mechanism (e.g., refractoriness, adaptation, recruitment of a common inhibitor) to initiate the interburst hiatus. Getting and Willows (1974) propose that electrically coupled TGN cells of *Tritonia* burst in this way; termination occurs when spikes become synchronous, and their afterhyperpolarization, which can then no longer be shunted onto more depolarized coupled cells, triggers an early G_K (see above) to retard repolarization (but see Friesen and Stent, 1978).

Cyclic inhibition produces alternating rather than simultaneous activity and is thus more appropriate to systems requiring more than one phase of excitation in the output cycle. The simplest cyclic inhibitory loop consists of two reciprocally inhibitory neurons (Fig. 10C); while one cell (A) fires, the other (B) is inhibited. To burst repetitively, this circuit requires (1) a source of excitation to make the cells fire and (2) a restorative mechanism to switch the activity between the cells. Interestingly, the single membrane property of post-inhibitory rebound can fill both requirements (see Section II,A,1,c,vi). Its delayed and transient excitatory effect causes cell A to fire rapidly when released from inhibition but then to slow down and stop, allowing a similar sequence in cell B. Theoretically capable of inducing alternating bursts even in normally silent cells (Perkel and Mulloney, 1974), and having been shown to occur in the oscillatory inhibitory circuits of the stomatogastric ganglion (Selverston et al., 1976), post-inhibitory rebound seems sure to be a very important element in network rhythm generators.

Other possible restorative mechanisms are spike adaptation and synaptic fatigue. As these have no excitatory component, tonic excitatory drive to cells A and B from a third cell C (Fig. 10C) may be required to maintain activity, unless A and B are tonically active. This requirement may be exploited by the nervous system to exert control over an oscillating subsystem. A rhythm can be turned on and modulated by the tonic activity of a central (command) neuron excitatory to both members of an antagonistic and reciprocally inhibitory pair (Friesen and Stent, 1978). Examples occur in the pyloric circuit (Fig. $10C_2$) and in the swimmeret system, where motoneurons again constitute at least part of the oscillator (Heitler, 1978). Paton (1978; personal communication) has shown here that a single command neuron produces time-locked excitatory effects on two antagonistic motoneurons.

Expanding the recurrent cyclic inhibitory loop to three or more neurons (Fig. 10D) again has the advantage of making more oscillation phases available for the motor sequence. Secondly, for an odd number of neurons, it eliminates the need for a restorative mechanism to ensure alternation. If A, B, and C are spontaneously active, and if A inhibits B, B inhibits C, and C inhibits A, then when A is active, B is suppressed allowing C to recover activity; when C then becomes active, A will be suppressed, and B can recover. Thus, activity automatically proceeds around the circuit in a direction opposite to that of the inhibitory transmission. (Examples occur in the pyloric circuit—see below and Fig. $10D_2$.) Cyclic inhibitory circuits with even numbers of neurons, however, require additional connections to maintain progression (see Friesen and Stent, 1978); otherwise, a stable state can develop in which every second cell is tonically active.

4. PYLORIC RHYTHMICITY

The pyloric system provides an instructive application of these ideas in that it embodies most of the oscillatory mechanisms described above. The

PD/AB neurons are endogenous bursters, and all the other pyloric neurons can produce plateau potentials. One could view the connectivity (Fig. 1) among these cells as acting primarily to impose the PD/AB rhythm on the other pyloric neurons: when the PD's fire, VD is recruited electrotonically while LP, IC, and the PY's are silenced by inhibition. When the PD's cease firing, LP and IC recover activity before the PY's because of the latter's delaying conductance. The P neurons reinforce PD's effect: they excite all the other pyloric neurons when PD/AB are silent, but they are in turn inhibited by AB's "efference copy" burst. (VD's timing is somewhat variable because of its inhibition by PD/AB, which becomes important when the PD's fire rapidly: Selverston et al., 1976.)

On the other view, the pyloric neurons' connectivity is itself capable of generating an emergent rhythm, independently of endogenous PD/AB bursting (Warshaw and Hartline, 1976). Indeed, Fig. $10D_2$ shows that several three-neuron loops of recurrent cyclic inhibition exist in the circuit, and that burst activity does progress around these loops in the direction opposite to that of the synaptic transmission, as predicted above. (This circumstance alone, for instance, predicts a delayed firing of the PY's relative to LP and IC.) In addition, reciprocal inhibition occurring between LP and each of PL, PE, VD, and PD should contribute to rhythmicity.

It is interesting to consider Selverston and Miller's (1980; see Section II,A,1,c,ix) selective inactivation experiments in this light. That pyloric rhythm persists when PD/AB are dead could be explained in two ways. First, "back-up" endogenous bursting may occur in VD (perhaps, e.g., plateau potentials are repeatedly triggered by tonic P cell excitation); or, secondly, the surviving network may produce emergent rhythm. Both mechanisms are probably at work. When LP (a key network element in that it participates in almost all pyloric cyclic inhibition loops) is inactivated along with PD/AB, VD continues to fire rhythmically as long as descending input to the ganglion is maintained. This would not be expected from VD's remaining known connectivity alone, and it implies an endogenous VD rhythm. On the other hand, if VD and PD/AB are killed, LP and the PY's can continue to fire in rough alternation, probably through reciprocal inhibition. (One could hardly expect a normal pyloric rhythm in the latter case, since all the neurons that normally fire in the second half of the cycle are now dead.)

It thus appears likely that both emergent and endogenous rhythmic tendencies interact in determining, somewhat redundantly, the pyloric oscillation.

5. RHYTHMICITY OF THE GASTRIC MILL

The gastric circuit lacks endogenously bursting cells, although several gastric neurons produce plateaus and several (LPG, GM, E, INT 1) are tonic pacemakers. The synaptic interaction of Interneuron 1 with the motoneurons driving the lateral teeth is of key importance. INT 1 (with LPG) alternates with the LG/MG pair through reciprocal inhibition; the duration of postinhibitory rebound in these neurons determines the period of their rhythm (B. Mulloney, personal communication). INT 1 in turn ensures properly phased alternation of the medial tooth neurons by simultaneously exciting DG (through "functional" connections: Selverston and Mulloney, 1974; see legend, Fig. 1) and inhibiting its antagonists, the GM's, as well as the E neurons of the commissural ganglion. These interpretations, while not detailed, probably capture the essence of the gastric mechanism, and provide an intuitive understanding of how a complex connectivity may produce rhythm.

6. CARDIAC GANGLION

The current picture of rhythmicity in the cardiac ganglion was outlined in Section II,A,2,c and need not be recapitulated here; suffice it to say that endogenous bursting of motoneurons (in stomatopods) or interneurons (in decapods) supplies the basic rhythm, and various synaptic interactions with motoneurons transmit it.

In closing, it may once more be remarked that motoneurons form a part of the oscillator, to some degree, in all of the crustacean rhythmic behaviors discussed here. The generality of this central influence exerted by motor output is becoming increasingly apparent as research on crustacean motor systems proceeds.

IV. TRENDS AND PROSPECTS

What have we learned from the crustaceans about simple nervous function, and what may we hope soon to learn? We now have some insight into the relationship between a neuron's structure and its integrative function. But before we can really know how that neuron integrates and passes on information, we will need to learn much more: e.g., (1) the detailed mapping of identified inputs and outputs onto that neuron's identified processes; (2) how the individual input and output synapses behave during different paradigms of individual or concerted activation; and (3) the conductance properties of the neuron, region by region. In this last regard—is integration purely electrotonic? Do specific dendrites spike? Where are spikes initiated? Does the membrane produce plateaus? If so, where? Is there long-term neurohumoral activation (unmasking) or suppression of specific conductances? The exciting prospect now is that these questions are answerable with today's anatomical, electrophysiological, and computational techniques applied to reproducibly identifiable neurons.

Given this kind of detailed knowledge about individual neurons, we may

7. Small Neural Systems and Control of Activity

be able to understand more deeply how circuits of such neurons behave; sufficiently sophisticated computer simulation will both aid and test our understanding by making verifiable predictions. More complete answers may then emerge to many long-standing questions: How are conflicting drives reconciled? Are behavioral priorities "hard-wired" into the circuit, and if so, how? How are diurnal, seasonal, and developmental changes in behavior mediated? Is a simple arthropod system's output modifiable by experience, as suggested by some experiments (e.g., Horridge, 1962; Stafstrom and Gerstein, 1977; Forman and Hoyle, 1978; Olson and Strandberg, 1979)? If so, what specific changes underlie this plasticity?

But there may be several catches in this optimistic program, aside from the imposing logistics. The whole progression assumes that the neurons involved are uniquely identifiable; identifiability, in turn, often rests on motor effect, spike patterns, and axonal location. We are now faced with a growing number of non-spiking interneurons that, in most cases, lack axons. Burrows (1979b), for instance, estimates that up to 100 such local interneurons may influence a single locust limb. Learning more about these and how to deal with them will likely be one of the major challenges facing neurobiologists in the next few years. Even, for example, if these cells are all unique and reproducible, current recording techniques can not gather enough simultaneous data to identify them all experimentally. How, then, can our circuit diagrams and computer models take account of these neurons? It may be that unique identifiability will not be necessary for understanding these neurons' roles—the vertebrate visual system may be a heartening example.

A related problem faces the identification and characterization of non-spike-mediated interactions: the lack of a presynaptic spike as a trigger or time reference must make the study of such interactions much more laborious. Similar problems in the vertebrate retina have been attacked with presynaptic injection of "white noise" current and postsynaptic analysis (e.g., Naka et al., 1979). A second technique that holds promise if it can be sufficiently refined is the optical recording method of Cohen (e.g., Salzberg et al., 1977), which permits the membrane voltages of a large number of cells to be monitored simultaneously. Coupled with computer cross-correlation analysis, such recording could reveal the existence and the effects of non-spike-mediated synaptic interactions. A third method now available to study any neuron's role in a circuit is to inactivate that neuron selectively (Miller and Selverston, 1979); this elegant technique promises to be invaluable in a variety of applications.

Thus, while large problems remain to be surmounted, we can expect solid systematic progress, a fruitful cross-fertilization with studies in other phyla, and quite possibly some major breakthroughs in our more universal understanding of neurophysiology.

ACKNOWLEDGMENTS

The author is grateful to Drs. J. A. Benson, I. M. Cooke, D. K. Hartline, C. Lingle, E. Marder, J. P. Miller, B. Mulloney, J. A. Paton, and A. I. Selverston for making available unpublished information, and to the National Research Council of Canada for support.

REFERENCES

Alexandrowicz, J. A. (1932). The innervation of the heart of the crustacea. I. Decapoda. *Q. J. Microsc. Sci.* **75**, 181-249.

Alexandrowicz, J. A. (1934). The innervation of the heart of the crustacea. II. Stomatopoda. *Q. J. Microsc. Sci.* **76**, 511-548.

Anderson, W. W., and Barker, D. L. (1977). Activation of a stomatogastric pattern generator by dopamine and L-dopa. *Soc. Neurosci. Abstr.* **3**, 171.

Atwood, H. L., and Wiersma, C. A. G. (1967). Command interneurons in the crayfish central nervous system. *J. Exp. Biol.* **46**, 249-261.

Ayers, J. L., and Selverston, A. I. (1979). Monosynaptic entrainment of an endogenous pacemaker network: A cellular mechanism for von Holst's magnet effect. *J. Comp. Physiol.* **129**, 5-17.

Barker, D. L., Kushner, P. D., and Hooper, N. K. (1979). Synthesis of dopamine and octopamine in the crustacean stomatogastric nervous system. *Brain Res.* **161**, 99-113.

Beltz, B., and Gelperin, A. (1978). The salivary burster of *Limax maximus:* A presumptive sensory-motor neuron. *Soc. Neurosci. Abstr.* **4**, 187.

Bennett, M. V. L., and Auerbach, A. A. (1969). Calculation of electrical coupling of cells separated by a gap. *Anat. Rec.* **163**, 152.

Berlind, A., and Cooke, I. M. (1970). Release of a neurosecretory hormone as peptide by electrical stimulation of crab pericardial organs. *J. Exp. Biol.* **53**, 679-686.

Bidaut, M., Russell, D. F., and Hartline, D. K. (1978). Distinguishing two types of inhibitory synapses from pacemaker neurons in the pyloric system of the lobster stomatogastric ganglion. *Soc. Neurosci. Abstr.* **4**, 188.

Biedermann, W. (1887). Beiträge zur allgemeinen Nerven—und Muskelphysiologie. Über die Innervation der Krebsschere. *Sitzungsber. Akad. Wiss. Wien, Math.-Naturwiss.Kl., Abt. 3* **97**, 49-82.

Blight, A. R., and Llinas, R. (1978). Depolarization—release coupling at a synapse lacking regenerative spikes. *Soc. Neurosci. Abstr.* **4**, 577.

Bullock, T. H., and Terzuolo, C. A. (1957). Diverse forms of activity in the somata of spontaneous and integrating ganglion cells. *J. Physiol. (London)* **138**, 341-364.

Bullock, T. H., Cohen, M. J., and Maynard, D. M. (1954). Integration and central synaptic properties of some receptors. *Fed. Proc., Fed. Am. Soc. Exp. Biol.* **13**, 20.

Burrows, M. (1967). Reflex withdrawal of the eyecup in the crab *Carcinus. Nature (London)* **215**, 56-57.

Burrows, M. (1975). Monosynaptic connexions between wing stretch receptors and flight motoneurones of the locust. *J. Exp. Biol.* **62**, 189-219.

Burrows, M. (1979a). Synaptic potentials effect the release of transmitter from locust nonspiking interneurons. *Science* **204**, 81-83.

Burrows, M. (1979b). Graded synaptic interactions between local premotor interneurons of the locust. *J. Neurophysiol.* **42**, 1108-1123.

Burrows, M., and Horridge, G. A. (1968). Eyecup withdrawal in the crab, Carcinus, and its interaction with the optokinetic response. *J. Exp. Biol.* **49**, 285-297.

Burrows, M., and Siegler, M. V. S. (1978). Graded synaptic transmission between local interneurones and motor neurones in the metathoracic ganglion of the locust. *J. Physiol. (London)* **285**, 231-255.

Bush, B. M. H. (1962). Peripheral reflex inhibition in the claw of the crab, Carcinus maenas (L.). *J. Exp. Biol.* **39**, 71-88.

Bush, B. M. H. (1963). A comparative study of certain limb reflexes in decapod crustaceans. *Comp. Biochem. Physiol.* **10**, 273-290.

Calabrese, R. L. (1976). Crayfish mechanoreceptive interneurons. I. The nature of ipsilateral excitatory inputs. *J. Comp. Physiol.* **105**, 83-102.

Carpenter, D., and Gunn, R. (1970). The dependence of pacemaker discharge of Aplysia neurons upon Na^+ and Ca^{++}. *J. Cell. Physiol.* **75**, 121-128.

Connor, J. A. (1969). Burst activity and cellular interaction in the pacemaker ganglion of the lobster heart. *J. Exp. Biol.* **50**, 275-295.

Connor, J. A., and Stevens, C. F. (1971a). Inward and delayed outward membrane currents in isolated neural somata under voltage clamp. *J. Physiol. (London)* **213**, 1-19.

Connor, J. A., and Stevens, C. F. (1971b). Voltage clamp studies of a transient outward membrane current in gastropod neural somata. *J. Physiol. (London)* **213**, 21-30.

Cooke, I. M. (1966). The sites of action of pericardial organ extract and 5-hydroxytryptamine in the decapod crustacean heart. *Am. Zool.* **6**, 107-121.

Cooke, I. M., and Goldstone, M. W. (1970). Fluorescence localization of monoamines in crab neurosecretory structures. *J. Exp. Biol.* **53**, 651-668.

Cooke, I. M., and Hartline, D. K. (1975). Neurohormonal alteration of integrative properties of the cardiac ganglion of the lobster Homarus americanus. *J. Exp. Biol.* **63**, 33-52.

Cooke, I. M., and Tazaki, K. (1979). Driver potentials isolated in crustacean cardiac ganglion cells by ligaturing. *Soc. Neurosci. Abstr.* **5**, 494.

Dando, M. R., and Maynard, D. M. (1974). The sensory innervation of the foregut of Panulirus argus (Decapoda Crustacea). *Mar. Behav. Physiol.* **2**, 283-305.

Dando, M. R., and Selverston, A. I. (1972). Command fibres from the supra-oesophageal ganglion to the stomatogastric ganglion in Panulirus argus. *J. Comp. Physiol.* **78**, 138-175.

Davis, W. J. (1971). Functional significance of motoneuron size and soma position in swimmeret system of the lobster. *J. Neurophysiol.* **34**, 274-288.

Davis, W. J., and Kennedy, D. (1972a). Command interneurons controlling swimmeret movements in the lobster. I. Types of effects on motoneurons. *J. Neurophysiol.* **35**, 1-12.

Davis, W. J., and Kennedy, D. (1972b). Command interneurons controlling swimmeret movements in the lobster. II. Interaction of effects on motoneurons. *J. Neurophysiol.* **35**, 13-19.

Davis, W. J., and Kennedy, D. (1972c). Command interneurons controlling swimmeret movements in the lobster. III. Temporal relationships among bursts in different motoneurons. *J. Neurophysiol.* **35**, 20-29.

Dudel, J., and Kuffler, S. W. (1961). Presynaptic inhibition at the crayfish neuromuscular junction. *J. Physiol. (London)* **155**, 543-562.

Eckert, B. (1959). Über das Zusammemwirken des erregenden und des hemmenden Neurons des M. Abduktor der Krebsschere beim Ablauf von Reflexen des myotatischen Typus. *Z. Vergl. Physiol.* **41**, 500-526.

Eckert, R., and Lux, H. D. (1976). A voltage-sensitive persistent calcium conductance in neuronal somata of Helix. *J. Physiol. (London)* **254**, 129-151.

Evoy, W. H., and Kennedy, D. (1967). The central nervous organization underlying control of antagonistic muscles in the crayfish. I. Types of command fibers. *J. Exp. Zool.* **165,** 223-238.

Evoy, W. H., Kennedy, D., and Wilson, D. M. (1967). Discharge patterns of neurones supplying tonic abdominal flexor muscles in the crayfish. *J. Exp. Biol.* **46,** 393-411.

Field, L. H. (1974). Neuromuscular correlates of rhythmical cheliped flexion behavior in hermit crabs. *J. Comp. Physiol.* **92,** 415-441.

Forman, R., and Hoyle, G. (1978). Position learning in behaviorally appropriate situations. *Soc. Neurosci. Abstr.* **4,** 193.

Friesen, W. O. (1975a). Physiological anatomy and burst pattern in the cardiac ganglion of the spiny lobster *Panulirus interruptus*. *J. Comp. Physiol.* **101,** 175-189.

Friesen, W. O. (1975b). Synaptic interactions in the cardiac ganglion of the spiny lobster *Panulirus interruptus*. *J. Comp. Physiol.* **101,** 191-205.

Friesen, W. O., and Stent, G. S. (1978). Neural circuits for generating rhythmic movements. *Annu. Rev. Biophys. Bioeng.* **7,** 37-61.

Getting, P. A., and Willows, A. O. D. (1974). Modification of neuron properties by electrotonic synapses. II. Burst formation by electrotonic synapses. *J. Neurophysiol.* **37,** 858-868.

Graubard, K. (1978). Synaptic transmission without action potentials: Input-output properties of a nonspiking presynaptic neuron. *J. Neurophysiol.* **41,** 1014-1025.

Graubard, K., Raper, J. A., and Hartline, D. K. (1977). Non-spiking synaptic transmission between spiking neurons. *Soc. Neurosci. Abstr.* **3,** 177.

Hagiwara, S. (1961). Nervous activities of the heart in crustacea. *Ergeb. Biol.* **24,** 288-311.

Hagiwara, S., and Bullock, T. H. (1957). Intracellular potentials in pacemaker and integrative neurons of the lobster cardiac ganglion. *J. Cell. Comp. Physiol.* **50,** 25-47.

Hagiwara, S., Watanabe, A., and Saito, N. (1959). Potential changes in syncytial neurons of lobster cardiac ganglion. *J. Neurophysiol.* **22,** 554-572.

Hartline, D. K. (1967). Impulse identification and axon mapping of the nine neurons in the cardiac ganglion of the lobster *Homarus americanus*. *J. Exp. Biol.* **47,** 327-340.

Hartline, D. K. (1979a). Integrative neurophysiology of the lobster cardiac ganglion. *Am. Zool.* **19,** 53-65.

Hartline, D. K. (1979b). Pattern generation in the lobster (*Panulirus*) stomatogastric ganglion. II. Pyloric network simulation. *Biol. Cybernet.* **33,** 223-236.

Hartline, D. K., and Cooke, I. M. (1969). Postsynaptic membrane response predicted from presynaptic input pattern in lobster cardiac ganglion. *Science* **164,** 1080-1082.

Hartline, D. K., and Gassie, D. V., Jr. (1979). Pattern generation in the lobster (*Panulirus*) stomatogastric ganglion. I. Pyloric neuron kinetics and synaptic interactions. *Biol. Cybernet.* **33,** 209-222.

Hartline, D. K., and Maynard, D. M. (1975). Motor patterns in the stomatogastric ganglion of the lobster *Panulirus argus*. *J. Exp. Biol.* **62,** 405-420.

Hartline, D. K., and Russell, D. F. (1978). Induction of regenerative properties in neurons of the lobster stomatogastric ganglion by identified neural inputs. *Soc. Neurosci. Abstr.* **4,** 195.

Hartline, D. K., Gassie, D. V., and Sirchia, C. D. (1979). Additions to the physiology of the stomatogastric ganglion (STG) of the spiny lobster *Panulirus interruptus*. *Soc. Neurosci. Abstr.* **5,** 248.

Hartline, H. K., Ratliff, F., and Miller, W. H. (1961). Inhibitory interaction in the retina and its significance in vision. *In* "Nervous Inhibition" (E. Florey, ed.), pp. 241-284. Pergamon, Oxford.

Heitler, W. J. (1978). Coupled motoneurones are part of the crayfish swimmeret central oscillator. *Nature* **275,** 231-234.

7. Small Neural Systems and Control of Activity

Henneman, E., Somjen, G., and Carpenter, D. O. (1965). Functional significance of cell size in spinal motoneurons. *J. Neurophysiol.* **28,** 560-580.

Hermann, A. (1979a). Generation of a fixed motor pattern. I. Details of synaptic interconnections of pyloric neurons in the stomatogastric ganglion of the crab, Cancer pagurus. *J. Comp. Physiol.* **130,** 221-228.

Hermann, A. (1979b). Generation of a fixed motor pattern. II. Electrical properties and synaptic characteristics of pyloric neurons in the stomatogastric ganglion of the crab, Cancer pagurus. *J. Comp. Physiol.* **130,** 229-239.

Hill, R. H., and Lang, F. (1979). A reinvestigation of the inhibitory innervation pattern of the thoracic limbs of crayfish and lobster. *J. Exp. Zool.* **208,** 129-135.

Hoffmann, P. (1914). Über die doppelte Innervation der Krebsmuskeln. Zugleich ein Beitrag zur Kenntnis nervöser Hemmungen. *Z. Biol. (Munich)* **63,** 30-44.

Horridge, G. A. (1962). Learning of leg position by the ventral nerve cord in headless insects. *Proc. R. Soc. London, Ser. B* **157,** 33-52.

Irisawa, A., and Hama, K. (1965). Contact of adjacent nerve fibers in the cardiac nerve of mantis shrimp. *Jpn. J. Physiol.* **15,** 323-330.

Kandel, E. R. (1976). "Cellular Basis of Behavior." Freeman, San Francisco, California.

Kennedy, D. (1971). Crayfish interneurons. *Physiologist* **14,** 5-30.

Kennedy, D., and Mellon, De F., Jr. (1964). Synaptic activation and receptive fields in crayfish interneurons. *Comp. Biochem. Physiol.* **13,** 275-300.

King, D. G. (1976a). Organization of crustacean neuropil. I. Patterns of synaptic connections in lobster stomatogastric ganglion. *J. Neurocytol.* **5,** 207-237.

King, D. G. (1976b). Organization of crustacean neuropil. II. Distribution of synaptic contacts on identified motor neurons in lobster stomatogastric ganglion. *J. Neurocytol.* **5,** 239-266.

Klein, M., and Kandel, E. (1978). Presynaptic modulation of voltage-dependent Ca^{++} current: Mechanism for behavioral sensitization in Aplysia californica. *Proc. Natl. Acad. Sci. U.S.A.* **75,** 3512-3516.

Kravitz, E. A., Battelle, B., Evans, P. D., Talamo, B. R., and Wallace, B. G. (1976). Octopamine neurons in lobsters. *Soc. Neurosci. Symp.* **1,** 67-81.

Larimer, J. L., and Kennedy, D. (1966). Visceral afferent signals in the crayfish stomatogastric ganglion. *J. Exp. Biol.* **44,** 345-354.

Lindsey, B. G., and Gerstein, G. L. (1977). Reflex control of a crayfish claw motor neuron during imposed dactylopodite movements. *Brain Res.* **130,** 348-353.

Lindsey, B. G., and Gerstein, G. L. (1979a). Proprioceptive fields of crayfish claw motor neurons. *J. Neurophysiol.* **42,** 368-382.

Lindsey, B. G., and Gerstein, G. L. (1979b). Interactions among an ensemble of chordotonal organ receptors and motor neurons of the crayfish claw. *J. Neurophysiol.* **42,** 383-399.

Lingle, C. (1980). The sensitivity of decapod foregut muscles to acetylcholine and glutamate. *J. Comp. Physiol.* **138,** 187-200.

Livengood, D. R., and Kusano, K. (1972). Evidence for an electrogenic sodium pump in follower cells of the lobster cardiac ganglion. *J. Neurophysiol.* **35,** 170-186.

Marder, E. (1974). Acetylcholine as an excitatory neuromuscular transmitter in the stomatogastric system of the lobster. *Nature (London)* **251,** 730-731.

Marder, E. (1976). Cholinergic motor neurones in the stomatogastric system of the lobster. *J. Physiol. (London)* **257,** 63-86.

Marder, E., and Paupardin-Tritsch, D. (1978a). The pharmacological properties of some crustacean neuronal acetylcholine, γ-aminobutyric acid and L-glutamate responses. *J. Physiol. (London)* **280,** 213-236.

Marder, E., and Paupardin-Tritsch, D. (1978b). Inhibitory responses evoked by cholinergic agonists in crustacean stomatogastric ganglion neurons. *Soc. Neurosci. Abstr.* **4,** 200.

Matsui, K., Kuwasawa, K., and Furamoto, T. (1977). Periodic bursts in large cell preparations of the lobster cardiac ganglion (*Panulirus japonicus*). *Comp. Biochem. Physiol. A* **56A,** 313-324.

Mayeri, E. (1973a). Functional organization of the cardiac ganglion of the lobster, *Homarus americanus*. *J. Gen. Physiol.* **62,** 448-472.

Mayeri, E. (1973b). A relaxation oscillator description of the burst-generating mechanism in the cardiac ganglion of the lobster, *Homarus americanus*. *J. Gen. Physiol.* **62,** 473-488.

Maynard, D. M. (1953). Activity in a crustacean ganglion. I. Cardio-inhibition and acceleration in *Panulirus argus*. *Biol. Bull. (Woods Hole, Mass.)* **104,** 156-170.

Maynard, D. (1955). Activity in a crustacean ganglion. II. Pattern and interaction in burst formation. *Biol. Bull. (Woods Hole, Mass.)* **109,** 420-436.

Maynard, D. M. (1960). Circulation and heart function. In "Physiology of Crustacea" (T. H. Waterman, ed.), Vol. 1, pp. 161-226. Academic Press, New York.

Maynard, D. M. (1961). Cardiac inhibition in decapod crustacea. In "Nervous Inhibition" (E. Florey, ed.), pp. 144-178. Pergamon, Oxford.

Maynard, D. M. (1966). Integration in crustacean ganglia. *Symp. Soc. Exp. Biol.* **20,** 111-149.

Maynard, D. M. (1967). Neural coordination in a simple ganglion. *Science* **158,** 531-532.

Maynard, D. M. (1972). Simpler networks. *Ann. N.Y. Acad. Sci.* **193,** 59-72.

Maynard, D. M., and Atwood, H. L. (1969). Divergent post-synaptic effects produced by single motor neurons of the lobster stomatogastric ganglion. *Am. Zool.* **9,** 248.

Maynard, D. M., and Burke, W. (1966). Electrotonic junctions and negative feedback in the stomatogastric ganglion of the mud crab, *Scylla serrata*. *Am. Zool.* **6,** 526.

Maynard, D. M., and Dando, M. R. (1974). The structure of the stomatogastric neuromuscular system in *Callinectes sapidus, Homarus americanus* and *Panulirus argus* (Decapoda Crustacea). *Philos. Trans. R. Soc. London, Ser. B* **268,** 161-220.

Maynard, D. M., and Selverston, A. I. (1975). Organization of the stomatogastric ganglion of the spiny lobster. IV. The pyloric system. *J. Comp. Physiol.* **100,** 161-182.

Maynard, D. M., and Walton, K. D. (1975). Effects of maintained depolarization of presynaptic neurons on inhibitory transmission in lobster neuropil. *J. Comp. Physiol.* **97,** 215-243.

Maynard, D. M., and Welsh, J. H. (1959). Neurohormones of the pericardial organs of brachyuran crustacea. *J. Physiol. (London)* **149,** 215-227.

Meech, R. W., and Standen, N. B. (1975). Potassium activation in *Helix aspersa* neurones under voltage clamp: A component mediated by calcium influx. *J. Physiol. (London)* **249,** 211-239.

Mellon, de F. (1969). The reflex control of rhythmic motor output during swimming in the scallop. *Z. Vergl. Physiol.* **62,** 318-336.

Mendelson, M. (1971). Oscillator neurons in crustacean ganglia. *Science* **171,** 1170-1173.

Mill, P. J., and Lowe, D. A. (1972). An analysis of the types of sensory unit present in the P-D proprioceptor of decapod crustaceans. *J. Exp. Biol.* **56,** 509-525.

Miller, J. P., and Selverston, A. I. (1979). Rapid killing of single neurons by irradiation of intracellularly injected dye. *Science* **206,** 702-704.

Mulloney, B. (1977). Organization of the stomatogastric ganglion of the spiny lobster. V. Coordination of the gastric and pyloric systems. *J. Comp. Physiol.* **122,** 227-240.

Mulloney, B., and Selverston, A. I. (1974a). Organization of the stomatogastric ganglion of the spiny lobster. I. Neurons driving the lateral teeth. *J. Comp. Physiol.* **91,** 1-32.

Mulloney, B., and Selverston, A. I. (1974b). Organization of the stomatogastric ganglion of the spiny lobster. III. Coordination of the two subsets of the gastric system. *J. Comp. Physiol.* **91,** 53-78.

Mulloney, B., and Sigvardt, K. A. (1978). Testing direct connections in the stomatogastric ganglion with TEA. *Soc. Neurosi. Abstr.* **4**, 384.

Mulloney, B., Skinner, K., and Edwards, D. H., Jr. (1979). Identified neurons of the stomatogastric ganglion have different, characteristic membrane time constants. *Soc. Neurosci. Abstr.* **5**, 256.

Naka, K. I., Chan, R. Y., and Yasui, S. (1979). Adaptation in catfish retina. *J. Neurophysiol.* **42**, 441-454.

Neher, E. (1971). Two fast transient current components during voltage clamp on snail neurons. *J. Gen. Physiol.* **58**, 36-53.

Olson, G. C., and Strandberg, R. (1979). Instrumental conditioning in crayfish: Lever pulling for food. *Soc. Neurosci. Abstr.* **5**, 257.

Otani, T., and Bullock, T. H. (1959). Effects of presetting the membrane potential of the soma of spontaneous and integrating ganglion cells. *Physiol. Zool.* **32**, 104-114.

Parnas, I., and Strumwasser, F. (1974). Mechanisms of long-lasting inhibition of a bursting pacemaker neuron. *J. Neurophysiol.* **37**, 609-620.

Paton, J. A. (1978). Command fiber to motor neuron connections in the lobster swimmeret system. *Soc. Neurosci. Abstr.* **4**, 384.

Paul, D. H. (1976). Role of proprioceptive feedback from nonspiking mechanosensory cells in the sand crab, *Emerita analoga*. *J. Exp. Biol.* **65**, 243-258.

Paul, D. H. (1981). An identified local interneuron in the neural oscillator of crayfish swimmerets. *Soc. Neurosci. Abstr.* **7**, 138.

Pearson, K. G., and Fourtner, C. R. (1975). Nonspiking interneurons in walking system of the cockroach. *J. Neurophysiol.* **38**, 33-52.

Perkel, D. H., and Mulloney, B. (1974). Motor pattern production in reciprocally inhibitory neurons exhibiting post-inhibitory rebound. *Science* **185**, 181-183.

Powers, L. (1973). Gastric mill rhythms in intact crabs. *Comp. Biochem. Physiol. A* **46A**, 767-783.

Raper, J. A. (1978). Graded synaptic transmission between spiking neurons during the generation of a central motor pattern. *Soc. Neurosci. Abstr.* **4**, 204.

Raper, J. A. (1979). Non impulse-mediated synaptic transmission during the generation of a cyclic motor program. *Science* **205**, 304-306.

Ripley, S. H., and Wiersma, C. A. G. (1953). The effect of spaced stimulation of the excitatory and inhibitory axons of the crayfish. *Physiol. Comp. Oecol.* **3**, 1-7.

Russell, D. F. (1976a). Rhythmic excitatory inputs to the lobster stomatogastric ganglion. *Brain Res.* **101**, 582-588.

Russell, D. F. (1976b). Making the stomatogastric ganglion work. *Soc. Neurosci. Abstr.* **2**, 354.

Russell, D. F. (1978). P cells: Distribution and coupling of pattern generators in the lobster stomatogastric nervous system. *Soc. Neurosci. Abstr.* **4**, 384.

Russell, D. F., and Hartline, D. K. (1978). Bursting neural networks: A re-examination. *Science* **200**, 543-546.

Salzberg, B. M., Grinvald, A., Cohen, L. B., Davila, H. V., and Ross, W. N. (1977). Optical recording of neuronal activity in an invertebrate central nervous system: simultaneous monitoring of several neurons. *J. Neurophysiol.* **40**, 1281-1291.

Sandeman, D. C. (1967). Excitation and inhibition of the reflex eye withdrawal of the crab *Carcinus*. *J. Exp. Biol.* **46**, 475-485.

Sandeman, D. C. (1969a). The synaptic link between the sensory and motoneurones in the eye-withdrawal reflex of the crab. *J. Exp. Biol.* **50**, 87-98.

Sandeman, D. C. (1969b). The site of synaptic activity and impulse initiation in an identified motoneurone in the crab brain. *J. Exp. Biol.* **50**, 771-784.

Sandeman, D. C. (1969c). Integrative properties of a reflex motoneuron in the brain of the crab *Carcinus maenas*. *Z. Vergl. Physiol.* **64**, 450–464.

Sandeman, D. C. (1971). The excitation and electrical coupling of four identified motoneurons in the brain of the Australian mud crab, *Scylla serrata*. *Z. Vergl. Physiol.* **72**, 111–130.

Selverston, A. I. (1974). Structural and functional basis of motor pattern generation in the stomatogastric ganglion of the lobster. *Am. Zool.* **14**, 957–972.

Selverston, A. I. (1977). Mechanisms for the production of rhythmic behavior in crustaceans. In "Identified Neurons and Behavior of Arthropods" (G. Hoyle, ed.), pp. 209–225. Plenum, New York.

Selverston, A. I., and Miller, J. P. (1980). Mechanisms underlying pattern generation in lobster stomatogastric ganglion as determined by selective inactivation of identified neurons. I. Pyloric system. *J. Neurophysiol.* **44**, 1102–1121.

Selverston, A. I., and Mulloney, B. (1974). Organization of the stomatogastric ganglion of the spiny lobster. II. Neurons driving the medial tooth. *J. Comp. Physiol.* **91**, 33–51.

Selverston, A. I., Russell, D. F., Miller, J. P., and King, D. G. (1976). The stomatogastric nervous system: Structure and function of a small neural network. *Prog. Neurobiol.* (Oxford) **7**, 215–290.

Simmers, A. J. (1979). Oscillatory potentials in crab ventilatory neurones. *J. Physiol.* (London) **287**, 39P–40P.

Smith, D. O. (1972). Central nervous control of presynaptic inhibition in the crayfish claw. *J. Neurophysiol.* **35**, 333–343.

Smith, D. O. (1974). Central nervous control of excitatory and inhibitory neurons of opener muscle of the crayfish claw. *J. Neurophysiol.* **37**, 108–118.

Smith, T. G., Jr., Barker, J. L., and Gainer, H. (1975). Requirements for bursting pacemaker activity in molluscan neurons. *Nature* (London) **253**, 450–452.

Spirito, C. P. (1970). Reflex control of the opener and stretcher muscles in the cheliped of the fiddler crab, *Uca pugnax*. *Z. Vergl. Physiol.* **68**, 211–228.

Stafstrom, C., and Gerstein, G. (1977). A paradigm for position learning in the crayfish. *Brain Res.* **134**, 185–190.

Strumwasser, F. (1971). The cellular basis of behavior in *Aplysia*. *J. Psychiatr. Res.* **8**, 237–257.

Takeda, K., and Kennedy, D. (1964). Soma potentials and modes of activation of crayfish motoneurons. *J. Cell. Comp. Physiol.* **64**, 165–180.

Tameyasu, T. (1976). Intracellular potentials in the small cells and cellular interaction in the cardiac ganglion of the lobster *Panulirus japonicus*. *Comp. Biochem. Physiol. A* **54A**, 191–196.

Tatton, W. G., and Sokolove, P. G. (1975). Analysis of postural motoneuron activity in crayfish abdomen. II. Coordination by excitatory and inhibitory connections between motoneurons. *J. Neurophysiol.* **38**, 332–346.

Tazaki, K. (1970). Slow potential changes during the burst in the cardiac ganglion of the crab, *Eriocheir japonicus*. *Annot. Zool. Jpn.* **43**, 63–69.

Tazaki, K. (1971a). Slow potentials in spontaneous ganglion cells of the lobster heart. *Bull. Nara Univ. Educ., Nat. Sci.* **20**, 15–27.

Tazaki, K. (1971b). The effects of tetrodotoxin on the slow potential and spikes in the cardiac ganglion of a crab, *Eriocheir japonicus*. *Jpn. J. Physiol.* **21**, 529–536.

Tazaki, K. (1971c). Small synaptic potentials in burst activity of large neurons in the lobster cardiac ganglion. *Jpn. J. Physiol.* **21**, 645–658.

Tazaki, K. (1972a). The burst activity of different cell regions and intercellular coordination in the cardiac ganglion of the crab, *Eriocheir japonicus*. *J. Exp. Biol.* **57**, 713–726.

Tazaki, K. (1972b). Electrical interaction among large cells in the cardiac ganglion of the lobster, *Panulirus japonicus*. *J. Exp. Zool.* **180**, 85–94.

Tazaki, K. (1972c). Adaptation in pacemaker neurons of the crab ganglion. *Bull. Nara Univ. Educ., Nat. Sci.* **21,** 31–38.

Tazaki, K. (1973). Impulse activity and pattern of large and small neurones in the cardiac ganglion of the lobster, *Panulirus japonicus. J. Exp. Biol.* **58,** 473–486.

Tazaki, K., and Cooke, I. M. (1979a). Spontaneous electrical activity and interaction of large and small cells in cardiac ganglion of the crab, *Portunus sanguinolentus. J. Neurophysiol.* **42,** 975–999.

Tazaki, K., and Cooke, I. M. (1979b). Isolation and characterization of slow, depolarizing responses of cardiac ganglion neurons in the crab, *Portunus sanguinolentus. J. Neurophysiol.* **42,** 1000–1021.

Tazaki, K., and Cooke, I. M. (1979c). Ionic bases of slow, depolarizing responses of cardiac ganglion neurons in the crab, *Portunus sanguinolentus. J. Neurophysiol.* **42,** 1022–1047.

Terzuolo, C. A., and Bullock, T. H. (1958). Acceleration and inhibition in crustacean ganglion cells. *Arch. Ital. Biol.* **96,** 117–134.

Thompson, S. H. (1977). Three pharmacologically distinct potassium channels in molluscan neurones. *J. Physiol. (London)* **265,** 465–488.

van Harreveld, A., and Wiersma, C. A. G. (1937). The triple innervation of crayfish muscle and its function in contraction and inhibition. *J. Exp. Biol.* **14,** 448–461.

Ware, R. W., Clark, D., Crossland, K., and Russell, R. L. (1975). The nerve ring of the nematode *Caenorhabditis elegans:* Sensory input and motor output. *J. Comp. Neurol.* **162,** 71–110.

Warshaw, H. S., and Hartline, D. K. (1976). Simulation of network activity in stomatogastric ganglion of the spiny lobster, *Panulirus. Brain Res.* **110,** 259–272.

Watanabe, A. (1958). The interaction of electrical activity among neurons of lobster cardiac ganglion. *Jpn. J. Physiol.* **8,** 305–318.

Watanabe, A., and Bullock, T. H. (1960). Modulation of activity of one neuron by subthreshold slow potentials in another in lobster cardiac ganglion. *J. Gen. Physiol.* **43,** 1031–1045.

Watanabe, A. I., and Takeda, K. (1963). The spread of excitation among neurons in the heart ganglion of the stomatopod, *Squilla oratoria. J. Gen. Physiol.* **46,** 773–801.

Watanabe, A., Obara, S. T., Akiyama, T., and Yumoto, K. (1967a). Electrical properties of the pacemaker neurons in the heart ganglion of a stomatopod, *Squilla oratoria. J. Gen. Physiol.* **50,** 813–838.

Watanabe, A., Obara, S., and Akiyama, T. (1967b). Pacemaker potentials for the periodic burst discharge in the heart ganglion of a stomatopod, *Squilla oratoria. J. Gen. Physiol.* **50,** 839–862.

Watanabe, A., Obara, S., and Akiyama, T. (1968). Inhibitory synapses on pacemaker neurons in the heart ganglion of a stomatopod, *Squilla oratoria. J. Gen. Physiol.* **52,** 908–924.

Watanabe, A., Obara, S., and Akiyama, T. (1969). Acceleratory synapses on pacemaker neurons in the heart ganglion of a stomatopod, *Squilla oratoria. J. Gen. Physiol.* **54,** 212–231.

Welsh, J. H., and Maynard, D. M. (1951). Electrical activity of a simple ganglion. *Fed. Proc., Fed. Am. Soc. Exp. Biol.* **10,** 145.

Wiens, T. J. (1976). Electrical and structural properties of crayfish claw motoneurons in an isolated claw-ganglion preparation. *J. Comp. Physiol.* **112,** 213–233.

Wiens, T. J., and Atwood, H. L. (1978). Motoneuron interactions in crayfish claw control: Evidence from intracellular recording. *J. Comp. Physiol.* **124,** 237–247.

Wiens, T. J., and Gerstein, G. L. (1975). Cross connections among crayfish claw efferents. *J. Neurophysiol.* **38,** 909–921.

Wiens, T. J., and Gerstein, G. L. (1976). Reflex pathways of the crayfish claw. *J. Comp. Physiol.* **107,** 309–326.

Wiersma, C. A. G. (1952). The neuron soma. *Cold Spring Harbor Symp. Quant. Biol.* **17**, 155-163.

Wiersma, C. A. G., and Boettiger, E. G. (1959). Unidirectional movement fibres from a proprioceptive organ of the crab, Carcinus maenas. *J. Exp. Biol.* **36**, 102-112.

Wiersma, C. A. G., and Ripley, S. H. (1952). Innervation patterns of crustacean limbs. *Physiol. Comp. Oecol.* **2**, 391-405.

Wilson, D. M. (1966). Central nervous mechanisms for the generation of rhythmic behaviour in arthropods. *Symp. Soc. Exp. Biol.* **20**, 199-228.

Wilson, D. M., and Davis, W. J. (1965). Nerve impulse patterns and reflex control in the motor system of the crayfish claw. *J. Exp. Biol.* **43**, 193-210.

Wilson, D. M., and Gettrup, E. (1963). A stretch reflex controlling wingbeat frequency in grasshoppers. *J. Exp. Biol.* **40**, 171-185.

Wilson, W. A., and Wachtel, H. (1978). Prolonged inhibition in burst firing neurons: Synaptic inactivation of the slow regenerative inward current. *Science* **202**, 772-775.

Wine, J. J. (1977). Crayfish escape behavior. II. Command-derived inhibition of abdominal extension. *J. Comp. Physiol.* **121**, 173-186.

Wine, J. J., Mittenthal, J. E., and Kennedy, D. (1974). The structure of tonic flexor motoneurons in crayfish abdominal ganglia. *J. Comp. Physiol.* **93**, 315-335.

8

The Cellular Organization of Crayfish Escape Behavior

JEFFREY J. WINE AND FRANKLIN B. KRASNE

I.	Introduction	242
II.	Multiple Systems for Escape: The Role of the Giant Axons	243
III.	Organization of the Afferent Pathways	246
	A. The LG System	246
	B. Habituation of LG-Mediated Escape Responses	251
	C. The MG and Non-Giant Systems	254
IV.	Motor Control: Excitatory Pathways for Flexion	255
	A. Tonic versus Phasic Systems	255
	B. Overview of the Fast Flexor Motor System	255
	C. The Fast Flexor Muscles	257
	D. The Fast Flexor Motoneurons: Direct and Indirect Pathways for Excitation	258
	E. Variations in Synaptic Connections Responsible for the Spatial Patterning of Tail Flips	263
	F. The Corollary Discharge	267
	G. Summary of the Flexor Motor Circuitry and a Hypothesis Concerning Giant and Non-Giant Systems	270
V.	Motor Control: The Extensor System	271
	A. Overview	271
	B. Extensor Muscles and Motoneurons: Sources of Excitation and Inhibition	271
	C. Analysis of Extension: Dual Modes of Control	274
	D. Premotor Circuits for Extension	275
VI.	Command-Derived Inhibition	275
	A. Multiple Levels of Inhibition	275

	B.	Functional and Timing Considerations	277
	C.	Inhibition of the Postural Motor Systems	280
	D.	Heterogeneity of Inhibition of Sensory Interneurons and Flexor Motoneurons	280
	E.	Inhibitory Control of Habituation	282
VII.	Modulation of Escape Tendencies		283
	A.	Restraint-Induced Inhibition	285
	B.	Other Modulating Factors	286
	C.	Functional Considerations	286
VIII.	Concluding Remarks		287
	Abbreviations		288
	References		289

I. INTRODUCTION

Crayfish possess two pairs of giant axons that travel the length of the central nervous system. The giant cells are command neurons, which produce, with a single impulse, a coordinated "tail flip" escape response. These conclusions were reached over 50 years ago by George Edwin Johnson, who carefully described the giant axons, provided anatomical evidence for connections between them and certain flexor motoneurons, and then, using a Harvard inductorium and a battery, showed that shocks to the region of the cord containing the giant axons could reliably initiate tail flips (Johnson, 1924, 1926). Conclusive evidence for the motor effects of giant axons was provided by the hallmark experiments of Wiersma (1947).

Our interest in escape behavior has been sustained by the ease with which these giant cells can be stimulated and recorded: they provide a dependable starting point for investigations into problems such as decision, command, coordination, and even simple forms of learning. Among the questions that intrigued us and others were: How can the animal produce such a wide range of escape maneuvers with just two kinds of giant axon? How and why does the animal "choose" to escape from a slight disturbance one moment, yet ignore a powerful stimulus the next? And how does a 1-msec impulse in a single cell give rise to a coordinated, spatiotemporal pattern involving most of the animal's muscles and lasting 100 times longer than the initiating event? We recently summarized the neural basis for escape behavior, with an emphasis on the many control systems that give the behavior variety and adaptiveness (Krasne et al., 1977). Much has been learned since then. In particular, new evidence about the extension phase, about indirect pathways to the flexor motoneurons, and about inhibition have enriched our understanding of the behavior. However, these and other findings have

8. The Cellular Organization of Crayfish Escape Behavior

raised a host of new problems, and one purpose of this chapter is to emphasize what we do not understand.

II. MULTIPLE SYSTEMS FOR ESCAPE: THE ROLE OF THE GIANT AXONS

Crayfish* can respond to a sudden visual or tactile stimulus by initiating a tail flip within 10-20 msec, and then swimming away, i.e., performing repeated tail flips at 10-20 Hz (Fig. 1). This rapid and adaptive performance is achieved by a division of labor among at least three neural systems (Table I). Two systems of giant axons produce the initial tail flip at short latency. The initial, giant-mediated tail flips are relatively stereotyped. One giant axon system is activated by sudden visual stimuli or taps to the front part of the animal and propels the animal backward through the water. A second giant axon system is activated by sudden mechanical stimuli to the abdomen and pitches the crayfish forward. A third system (probably a set of systems) is called the non-giant system and mediates all other forms of tail flips, including the repetitive tail flips during swimming (Fig. 1), tail flips seen during struggles to escape, tail flips during righting, and so on. Any gradual stimulus, such as a pinch or a slowly looming object, tends to activate the non-giant system, and the resulting tail flips and trajectories can assume many forms. The gain in flexibility comes at the expense of speed, as is shown when the non-giant systems occasionally respond to discrete stimuli that permit their latencies to be measured. They are then seen to be delayed at least ten times longer than giant-mediated tail flips. The mean latency for 66 tail flips mediated by the giant systems, measured from the stimulus to the beginning of the flexor EMG, was 8.12 ± 5.0 msec, while for 12 tail flips initiated by the non-giant system it was 217 ± 97 msec (Wine and Krasne, 1972).

The lateral and medial giant axons are command neurons according to all criteria usually deemed important (see Kupfermann and Weiss, 1978). The evidence for activity of the giant axons during normal behavior is easy to record (Fig. 1B). It is also possible to fire the giants via discrete electrical

---

Procambarus clarkii Girard was used for most experiments reported in the chapter. Because of the difficulties inherent to tracing neural circuitry and explaining behavior in cellular terms, it is desirable to focus attention on a single behavior in a single species, at least until enough is learned to form a basis for comparative work. The intrinsic advantages of the crayfish for such investigations should become apparent to readers of this chapter. To these advantages must be added the considerable knowledge that has been accumulated about the anatomy and physiology of crayfish escape behavior.

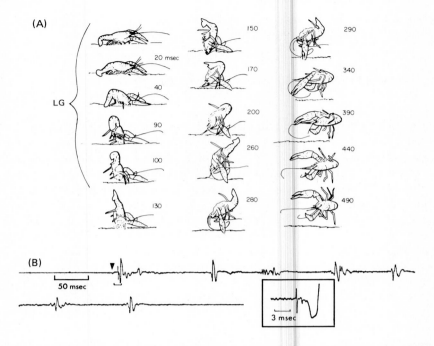

Fig. 1. (A) An escape response produced by a tap to the animal's abdomen. Three tail flips are shown. The first flexion-extension cycle (brackets) was mediated by LG axons; the following two tail flips are mediated by non-giant systems. Traces are from high speed cinematographs; numbers indicate elapsed time from the stimulus. (B) Nerve and muscle recordings from an unrestrained crayfish during an escape response like the one shown in A. Electrodes were positioned near the nerve cord where they recorded giant axon spikes with a large signal-to-noise ratio. Area in the small bracket is enlarged in inset. Arrow point shows stimulus onset-time, obtained from a transducer on the rod used to tap the animal. Note that only the first response involves giant axons. (Wine and Krasne, 1972.)

shocks and thereby produce an escape tail flip. This has been done repeatedly (Johnson, 1926; Wiersma, 1947; Roberts, 1968; Larimer et al., 1971). Figure 2A shows a tail flip sequence produced by a direct electrical shock to a medial giant (MG)* axon; a naturally evoked MG flip is shown in Fig. 2B (G. Hagiwara, unpublished). Finally, it has been possible to eliminate tail flips evoked by electrical shocks to afferent nerves by hyperpolarizing the giant axon to prevent its firing (Olson and Krasne, 1978).

Our knowledge of the neural basis for these systems is inversely proportional to their complexity. Least is known about the multimodal, flexible, non-giant systems, precisely because they consist of smaller and more nu-

*A full list of abbreviations and symbols is given at the end of this chapter.

TABLE I
Stimulus–Response Relationships for Unrestrained Crayfish[a]

Stimulus	Number of responses obtained			Number of animals observed
	LG	MG	Non-G	
Tap abdomen	81	0	7	20
Touch antenna or tap rostrum	0	56	5	16
Tap thorax, claws or rostral legs	0	10	4	2
Very rapidly approaching object (visual)	0	10	0	2
Pinch antenna or leg	0	2	17	7
Pinch uropods	1	0	12	3
"Ordinary" visual	0	1	15	3
Non-specific: pull leads, handle, etc.	0	1	8	3
"Spontaneous"	0	0	15	5

[a] From Wine and Krasne, 1972.

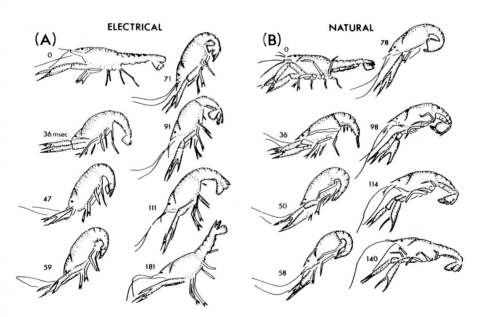

Fig. 2. Comparison of natural and centrally evoked tail flips. (A) Escape response produced by a single, 0.1-msec shock to a medial giant axon, via chronically implanted electrodes in circumesophageal connective. (B) An escape response produced in the same animal by a tap to the rostrum. Variation between the two is no greater than that seen among naturally elicited tail flips. (G. Hagiwara and J. Wine, unpublished.)

merous neurons. The MG system integrates both visual and tactile stimuli, but analysis of its afferent circuitry has proceeded slowly because of difficulties in recording from the brain. The stereotyped lateral giant (LG) system, which responds only to mechanical stimulation, is the most thoroughly described, since its afferent pathways are located in the accessible abdominal nervous system. The three systems are intimately related; they share most of their motor and inhibitory circuitry, and we suspect that much of the neural complexity that we have encountered in the giant axon circuits actually evolved primarily to mediate non-giant responses—the giant axons having usurped certain portions of the non-giant circuitry.

III. ORGANIZATION OF THE AFFERENT PATHWAYS

A. The LG System

The command interneurons in this system are the lateral giants (LG's), named because of the size and position of their axons. The LG axon network, although represented by a single symbol in our circuit diagrams, is actually a ladder network of at least 12 pairs of neurons, electrically coupled via junctions that transmit impulses 1:1 (Fig. 4C). The longitudinal, nonrectifying coupling among segments causes them to conduct as a single axon (Wiersma, 1947; Kao, 1960; Watanabe and Grundfest, 1961); the extensive, nonrectifying cross-coupling (Watanabe and Grundfest, 1961; Kusano and Grundfest, 1965) ensures that both axons always fire. These features confer a very high safety factor for transmission within the network.

How are the LG axons normally activated? Behavioral experiments show that a sharp tap to the abdomen or tailfan is the best stimulus. Neither water jets (from a Water-pic), nor pinches severe enough to do damage, fire the LG's (Table I). Furthermore, repeated taps of an intensity just above threshold for responses in fresh animals quickly cause the response to habituate (Wine et al., 1975; and see below). What mechanisms in the afferent circuit to the LG axons account for its response features?

One portion of the afferent pathway that excites the LG axons is shown schematically in Fig. 3. Anatomy of representative elements is shown in Fig. 4, and key aspects of circuit electrophysiology are shown in Fig. 5. The circuit diagram for the afferent pathway and most of the following description is based on work by Krasne (1969) and Zucker (1972a).

The major input to the LG axon is a disynaptic pathway from cuticular hair receptors via sensory interneurons to the LG. Approximately 1000 hairs cover the abdomen, each containing two bipolar receptors that are directionally sensitive to movement (Figs. 4A, 5A; Wiese, 1976). Many of these

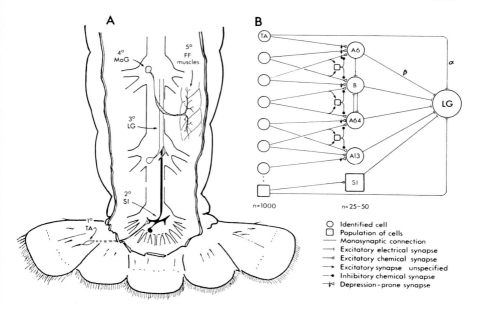

Fig. 3. Wiring diagram for afferent pathways to LG interneuron. (A) Semi-schematic representation of shortest major pathway from receptors to muscles; this pathway involves five stages. The nervous system and elements are not drawn to scale, and only one LG segment is shown. (B) Schematic of afferent pathways to LG interneuron, emphasizing considerable convergence. Interneurons causing presynaptic inhibition of afferents are not identified, nor is it known for sure that the pathway is as direct as shown. (From Krasne, 1969; Zucker, 1972a; Kennedy et al., 1974.)

afferents make direct electrical synapses onto the LG, but not in sufficient numbers to fire it. The same afferents also make chemical, presumably cholinergic (Barker et al., 1972), synapses onto sensory interneurons, which in turn synapse electrically with the LG.

The circuit (Fig. 3B) displays a large amount of convergence, the population of relevant sensory interneurons numbering perhaps 25 to 50. Divergence also occurs, with a single afferent synapsing upon many interneurons (Calabrese, 1976a,b). Mechanosensory interneurons constitute the main synaptic drive to the command cell. Three examples of sensory interneurons that synapse onto the LG's are shown in Fig. 4B; some of the interneurons have inputs in several ganglia, all of them seem to have multisegmental outputs to the LG's, and many of them are interconnected with one another.

Two distinct classes of interneurons synapse with the LG. One class of interneuron, exemplified by interneuron A6, responds to a root shock with

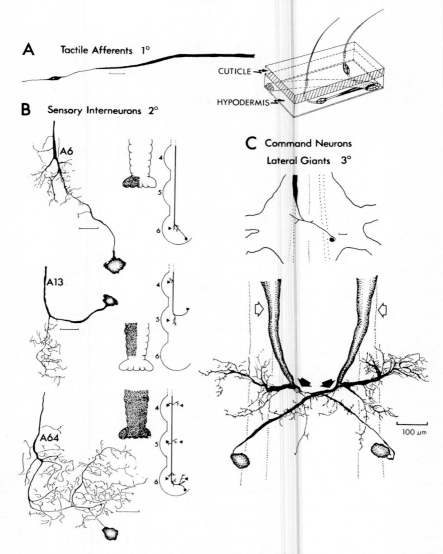

Fig. 4. Structures of neural elements in afferent pathway to LG axons. (A) Bipolar mechanoreceptors. Two receptors are associated with each hair; their dendrites extend into the hollow hair shaft as shown in inset for hairs on cephalothorax (Mellon and Kennedy, 1964; Mellon and Kaars, 1974). (B) Sensory interneurons. These receive direct and polysynaptic input from afferents. Three of four interneurons known to be presynaptic to LG axons are shown; insets show receptive fields and schematic locations in cord. Input sites are shown as solid arrows. A6, A13, and A64 are numbers assigned by Wiersma and Hughes (1961) to axons with same properties as for these interneurons. Axons of all three cells extend at least into the thoracic nervous system (Sigvardt et al., 1982). (C) LG axon segments in second abdominal ganglion. Top figure shows one segment in relation to ganglion, bottom shows details of both

8. The Cellular Organization of Crayfish Escape Behavior

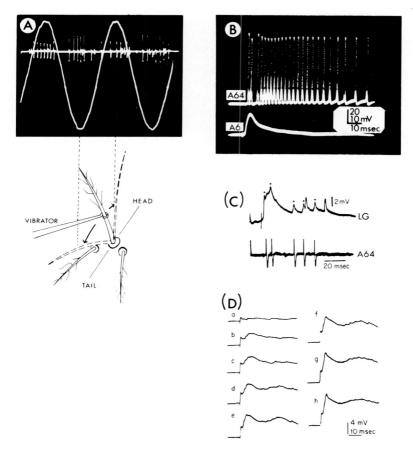

Fig. 5. Examples of electrophysiological responses in afferent circuit to LG's. For schematic of circuit, see Fig. 3B. (A) Response of mechanoreceptors. Evidence for two sensory cells sensitive to opposite directions of movement of a single hair. Top trace shows two spike amplitudes in sensory nerve from hair in response to a 5 Hz sinusoidal movement of magnitude indicated in drawing. (B) Simultaneous comparison of responses in A6, a class I interneuron (bottom trace), and in A64, a class II interneuron, to same sensory volley: a 0.1-msec shock to nerve from ipsilateral endopodite. Soma sites were used for both recordings. Note that the gain for nonelectrogenic class I cell is twice the gain for A64. (C) Simultaneous recordings in LG dendrite of third ganglion (top trace) and from axons of A64 in 2–3 connective (bottom trace) showing 1:1 EPSP (dots) associated with interneuron spikes. (D) Compound EPSP recorded in LG axon just rostral to septal junction, following series of 0.1-msec shocks to ipsilateral second root at increasing intensities (a–h). Early component (α) is monosynaptic excitation from afferents (A: Wiese, 1976; B: G. Hagiwara, unpublished; C: Zucker, 1972a; D: Krasne, 1969).

---

segments. Open arrows indicate septal junctions; closed arrows are presumed sites of commissural synapses, dotted lines show rostral portions of third ganglion LG segments (G. Hagiwara, unpublished).

only a brief burst of impulses and has a passive soma membrane. The second class, exemplified by interneurons A13 and A64, responds to a brief root shock with a prolonged train of impulses and has electrogenic soma membrane (Calabrese, 1976a; Wine, 1975). Responses of class I and II sensory interneurons are shown in Fig. 5B.

How many sensory interneurons must fire to trigger the LG? Recordings from near the spike-initiating zone of the LG (Fig. 5D) showed unitary EPSP's that were rarely as large as 0.75 mV (Krasne, 1969), while the threshold for spike initiation in these same regions was about 8 mV. If the electrode was either in the region of spike initiation or distal to it, and if all inputs were equal to the largest, it would require about 10 to 12 simultaneous inputs to trigger the LG (summation is linear because the synapses are electrical). Actually, most interneurons probably have smaller effects, and any asynchrony in the input will further increase the number of interneurons needed to fire the LG. On the other hand, the feedforward connections of afferents produce a depolarization of about 2 mV (Fig. 5D), equivalent to input from several of the most effective interneurons; and temporal summation from repetitively firing interneurons enhances their effectiveness. In one experiment, electrical stimulation of a segmental nerve was used to produce a synchronous volley of afferent input; and the disynaptic component, which typically fires the cell, was graded into about 14 steps. Eight steps are shown in Fig. 5D (Krasne, 1969). While these results only estimate the convergence required, they do make the point that the few interneurons so far identified as presynaptic to the LG (Zucker, 1972a) are only a small fraction of the number required to trigger LG firing.

What features of the afferent circuit account for its selectivity for abrupt stimuli? Three features are potentially important. (1) Chemical synapses between mechanoreceptors and sensory interneurons show rapid homosynaptic depression (Zucker, 1972b). (2) Synapses of interneurons onto the LG are electrical, and the LG has a high threshold and short time constant, so that summation requires closely spaced inputs (Zucker, 1972b). (3) Activity in afferents evokes recurrent, presynaptic inhibition onto afferent terminals (Kennedy et al., 1974).

When a stimulus is gradually increased in intensity, causing more and more afferents to fire, many afferents will be firing at peak stimulus intensity; but presumably the synapses of those with lowest thresholds would by that time have become severely depressed. In addition, as inhibition is recruited, the peak response is presumably depressed even further. These effects are avoided when all of the afferents are recruited synchronously. Thus, in the system shown in Fig. 3B, presynaptic and postsynaptic inhibition at the first synapse, plus homosynaptic depression of that synapse, act synergistically to severely attenuate sustained stimuli, while abrupt volleys are transmitted. Thus, the afferent circuit is a high pass filter.

UNRESOLVED QUESTIONS ABOUT THE AFFERENT CIRCUIT TO THE LG's

We do not yet know what proportion of the abdominal mechanosensory interneurons project to the LG's. Many do not. For example, one type of interneuron is excited by only one of the two bipolar cells in each hair, and as a consequence it displays beautiful directional sensitivity to water waves (Wilkens and Larimer, 1972; Wiese, 1976; Wiese et al., 1976). These interneurons do not contribute much to the LG's, although some directional selectivity can be detected in recordings from the LG (Fricke, unpublished). Furthermore, the four interneurons so far identified as presynaptic to the LG (Fig. 3B) are all nondirectional (K. A. Sigvardt, G. Hagiwara, K. Wiese and J. J. Wine, unpublished; R. A. Fricke, unpublished).

Another problem is that we do not yet understand the significance of the two classes of interneurons that synapse with LG. Those which show a prolonged afterdischarge to a discrete stimulus (class II) are especially puzzling in view of the phasic response of the LG command cell.

Finally, we have evidence that interneurons in the LG circuit also excite the postural circuits discussed by Page in Chapter 2 of this volume (Kuwada and Wine, 1979; R. A. Goldberg, C. L. Tillotson, and J. J. Wine, unpublished). This raises the general question of the organizational plan used to connect the receptor surface to the animal's motor circuits.

B. Habituation of LG-Mediated Escape Responses

If a crayfish is tapped on the abdomen repeatedly, the probability of a response diminishes. Stimulation at intervals of 1/min can drive responsiveness to zero, with little recovery seen after 3 hr. This occurs even if the abdominal nervous system is isolated from the rest of the CNS by transecting the connectives between thorax and abdomen (Wine et al., 1975).

All available evidence is consistent with the interpretation that habituation occurs as a result of homosynaptic depression of synapses between afferents and interneurons. The analysis of habituation, which became straightforward after the circuit had been mapped, is shown in schematic form in Fig. 6A (Wine and Krasne, 1978; based on Krasne, 1969, and Zucker, 1972a,b). In brief, the known pathway has ten potential sites for lability that could lead to habituation. Sites 6–10, the efferent limb, can be tested by shocking the giant axons directly and recording muscle tension or movements. When this is done, flexion continues for over 30 trials at 1/sec, with some waning in flexor magnitude; this is unlike normal habituation, in which the response quickly fails in an all-or-none fashion (Krasne and Woodsmall, 1969; Wine et al., 1975). When impulses in LG axons are recorded during an habituation session, they are seen to drop out exactly when the behavior does—a spike

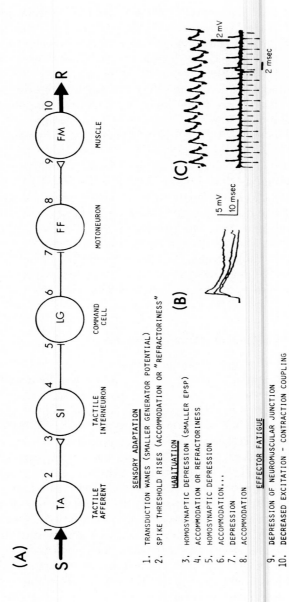

Fig. 6. Analysis of habituation of escape response in crayfish. (A) Simplified version of LG escape circuit, showing ten possible sites of lability. Sites are numbered from 1 to 10 and listed below diagram. (B) Decline in EPSP in LG command cell due to stimulation of axons of tactile afferents at 1/5.7 sec. First, second, and fifteenth responses are shown. (C) Stability of EPSP's in command cell (top trace) when sensory interneurons are stimulated directly (bottom trace). No change is seen at frequencies of 400/sec (B: from Krasne, 1969; C: from Zucker, 1972a).

8. The Cellular Organization of Crayfish Escape Behavior

is never seen without a concomitant tail flip. The weakest link is evidently in the afferent and not the efferent limb.

The afferent limb can be tested directly by recording intracellularly in the LG axon. Afferent volleys cause a large, compound excitatory postsynaptic potential (EPSP), which rapidly declines during repeated stimuli at rates of 1/min or faster (Krasne, 1969, and Fig. 6B). This test bypasses sites 1 and 2, which were independently tested by Zucker (1972b) and found to be stable. Further work by Zucker proved that the labile site is the synapse between the tactile afferents and the sensory interneurons (Site 3). No sign of long-lasting accommodation was detected in sensory interneurons, while synaptic transmission between sensory interneurons and the LG command cell was shown to be electrical and hence extremely stable (Fig. 6C). However, synapses made by terminals of tactile afferents onto the sensory interneurons are chemical and these synapses are labile. A unitary EPSP in a tactile interneuron was reduced to 50% of its initial amplitude by five stimuli at 1/sec, while stimuli at 1/min caused a similar rate of decline in the number of spikes recorded in tactile interneurons (Zucker, 1972b). Although some discrepancies remain between the time course of recovery in physiological and behavioral experiments, the agreement seems acceptable in view of the many differences between the two kinds of experiment. In any event, while it is not yet possible to say how the mechanism of habituation might be influenced by various experimental conditions, the evidence does suggest that no other site within the pathway has the necessary properties to account for the behavioral change. Hence, behavioral habituation appears to be the result of a decrement in synaptic transmission between the tactile afferents and the sensory interneurons.

What is the mechanism for homosynaptic depression at this synapse? Zucker (1972b) described experiments in which the coefficient of variation (i.e., the ratio of the standard deviation to the mean) of EPSP's in one of the first order interneurons of the LG pathway increased during depression. Assuming (1) that transmitter release occurs according to Poisson statistics, (2) that homosynaptic depression occurs uniformly at all the synapses activated by the afferent volley, and (3) that the volley itself is constant, Zucker could conclude that depression is due, at least in part, to diminished release of quanta from presynaptic terminals. Unfortunately, the conditions of the experiment made the validity of these assumptions somewhat doubtful; but a subsequent experiment done under more favorable circumstances and not assuming Poisson release led to similar conclusions (see Krasne, 1976). Quantal analyses on habituation-producing synaptic depression in the *Aplysia* gill withdrawal reflex suggest the same mechanism (see Kandel, 1976). The conclusion that depression is associated with diminished release from primary afferents at their synapses onto first order interneurons is well founded.

What are the functional consequences of homosynaptic depression at afferent terminals? We have focused on the resulting habituation of the LG escape response, but this is only one of several consequences. Earlier, we already noted the potential usefulness of synaptic depression in filtering out sustained stimuli. Furthermore, evidence exists that the depression-prone sensory pathways are not private lines to the LG axons, but rather contribute to other behavioral systems (Kuwada and Wine, 1979; Goldberg et al., 1982). Hence, the use-induced decreases are not response-specific and are better viewed as sensory adjustments. They have much in common with receptor adaptation, but they differ in one striking respect—the animal can prevent depression from occurring by inhibiting the synapses (see below).

C. The MG and Non-Giant Systems

Our knowledge of the MG system is based on intracellular recordings from the MG axons of heavily dissected preparations in poor physiological condition, and on behavioral observations. Stimulation of antennal or optic nerves produces compound depolarizations in the MG axon. Tactile stimulation also excites the MG's, but "sustaining fiber" responses (Wiersma, 1967) to onset or offset of light have no effect. Electrically evoked EPSP's show an early, stable component and a later, decrementing component, suggesting the same basic organization as in the LG system (F. B. Krasne and J. J. Wine, unpublished results). An interesting difference between the MG and LG systems is that the MG interneurons have a single input region and spike-initiating zone. Since the MG axon can be excited via tactile input anywhere on legs or thoracic carapace, central afferent pathways to the MG's must be long.

The non-giant system is known chiefly by its effects. No decision neurons are known for the non-giant system; perhaps single command neurons for non-giant swimming do not exist. It is possible to find axons in the circumesophageal connectives that provoke swimming or single, non-giant tail flips, but they do so unreliably and often elicit other behaviors at stimulation frequencies lower than those needed for tail flips (Atwood and Wiersma, 1967; Bowerman and Larimer, 1974). Similar responses have not been reported with axons dissected out at more caudal levels, and swimming and non-giant tail flips are abolished by transections of the nerve cord below the subesophageal ganglion (Wine and Krasne, 1972). Thus, high-frequency stimulation of circumesophageal axons may elicit tail flips indirectly, as might any strong sensory stimulus.

Since non-giant decision neurons are unknown, a detailed analysis of afferent circuits is not yet possible for this system. Behavioral observations suggest that most modalities will be involved, and we think of non-giant

8. The Cellular Organization of Crayfish Escape Behavior

responses as being "voluntary," by which we mean that they are not tightly linked to external stimulation, and they are subject to mediation by complex central processes. To clarify this distinction, we think a crayfish might be taught to flip its tail with the non-giant system, but not with the giant one. At present our best guess, based on work with lesions, is that non-giant swimming is initiated and perhaps controlled in the subesophageal ganglion (Wine and Krasne, 1972). However, we have not eliminated the possibility that swimming oscillators are located elsewhere and require the subesophageal ganglion only for tonic excitations.

One avenue of promise is that many identified motoneurons are shared by the giant and non-giant systems. A few premotor interneurons common to both systems have also been identified (Kramer, 1978), so it may be possible to work backward through the non-giant system (see below).

IV. MOTOR CONTROL: EXCITATORY PATHWAYS FOR FLEXION

A. Tonic versus Phasic Systems

A significant experimental advantage of the crayfish abdomen is that slow and fast axial flexor muscles are divided into separate groups, each innervated by a separate set of motoneurons running in separate branches of purely motor roots. The action of either system can thus be displayed in isolation by cutting one or the other motor roots. Postural control is unaffected by loss of the fast (phasic, deep) flexors, while rapid tail flips are unaffected by loss of the slow (tonic, superficial) flexors (Kennedy and Takeda, 1965a,b). The slow motor system is dealt with in Chapter 2 of this volume; characteristics of fast and slow crustacean muscles are covered in Chapters 2 and 4 of Volume 3. The following account deals exclusively with the fast flexor muscles, except in the tailfan, where the separation of phasic and tonic systems is less distinct.

B. Overview of the Fast Flexor Motor System

The fast flexor muscles are innervated, in parallel, by a system of giant motoneurons (MoG's) and non-giant fast flexor (FF) motoneurons. The simplest pathway for production of tail flips is from giant axons directly to the MoG's. This pathway is sufficient for production of strong, rapid flexor contractions and a consequent tail flip due to a single giant axon impulse. The MG axons excite the whole population of giant motoneurons, while the LG axons excite only a subset. In this way, different patterns of flexion are produced by the two kinds of giant axons. A parallel but less direct pathway exists from the giant axons to a giant, intraganglionic premotor neuron (the

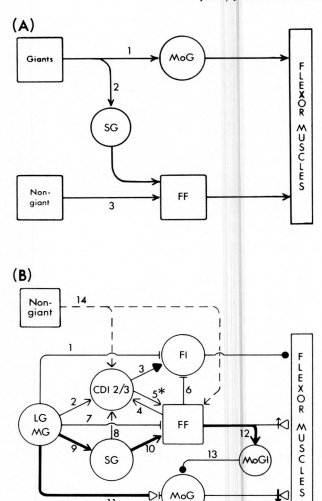

Fig. 7. (A) Pathways for exciting flexor muscles. Pathway 1 is disynaptic from giants via motor giant. Pathway 3 is from non-giant premotor elements to fast flexor motoneurons. Pathway 2 uses segmental giant to link giants to fast flexor motoneurons. (B) Details of flexion circuit for one side of segment 2 or 3. Direct LG/MG input via CDI's and SG cells (note, however, that CDI output pathway 5* does not exist in segment where CDI receives input). FF motoneurons in turn excite at least CDI's and FI, and inhibit MoG via MoGI. (Pathways 1, 3, and 6: Wine and Mistick, 1977; pathways 2, 4, 5, and 8–10; Kramer, 1978; Kramer et al., 1981; pathway 7: Takeda and Kennedy, 1964; pathway 11: Furshpan and Potter, 1959; pathways 12 and 13: Wine, 1977a; pathway 14: Kramer, 1978). Pathway 9/10 is main route from giant axons to FF motoneurons (Roberts et al., 1982). No claim is made that this circuit is complete. In particular, many other CDI's exist.

8. The Cellular Organization of Crayfish Escape Behavior

segmental giant) and then to the FF motoneurons. The role of this pathway is not yet clear. It is effective in anterior ganglia where it reinforces the MoG pathway; but in the caudal ganglia, the pathway often does not fire FF motoneurons. A third pathway consists of poorly understood non-giant premotor elements, which can fire FF motoneurons while bypassing both motor giants and segmental giants. This third pathway, as was stated earlier, mediates all repetitive tail flips (swimming) and all initial tail flips, except those caused by very abrupt stimuli; portions of the third pathway are also driven by the giants. An overview of the various pathways is shown in Fig. 7A; details of the giant-mediated pathways to the flexor muscles are shown in Fig. 7B.

C. The Fast Flexor Muscles

The fast flexor muscles of the abdomen employ a fascinating mechanical arrangement to increase the power and speed of the tail flip (Rayner and Wiersma, 1967). As shown in Fig. 8, the massive anterior oblique muscles

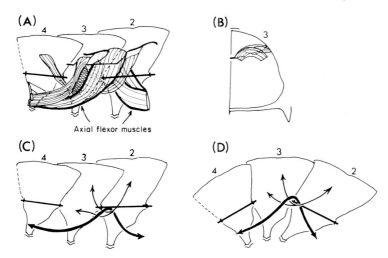

Fig. 8. "Pulley-action" of the fast flexor muscles in crayfish. (A and B) Internal view of abdominal flexor musculature in extended position. Diagrammatic reconstruction prepared from camera lucida drawings. (A) Medial view of second, third, and fourth abdominal segments. Lines across the second and fourth segments have been drawn through intersegmental hinges, the positions of the hinges marked by short vertical bars. Parts of transverse membrane and all inscriptions are shown stippled, except that midline inscription between lateral halves of transverse-running muscles is cross-hatched. (B) Anterior view of third abdominal segment (only medial part of transverse membrane is shown in this figure). (C and D) Diagrammatic representation of pulley-action occurring in flexor musculature of abdomen. (C) Second, third, and fourth abdominal segments in extended position. (D) Same segments in fully flexed position. (Rayner and Wiersma, 1967.)

loop over the central muscles. The central muscles are attached to a structure called the transverse membrane, which is also the point of attachment for the dorsolateral and transverse muscles. These latter muscles act to raise the transverse membrane, causing it to pull up both the central muscle and the anterior oblique. If both of the latter muscles are also contracting, the loop acts as a mutual pulley to amplify their effects (see Fig. 8C,D). Rayner and Wiersma calculated that this trick reduces by half the change in length of the muscle needed to produce flexion, with a consequent increase in velocity of the tail flip.

Unfortunately, the brief accounts by Rayner and Wiersma (1964, 1967) have never been followed by a detailed exposition of the deep flexor muscles. These muscles are treated briefly by Pilgrim and Wiersma (1963), but otherwise only classical accounts are available. The best of these is by Daniel (1931); however, this has never been referred to by workers interested in the innervation of the deep flexors, so that nothing is known about the innervation of the dorsolateral and central muscles. Also, serial homology of the muscular system—crucial for an understanding of different modes of swimming—has not been established definitively. Given the completeness with which we understand the structure of the central pool of flexor motoneurons, our limited knowledge about the muscles has become a bottleneck in our program to explain the behavior.

D. The Fast Flexor Motoneurons: Direct and Indirect Pathways for Excitation

A pool of 96 flexor efferents supplies the fast flexor muscles innervated by the first five abdominal ganglia. Soma maps for each ganglion show a basic plan that undergoes some modification in the anterior and posterior ganglia (Fig. 9A,B). Three functional categories are represented in every half segment: (1) the motor giant (MoG), (2) the peripheral inhibitor (FI), and (3) the fast flexor (FF) motoneurons. Each half segment has one MoG, one FI, and from five to nine FF motoneurons. The somata of the FF motoneurons are arranged in three groups (Fig. 9A).

The motor giant and flexor inhibitor branch extensively in the periphery: all muscle fibers receive input from FI, and all except the transverse muscles receive input from MoG (Selverston and Remler, 1972). As a group, the FF motoneurons innervate all flexor muscles, but a given FF motoneuron distributes excitation to a fairly restricted portion of the segmental muscles (Table II).

The structure and properties of the MoG and FF motoneurons differ strikingly (Fig. 9 and Table III). Evidence suggests that the MoG is exclusively activated by the giant axons, and it is strongly inhibited prior to and during non-giant tail flips (Fig. 10, F. B. Krasne, unpublished). This restriction of MoG activity to the initial, giant-mediated, stereotyped tail flip is consistent with its extremely depression-prone peripheral junction (Bruner and Kennedy, 1970; Zucker and Bruner, 1977), its extensive and nonselective

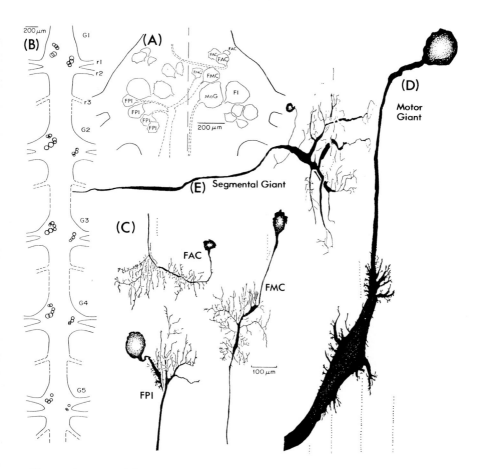

Fig. 9. Anatomy of flexor motor elements. (A) A soma map of all flexor efferents in the second ganglion. Cells with axons leaving by anterior and posterior roots of left side are labeled, and approximate position of their axons are indicated with dotted lines. (B) Variations in soma map in ganglia 1–5. Cells leaving by roots to right are shown. (C) Representative FF motoneurons from three clusters: anterior contralateral cluster (FAC); posterior ipsilateral cluster (FPI); and medial contralateral cluster (FMC). (D) Motor giant. Outlines of giant axons are indicated with dotted lines. (E) Segmental giant. Axon of segmental giant exits via ipsilateral first root (A: Wine and Mistick, 1977; B: Mittenthal and Wine, 1978; C–E: G. Hagiwara and J. J. Wine, unpublished).

TABLE II

Motoneurons to the Fast Flexor Muscles

Nomenclature references			Axon branching pattern	Dendritic domain	Muscles innervated
(1)[a]	(2)[b]	(3)[c]	(4[d], except as noted)	relative to axon	
MoG	F-1	—	All[b]	Partial Bilateral	All except T[b]
Fl	F-10	I2	All[b,d]	Bilateral	All[b]; A_{2nd}, caudal[c]
FAC-1	F-8	M3	Lateral	Partial Bilateral	T[b,c]
FAC-2	F-9	M2	Lateral and Dorsal	Partial Bilateral	A_{ii}XVII[b], A, P[c]
FAC-3	—	M1	Lateral and Dorsal	Partial Bilateral	P[c]
FMC-1	F-5	M4	Lateral and Dorsal	Bilateral[d] Ipsilateral[b]	P[b]; P, A_{2nd}, A_{4th}, T_{3rd}[c]
FMC-2	—	M5	Lateral and Dorsal	Ipsilateral	A_{3rd}[c]
FPl-1	F-2	M6	Lateral	Mainly Ipsilateral	T[b,c]
FPl-2	F-3	M7	Superficial[e]	Mainly Ipsilateral	A_iXVII[b], A_{2nd}[c]
FPl-3	F-7	M9	Dorsal	Mainly Ipsilateral	A_i ?, P[b], A_{4th}[c]
FPl-4	F-6	M10	Dorsal	Mainly Ipsilateral	A_iXVIII[b], A_{4th}[c]

[a] (1) = Mittenthal and Wine, 1978.
[b] (2) = Selverston and Remler, 1972; based on the third ganglion of crayfish.
[c] (3) = Otsuka et al., 1967; based on the second ganglion of lobster.
[d] (4) = Stretton and Kravitz, 1973; based on the second ganglion of lobster.
[e] In the lobster the superficial branch contains the axons of two fast flexor motoneurons as well as the tonic flexors.

TABLE III

Contrasting Properties of the Giant and Non-Giant Fast Flexor Motoneurons

Motor giant (MoG)	Non-giant (FF)
One per hemiganglion	Five to nine per hemiganglion
Passive soma	Active soma
No dendrites in ganglion	Extensive dendrites in ganglion
Extensive peripheral branching	Restricted peripheral branching
Synapses in periphery are extremely depression-prone	Synapses in periphery show mild facilitation
Inhibited by recurrent Renshaw-like pathway from other motoneurons	Not inhibited
No known central outputs	Many central outputs (possibly indirect, via SG)

8. The Cellular Organization of Crayfish Escape Behavior

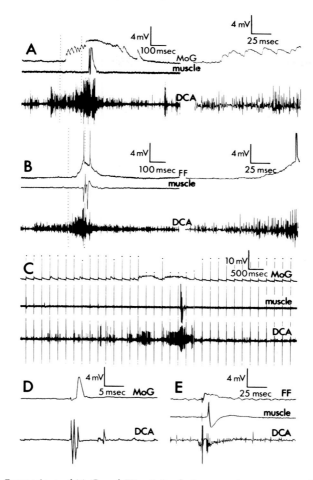

Fig. 10. Comparison of MoG and FF activity during non-giant responses. Traces labeled "muscle" are thoracic flexor EMG's; MoG and FF activity was recorded intracellularly at origin of third roots; traces labeled DCA (dorsal cord axons) were recorded with a large suction electrode. (A) Depolarizing IPSP's in a fourth ganglion MoG preceding and during a non-giant tail flip produced by manipulating exopodite of tailfan. Note that both dorsal cord activity and IPSP start well before motor response. First part of response (between dotted lines) is shown expanded on right. (B) EPSP leading to spikes in a FF motoneuron during a non-giant response evoked by brushing ventrum between abdomen and thorax. Again, note gradual development of EPSP and DCA activity. (C) In a different preparation a non-giant response was evoked in a fifth ganglion MoG by pinching an appendage of the tailfan during 4 Hz stimulation of cord giants in order to examine the extent of inhibition of MG-produced EPSP's. Dots indicate tops of EPSP's in MoG. (D) A giant-evoked EPSP in the above MoG neuron at fast sweep. (E) A giant-evoked EPSP in the FF motoneuron shown in B. Note much more rapid rise and short duration of the EPSP. Duration of such EPSP's is often augmented by late-arriving CDI input and sometimes shows a second rise from such input.

peripheral branching pattern (Kennedy and Takeda, 1965a; Selverston and Remler, 1972), and its structure of short central branches restricted to the area of the giant axons (Mittenthal and Wine, 1973, 1978). Further evidence is the failure to observe any excitatory input to the MoG, other than that from the giant axons, when stimulating the cord with strong electrical shocks (Wine, 1977a). The role of the MoG in shaping escape behavior is discussed in Section IV,E.

PATHWAYS TO THE FF MOTONEURONS

The giants also excite the entire pool of fast flexor motoneurons, though not always sufficiently to cause firing. The pathway from the giant axons to the fast flexor motoneurons, unlike that to the MoG's, and in contrast with prior interpretations, is disynaptic, with only minor direct connections from giant axons to motoneurons. Anatomical and electrophysiological studies reveal a striking neuron, termed the segmental giant, interposed between giant axons and FF motoneurons in each hemiganglion (Fig. 9E; Kramer et al., 1981; Roberts et al., in press). Coupling is electrical and suprathreshold, so that the segmental giant (SG) introduces little transmission delay. The SG has an axon in the ipsilateral first root; antidromic stimulation of the SG axon accounts for the activation of ipsilateral FF motoneurons noticed in previous studies (Wiersma, 1949; Takeda and Kennedy, 1964). The peripheral destination of the SG axon is not yet known; it has no discernible direct effect on muscles of the swimmerets, but its central effects include excitation of swimmeret powerstroke motoneurons (Kramer et al., 1981). While the giant axons fire most of the motoneurons bilaterally, a segmental giant tends to fire only FF motoneurons with ipsilateral dendrites, although it produces subthreshold EPSP's in motoneurons with contralateral dendrites. This activation pattern is consistent with anatomical studies showing that most FF motoneurons have asymmetric dendritic domains, with the heaviest concentration of processes ipsilateral to the axon (Stretton and Kravitz, 1973; Table II; Fig. 9C). The heterogeneity of peripheral branching patterns of different FF motoneurons in middle abdominal segments, and the variable movements seen during non-giant escape, suggest that non-giant inputs might be able to activate portions of the FF pool within a segment selectively. However, evidence for such selectivity is not presently available.

The FF motoneurons are also excited, often to firing level, during non-giant tail flips. The pattern of excitation of the FF motoneurons is very different during giant and non-giant flips. A giant axon impulse causes a short-latency, rapidly rising depolarization which peaks within a few milliseconds and seldom causes more than a single spike. In contrast, during non-giant tail flips, the FF's often fire in bursts, and each burst is preceded by about 60 msec of gradually increasing depolarization (compare Fig. 10B and E).

The premotor interneurons responsible for FF excitation during non-giant flips have only begun to be identified, but a large population of dorsally located axons fires concomitantly with depolarizations of the FF's seen during non-giant mediated tail flips (Fig. 10, dorsal cord interneurons). The two largest axons have been identified and called CDI 2 and CDI 3. Both interneurons have the interesting properties of (1) being contained entirely within the abdomen, even though control and initiation of non-giant flips is believed to depend upon the subesophageal ganglion, and (2) being reliably fired by the giants (and/or the SG's) in their ganglia of origin. Both provide to the FF's, along the route of their axons, excitation that is (for the most part) weak compared to that from the giants and thus requires summation with other premotor elements to fire the FF's.

E. Variations in Synaptic Connections Responsible for the Spatial Patterning of Tail Flips

The different forms of abdominal flexion that produce backward and upward escape responses can be explained, in part, by the pattern of synaptic connections between giant axons and motor giants. The pattern of these connections has been demonstrated both anatomically and physiologically (Mittenthal and Wine, 1973, 1978). The central finding is that the LG axons do not synapse with MoG neurons in the fourth and fifth abdominal segments, which thus remain unflexed during the tail flip, resulting in a downward propulsive force that pitches the animal forward (Fig. 11). In contrast, the medial giant (MG) axons synapse with the MoG's in all segments, so that the abdomen forms a tight tuck, producing the main propulsive force to drive the animal backward.

What role do the FF motoneurons play in different motor patterns produced by the LG and MG axons? Investigations by L. Miller (in preparation) show that LG and MG impulses produce identical amounts of depolarization in FF motoneurons of the fifth abdominal ganglion. Since the depolarizations are usually subthreshold, the lack of selectivity does not interfere with the LG-induced pattern of flexion which depends on the absence of flexor drive in the caudal segments. In other words, the giant axons usually fire the FF motoneurons in the second and third ganglia and rarely fire them in the fifth ganglia. In the fourth ganglion, the giant input is again subthreshold but larger than in the fifth. What is the basis of this rostrocaudal gradient of excitatory drive? Miller tested the segmental giant's excitation of FF motoneurons in each abdominal ganglion from the second through the fifth and found a clear decline both in absolute amplitude of the SG EPSP and in the percentage of times that it fired the motoneurons either alone or in combination with other inputs. Hence, during MG-mediated tail flips, flexor

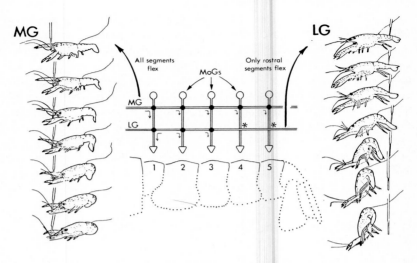

Fig. 11. Connections of giant axons with MoG's explain form of tail flips. When MG's fire, all MoG's are excited, all segments flex, and the abdomen curls and propels the animal backward. When LG's fire, there is no output to caudal segments, which remain straight and cause the thrust to be directed mainly down, thus pitching animal forward. Since MG's respond to rostral inputs and LG's to caudal ones, tail flips always remove the animal from the source of stimulus. (Based on Wine and Krasne, 1972; Mittenthal and Wine, 1973.)

drive in caudal segments must often be exclusively the result of the motor giant pathway (see Fig. 14). Since peripheral junctions of the motor giants are extremely labile (Bruner and Kennedy, 1970; Zucker and Bruner, 1977), use-dependent variations in the form of the tail flip might be expected. Variation in the extent of caudal flexion is, in fact, seen (compare Figs. 2 and 11), with consequent changes in the escape trajectory; recent evidence has established that at least some changes are use-dependent (Miller, 1981).

On occasion, the giant axons fire multiple spikes that can drive the FF motoneurons to discharge. However, repetitive firing of giant axons also causes the peripheral inhibitor to burst (Wine and Mistick, 1977), with the shortest latencies occurring in the fourth and fifth ganglia (Uyama and Matsuyama, 1980). The peripheral inhibitor may cancel discharge of the FF motoneurons and thus preserve the LG pattern, but it is also possible that in some cases partial activation of FF motoneurons in caudal ganglia occurs, with a consequent change in the animal's trajectory.

The uropods and telson are also differentially affected by the LG and MG axons (Larimer and Kennedy, 1969). The complex tailfan system includes homologs of both axial muscles and muscles of the appendages, and has three degrees of freedom of movement. The muscles are innervated by five

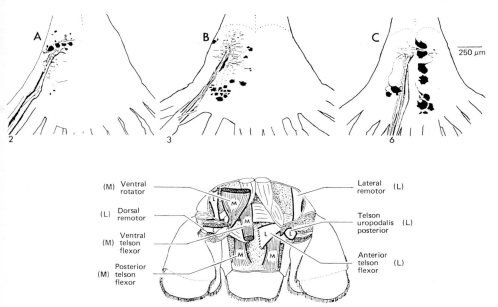

Fig. 12. Motor system of the tailfan. (A–C) Neuron pools serving muscles of the exopodite, endopodite, and telson, respectively. These pools are based on backfills of entire roots. Most and possibly all neurons are efferents, but sensory elements with central somata have been described in homologous roots of other Crustacea (Paul, 1972). (D) Muscles of tailfan. Striped muscles (M) are activated by medial giant axons, clear muscles (L) are activated by lateral giant axons (D: From Larimer and Kennedy, 1969).

roots from two ganglia; about 55 motoneurons were detected electrophysiologically (Larimer and Kennedy, 1969) and about 75 somata were seen with backfills of Roots 2, 3, and 6 of the sixth ganglion (Fig. 12A–C).

A partial analysis of phasic muscles of this system yielded the data shown in Fig. 12D and Table IV (Larimer and Kennedy, 1969). EMG recordings showed that seven muscles were differentially activated by the LG and MG axons. Results for MG impulses are consistent with observed behavior, but the observed activation of the lateral and dorsal remotors (which act on the uropods) by LG axons is opposite to the promotion observed in behavioral experiments both by Larimer et al. (1971) and by us (Wine and Krasne, 1972, and unpublished observations).

We have not yet resolved this discrepancy, but recent results indicate a direction for future research. Motoneurons controlling flexion of the telson were recorded intracellularly while the MG and LG axons were stimulated. The LG's excite, at short latency, the anterior telson flexor motoneuron, and the MG's similarly excite ventral and posterior telson motoneurons. How-

TABLE IV

Actions and Innervation of Uropod Muscles[a]

Muscles	LG or MG Escape	Action	Type	Innervation	Axons/muscle	Max. number axons/fiber
Flexors						
Ventral telson flexor	MG	Telson flexion	Phasic	R6 G6	3E, 11	3
Posterior telson flexor	MG	Telson flexion	Phasic	R6 G6	5E, 11	5
Anterior telson flexor	LG	Telson flexion	Phasic	R6 G6	1E	1
Telson uropodalis anterior			Phasic	R2 G6	2+ E	2
Telson uropodalis posterior	LG	All flex, and close uropods on telson	Phasic	R2 G6	2+ E	2
Telson uropodalis lateral			Phasic	R2 G6	2+ E	2
Rotators						
Ventral rotator	MG	Rotation, some adduction and flexion	Mixed	R3 G5	2+ tonic 3+ phasic	—
Dorsal rotator		Rotation and flexion	Mixed	R1 G6	2E, 11 tonic 4+ E phasic	3
Extensors						
Lateral remotor		Remotion and extension of uropod	Phasic	R3 G6	2E	2
Medial remotor		Slight remotion and extension of uropod	Phasic	R3 G6	3E	2
Promotors						
Lateral promotor		Promotion	Phasic	R3 G6	2E, 11	3
Dorsal promotor		Promotion	Phasic	R3 G6	2+ E, 11	3
Remotors						
Lateral remotor	LG	Remotion, closing exopodite upon endopodite	Phasic	R2 G6	1E	1
Dorsal remotor	LG	Remotion, closing exopodite upon endopodite	Phasic	R2 G6	1E	1

[a] From Larimer and Kennedy, 1969.

8. The Cellular Organization of Crayfish Escape Behavior

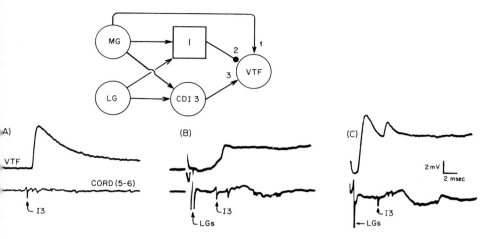

Fig. 13. Potential switching mechanism in CNS of the crayfish. Schematic shows circuitry between giant axons and a ventral telson flexor (VTF) motoneuron. Both giant axons excite VTF via CDI 3 and inhibit VTF via an unknown pathway (I). MG cell also excites VTF directly. Numbers indicate order of arriving information: (1) MG excitation is first, (2) Inhibition from either cell is second (a common pathway is probably used), and (3) CDI 3 excitation is last to arrive. (A–C) Intracellular responses in a VTF soma. (A) Response to CDI 3 fired alone: VTF fires. (B) Response to LG stimulation: since LG has no direct connection, the first event is the IPSP, followed by shunted EPSP from CDI 3. Note that if pathway 2 were suppressed, LG would fire VTF. (C) Response to MG stimulation: direct connection spikes VTF, then IPSP begins and shunts final EPSP from CDI 3. (From Kramer, 1978; Kramer et al., 1981.)

ever, this straightforward difference in direct connections is complicated by an odd arrangement of supplementary connections. These are illustrated for a given ventral telson flexor (VTF) motoneuron in Fig. 13 (Kramer, 1978; Kramer et al., 1981). It turns out that both MG and LG axons recruit an interneuron (CDI 3) that powerfully excites the VTF, yet both also recruit inhibition that begins soon enough to shunt the EPSP from CDI 3. If CDI 3 is fired alone, it excites the VTF, but when it is recruited by the giants, its EPSP is *inhibited*. This circuit constitutes a potential switching mechanism in which the LG axon's access to the VTF depends on the state of the inhibitory pathway. Kramer (1978) has some evidence that modulation does occur in semi-intact animals. (The MG axons always recruit the VTF via a direct connection that operates prior to the onset of inhibition.)

F. The Corollary Discharge

CDI 3 is not the only interneuron fired by the giants. In fact, the giant axons recruit a large population of interganglionic interneurons, whose spikes arise in every abdominal ganglion and travel various distances up and down the cord (Wine, 1971b). The interneurons are excited centrally, for they can be driven by the giants in an isolated abdominal nervous system. The burst of "corollary discharge interneurons," which may last for 40-50 msec following a single giant axon impulse, is partly responsible for a host of auxiliary effects. These include massive and widespread inhibition. For example, the postural systems are inhibited centrally and peripherally, and elements in every level of the escape circuit are also inhibited. These effects will be discussed in Section VI.

Two of the best characterized corollary discharge interneurons produce only excitation (at least directly) and have not yet been shown to have any functions critical to giant-mediated responses. These interneurons, termed CDI 2 and the previously mentioned CDI 3, originate in the second and third ganglia, respectively, and send large axons to the last ganglion. CDI 2 and 3 are excited by the giant axons, by the segmental giant, and (weakly) by the fast flexor motoneurons (Fig. 7B). Their output effects include: weak or moderate excitation of FF motoneurons in ganglia 3-5, strong excitation or firing of telson flexor motoneurons (seen only for CDI 3), weak excitation of the peripheral inhibitors of the flexor muscles, and (indirectly) presynaptic inhibition of tactile afferents (Kramer, 1978; Kramer et al., 1981; Krasne et al., 1977; Wine and Mistick, 1977). These effects parallel those of the giant axons (CDI 2 resembles the LG's and CDI 3 the MG's) and follow them so closely in time that their usual contribution during a giant-mediated response would seem inconsequential. In cases where they are known to excite motoneurons strongly, excitation is cancelled by inhibition (Fig. 13). These interneurons are also activated during some non-giant tail flips (Kramer, 1978), and it may be then that their chief function is expressed. Since we cannot be certain that we have discovered all of their outputs or inputs, we reserve judgment on their functions.

Fig. 14. Flexor motor circuitry of abdomen. This schematic shows the major connections of MG, LG, and non-giant pathways to all 86 fast flexor motoneurons in abdominal ganglia 1-5. Important features are: (1) All non-giant premotor elements, which are lumped together for clarity, bypass MoG and SG to excite FF motoneurons directly. (2) Giant axons make strong connections only to MoG and SG. In fourth and fifth ganglia, connections from LG to MoG are omitted, and connections from SG to FF motoneurons are weak (dotted lines). (3) MoG has extensive, depression-prone synapses with muscles; FF motoneurons have more restricted, facilitating synapses. Bold lines from SG to FF neurons in ganglia 1-3 emphasize that these connections can be 1:1. Various connections have been omitted for clarity: LG's and MG's are cross-connected; SG's receive input from both MG's; giant axons make weak, feed-forward connections to FF motoneurons; and finally, inhibitory neurons and connections of identified CDI's are not shown. For these details see Fig. 7B.

8. The Cellular Organization of Crayfish Escape Behavior

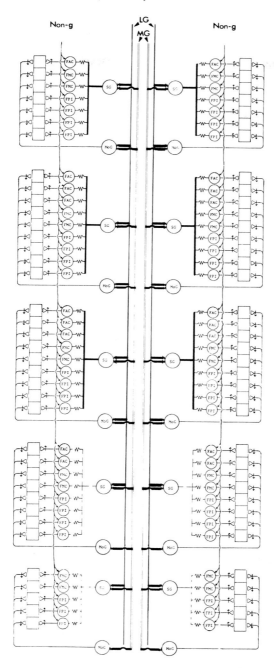

G. Summary of the Flexor Motor Circuitry and a Hypothesis Concerning Giant and Non-Giant Systems

Even an attentive reader must find the foregoing account somewhat overwhelming. In this summary, we remove all caveats and present our best guess about how the system works. For circuitry, see Fig. 14.

The basic system is the non-giant one, which we believe evolved first and has been elaborated to provide a range of movements during steered swimming. The FF motoneurons are the efferents of the non-giant system.

The giant axon system evolved later. Three basic changes occurred: (1) The giant axons evolved from smaller flexor premotor elements. (2) The motor giants evolved, probably from FF motoneurons, as the primary efferents of the giant system. (3) The segmental giants evolved, probably from limb motoneurons, as the primary "central driver" neurons of the system. The giant axons are thus known to make only two strong electrical synapses in each segment. This small number and the spatial separation of the synapses ensure that the axons are not unduly loaded, thus contributing to a higher safety factor for conduction and transmission. (Spatial separation occurs because the MoG synapse has migrated into the connectives; it is a rare example of a central, extraganglionic synapse.)

The giant and non-giant systems have retained or established with each other all synaptic contacts that are not maladaptive. This imparts a good deal of apparent redundancy to the system, but many synapses are probably useful in circumstances that we have not tested. The extent to which interconnections among the systems are useful must vary greatly from case to case—the main point is that synapses will not be eliminated unless they are maladaptive. When synapses are maladaptive, they are eliminated, weakened, or inhibited (inhibition connotes the possibility of disinhibition and underscores the potential adaptiveness of the pathway under some circumstances). We have detected evidence for the latter two solutions. In principle, this is all we can do; evidence for eliminated synapses must come from developmental or comparative studies.

The system is clearly asymmetric. The giants, which act in a massive fashion, activate large portions of the non-giant circuitry. However, the non-giant system, which operates with finesse, would be disrupted by participation of the powerful giant elements. For that reason the giant axons, the motor giants, and the segmental giants are all powerfully inhibited during non-giant escape.

Although these conclusions will lack force until developmental and comparative evidence is gathered, we have arrived at them by applying a modified comparative approach, in which different ganglia within this single species are examined relative to one another. For example, coactivation of giant and non-giant flexor motoneurons in the anterior abdominal ganglia gives no clue about which motor system is tied most closely to the giants; but the weak excitation by the giants of the FF motoneurons in caudal ganglia provides evidence that the giant-FF pathway is the secondary one.

V. MOTOR CONTROL: THE EXTENSOR SYSTEM

A. Overview

Abdominal flexion is the power stroke of escape in crayfish; the return stroke is a more complicated act in which the abdomen is rapidly re-extended in such a way that minimal counterthrust is produced. Extensors, like flexors, are controlled by different mechanisms during giant and non-giant tail flips. However, the extensors, unlike the flexors, are not centrally driven by the giant axons. Instead, short-latency re-extension following giant-mediated tail flips is the result of sensory input produced both by the initiating stimulus and by feedback from the tail flip: sensory feedback for this form of re-extension is both necessary and sufficient. In contrast, during non-giant-mediated swimming, sensory input to the extensor motoneurons is inhibited, and a central oscillator drives the extensors. These conclusions are documented by Reichert et al. (1981).

B. Extensor Muscles and Motoneurons: Sources of Excitation and Inhibition

The phasic, fast, or deep extensor muscles run from the thorax to the sixth abdominal segment and produce rapid extensions during escape. The fast extensors are innervated via the middle of three roots (Root 2) in each segment. The innervation shows a pronounced posterior slant, so that efferents from a given ganglion innervate the next two caudal segments (Fig. 15A). The fast extensor muscles consist of a spiralled, medial group and two lateral groups, innervated by five excitatory and one inhibitory axon. As in the case of the flexors, the fast muscles are paralleled by slow, tonically active ones which control posture. The tonic extensor muscles form thin sheets superficial to the phasic extensors and are also innervated by one inhibitory and five excitatory axons, which are distinguished from the fast axons by their much smaller diameter and their tonic activity (Fields and Kennedy, 1965; Fields, 1966; Kennedy et al., 1966).

Four of the six physiologically identified phasic extensor efferents have been anatomically defined (Triestman and Remler, 1975; Wine and Hagi-

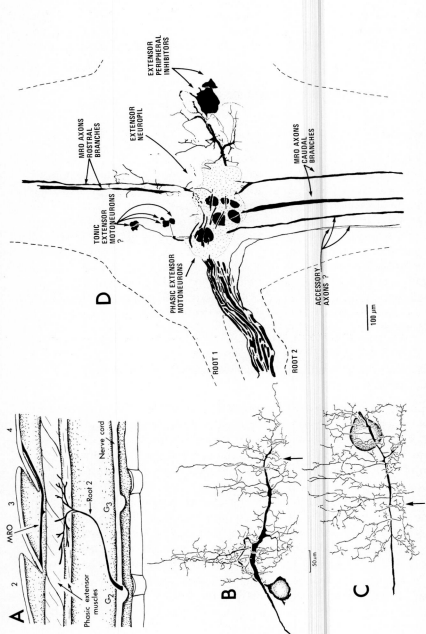

Fig. 15. Anatomy of phasic extensor system. (A) Details of muscle innervation in Segments 3 and 4 and location of muscle receptor organ (MRO). (B) One of larger phasic extensor efferents in G2. Arrow indicates midline. (C) Phasic extensor peripheral inhibitor (EI). (D) Pool of extensor efferents. A total of 16 (sometimes 17) neurons exit via a given second root. These include five classes of cells: phasic and tonic extensor motoneurons, phasic and tonic peripheral inhibitors, and peripheral inhibitors of the muscle receptor organs. Four neurons originate

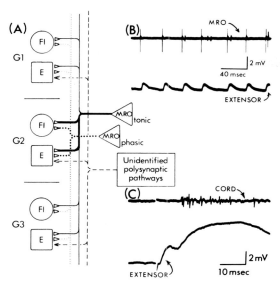

Fig. 16. (A) Sensory pathways to extensor efferents and flexor inhibitor. Both tonic and phasic MRO's synapse with flexor inhibitor and extensor motoneurons in their own and adjacent ganglia; synaptic efficacy diminishes in adjacent ganglia, as indicated by thinner lines. Polysynaptic input from unidentified receptors also occurs. (B) EPSP's produced in a phasic extensor motoneuron by tonic MRO. (C) Mono- and polysynaptic excitation produced by electrical stimulation of an ipsilateral second root. Early, stable component consists of two steps and is caused by MRO's; it is followed by a labile, polysynaptic component correlated with activation of sensory interneurons. (From Wine, 1977c.)

wara, 1977; Fig. 15). Recordings from cell bodies of these neurons in isolated abdomens reveal one source of inhibition and three sources of excitation (Triestman and Remler, 1975; Wine, 1977b,c; see Fig. 16).

(1) An impulse in any giant axon causes a short latency, short duration IPSP in the extensor motoneurons (Wine, 1977b).

(2) The phasic and tonic stretch receptors or muscle receptor organs (Fig. 15), which span the dorsal segments and are stretched by flexion, produce large, stable, monosynaptic, chemical EPSP's in the extensor motoneurons of their own ganglion, and smaller EPSP's in extensor motoneurons of neighboring ganglia (Wine, 1977c). The transmitter is probably acetylcholine, based on results in the lobster *Homarus americanus* (Barker et al., 1972).

(3) Sensory axons in any abdominal sensory root excite the extensors via polysynaptic pathways. Latency to onset of depolarization is up to 10 msec, and the peak of the EPSP is 30 msec or greater (Wine, 1977c). These path-

ways are quite depression-prone, a characteristic so far found only for mechanosensory afferent pathways (Zucker, 1972b).

(4) Stimulation of unidentified axons in the connectives cause stable EPSP's that can summate to fire the extensor motoneurons (Triestman and Remler, 1975; J. J. Wine, unpublished).

C. Analysis of Extension: Dual Modes of Control

As the above account shows, in the isolated abdominal CNS the only effect of a giant axon impulse on the extensors is to inhibit them. Unlike the flexor inhibitor (Wine and Mistick, 1977), the extensors do not receive delayed central excitation. Instead, excitation occurs from sensory feedback. The latter conclusion is based on an analysis of extensor EMG activity under various conditions (Reichert et al., 1981).

(1) If, in unrestrained animals, the giants are fired electrically so that the only sensory input to the extensors is from feedback, phasic re-extension is variable and labile, often disappearing completely after several trials at 1-min intervals.

(2) Phasic extension following electrically elicited giant-axon tail flips is also reduced or abolished by any form of interference with the animal—it never occurs if the abdomen is restrained, if the abdomen is initially flexed, or if the animal is in air; these conditions all reduce sensory influx.

(3) In all cases, if phasic extensor EMG's occur during a single, giant-mediated tail flip, they follow the flexor EMG's at short latency, as may be expected if they are triggered by sensory feedback.

These findings are in sharp contrast to the behavior of the extensors during non-giant swimming (Reichert et al., 1981; Schrameck, 1970).

(1) Extensor EMG's during non-giant swimming are not labile. They continue to occur even after repetitive, electrical stimulation of the giants has reduced post-giant extension activity to zero.

(2) Extensor EMG's during non-giant mediated behavior still occur when the abdomen is restrained in either flexed or extended positions or when all flexor motor roots are cut.

(3) During swimming, the extensors lead the flexors. The flexors follow extensors at near constant latencies of 20-30 msec over a wide range of swimming frequencies, whereas the intervals between flexion and subsequent extension are typically long and lengthen further as the swimming frequency declines. This must mean that sensory feedback to the extensors is inhibited during swimming.

8. The Cellular Organization of Crayfish Escape Behavior

D. Premotor Circuits for Extension

No interneurons of the fast extensor system have yet been identified. Transections of the ventral nerve cord anywhere below the subesophageal ganglion eliminate swimming (Wine and Krasne, 1972), and we have hypothesized that the swimming oscillator is located in the subesophageal ganglion. However, non-giant fast flexor activity eventually returns in cord-cut animals, which become hyper-reflexic (Wine, 1971a), and recent work shows that fast extensor activity also returns and may alternate with flexor outflow (H. Reichert, unpublished). Analysis of the premotor extensor circuitry is an important unsolved problem.

VI. COMMAND-DERIVED INHIBITION

A. Multiple Levels of Inhibition

An impulse in either pair of command axons unleashes a massive, widely distributed, and heterogeneous array of inhibitory effects. Thirteen classes of elements receiving inhibition have been found to date, including receptors,

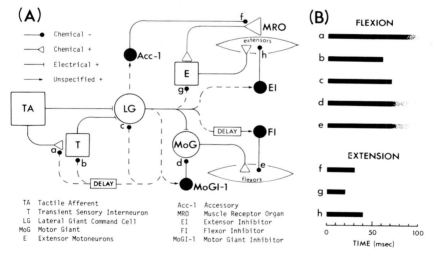

Fig. 17. (A) Pathways for command-derived inhibition. Simplified version of escape circuit is shown with eight inhibitory actions. Pathways a through e inhibit five levels of flexion circuit. Two inhibitory neurons have been identified: MoGI inhibits motor giant and FI inhibits flexor muscles. Pathways f, g, and h inhibit three levels of extension circuit; two identified inhibitory neurons are both peripheral inhibitors: accessory neuron inhibits muscle receptor organ; EI inhibits phasic extensor muscles. (B) Durations of unitary IPSP's are matched to circuits in which they occur. Inhibition of flexor elements is always long (greater than 50 msec in duration); inhibition of extensor elements is short (less than 40 msec). (From Wine, 1977b.)

TABLE V

Targets of Command-Derived Inhibition

Inhibited structures	Inhibitory neurons	Hyperpolarizing or depolarizing	Duration	Latency	Transmitter	Ionic basis
1. Mechanoreceptor terminals	unidentified	D	long	long	—	—
2. Mechanosensory interneurons	unidentified	H	short	long	—	—
3. LG command neuron	unidentified	D	long	short	—	Cl⁻ (?)
4. MG command neuron	unidentified	D	long	short	—	—
5. Motor giant	MoG1-1	D	long	short	GABA	Cl⁻
6. Telson flexor motoneurons	unidentified	D	long	short	—	—
7. Fast flexor muscles	FI	D	long	long	GABA	Cl⁻
8. Muscle receptor organs	Accessory Neurons	H	short	short	GABA	Cl⁻
9. Fast extensor motoneurons	unidentified	H	short	short	—	—
10. Fast extensor muscle fibers	EI	D	short	short	—	—
11. Tonic flexor motoneurons	unidentified	H	long	short	—	—
12. Tonic flexor muscles	fI (f5)	H	long	short	—	—
13. Tonic extensor muscles	eI	H	long	short	—	—

8. The Cellular Organization of Crayfish Escape Behavior

interneurons, motoneurons, and muscles that mediate both postural and escape flexions and extensions (see Figs. 17 and 20 and Table V). Six kinds of inhibitory neurons have been identified. All but one of these are peripheral inhibitors; the one central inhibitory neuron so far identified appears to inhibit only the giant flexor motoneurons (Wine, 1977a). This finding and the differences in the time of onset of inhibition (see below) mean that much of the central inhibition must be produced by separate pathways. All known inhibitors are activated centrally, as can be demonstrated in completely isolated abdominal nervous systems, but at least two of them are supplemented by sensory feedback (see below).

B. Functional and Timing Considerations

Command-derived inhibitory potentials within the circuitry differ from one another in delay of onset and in duration. In some instances, the specific temporal characteristics of particular actions can be understood in adaptive terms, and the timing mechanisms can also be analyzed to some extent.

1. DELAY

With three exceptions, all inhibitory actions begin within a few milliseconds of the giant-axon impulse; for example, the IPSP's in extensor motoneurons have latencies just as short as the EPSP's in FF motoneurons (Wine, 1977b). In contrast, both pre- and postsynaptic inhibition of the first central synapse is delayed by about 15 msec (Krasne and Bryan, 1973; Bryan and Krasne, 1977a,b), and peripheral inhibition of the flexor muscles is delayed at least as much.

Early inhibition of extensor motoneurons presumably clears the way for flexor activity, whereas late inhibition of the flexor muscles, themselves, clears the way for the re-extension that usually follows flexion (Wine and Mistick, 1977). The onset of inhibition at the first central synapse correlates with the actual initiation of movement; this is consistent with its having a role in canceling reafference from the flip movement. Adaptive reasons for the existence and for the early onset of LG and MoG inhibition are less easy to envisage, but we have speculated (Wine, 1977a; Bryan and Krasne, 1977b; also see Kandel et al., 1979) that they serve to limit the length of the LG and MoG spike trains.

The major factor causing delays seems to be the interposition of polysynaptic pathways of corollary discharge interneurons between the giant command axon and the fast flexor and (unidentified) first central synapse inhibitors (Wine and Mistick, 1977; Kramer et al., 1981).

The flexor inhibitor (FI; see Fig. 18) has been most intensively investigated. The giant axons, FF motoneurons, and CDI 2 and CDI 3 all synapse

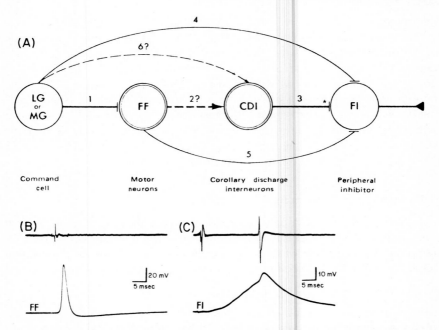

Fig. 18. (A) Minimal circuit responsible for delayed burst in flexor inhibitor. Command cells (LG or MG) are postulated to activate FI via a trisynaptic pathway involving motoneurons and corollary discharge interneurons, while direct, feed-forward links from command cells and motoneurons contribute to early depolarization in FI but do not fire it. Bold links indicate pathways postulated to be more powerful. However, evidence for links 2 and 6 is circumstantial, and their relative strengths are purely conjectural. This circuit was worked out before role of the segmental giant was known. Since SG is tightly coupled to both giants and FF motoneurons, pathways 4 and 5 may either originate from SG or parallel an SG pathway. In either case, all pathways other than pathway 3 are minor and of relatively short latency. (B and C) Comparison of responses in FF and FI to a giant axon impulse. Separate but typical experiments. Inset shows recording sites. Top trace in B shows giant axon impulse, in C it shows both giant and third root outflow. (From Wine and Mistick, 1977.)

with FI to contribute to the early depolarization. However, peak depolarization is reached only after other CDI's are recruited in distant ganglia and travel to FI. It also seems likely that several more synaptic steps intervene—i.e., CDI's probably excite other CDI's. Recent investigations indicate that the fourth abdominal ganglion (G4) is a particularly potent source of excitatory drive to the FI neurons (Uyama and Matsuyama, 1980). Extracellular records showed an intersegmental gradient of latencies for FI spikes following giant axon impulses, with the shortest mean latencies (about 12 msec) seen for G4 and G5. In addition, stepwise removal of ganglia from the caudal end had little effect on the FI discharges of more anterior ganglia until G4 was removed; this procedure suddenly eliminated FI discharges.

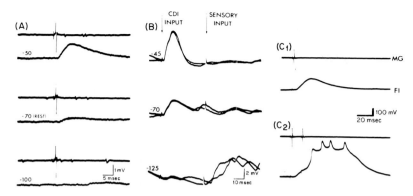

Fig. 19. Anomalous behavior of EPSP's produced by CDI's onto FI. (A) Depolarization increases and hyperpolarization decreases to zero but does not reverse amplitude of PSP's produced in FI by CDI 2 and 3. (B) Opposite effect of polarization on sensory-evoked and CDI-evoked EPSP's. (C) Dramatic augmentation of response in FI to multiple giant axon impulses. C_1, response to a single MG impulse; C_2, response to two MG impulses. (From Wine and Mistick, 1977).

In addition, "anomalous" synapses are made between CDI 2 and 3 and FI. These synapses, unlike most chemical synapses, produce EPSP's that increase with depolarization of the postsynaptic neuron and decrease (but do not reverse) with hyperpolarization of the postsynaptic neuron (Wine and Mistick, 1977). This behavior is specific to the CDI inputs; sensory pathways to FI produce PSP's that respond conventionally (Fig. 19A,B). Functionally, such anomalous synapses should augment temporal and spatial summation of events in FI. In fact, the FI neurons often do show enormously augmented discharges to paired pulses in the giant axons (Fig. 19C); this may result from actions of the anomalous synapses.

In addition to centrally originating excitation, FI receives both monosynaptic and polysynaptic excitations from sensory sources. The monosynaptic excitation is powerful and comes from the muscle receptor organs, which also excite the extensors, thus helping to ensure that FI firing will accompany extensor activity.

2. DURATION

The duration of inhibition is also found to vary in a functionally significant manner (Wine and Hagiwara, 1978; see Fig. 17B). IPSP's are either of long duration (20 msec or longer from beginning until decay to ½ amplitude) or of short duration (15 msec or less to reach ½ amplitude). Long-duration IPSP's occur exclusively in the flexion circuit and function to prevent flexions from re-occurring after the initial flexion. Thus, long-duration IPSP's prevent interference with re-extension. Short-duration IPSP's, in contrast, occur exclusively in the extension circuit. Again this makes sense, since the extensor elements must be released from inhibition and allowed to fire about 20–30 msec after the flexion command. The differences in IPSP durations result from the differences in duration of the conductance increases, with only minor distortions due to the membrane time constants, which are typically rather short. Identification of two inhibitory neurons (MoGI and FI) producing inhibition of long duration, and two others (the accessory neuron to the stretch receptor and EI) producing inhibition of short duration has made it possible to demonstrate that the IPSP durations are due to intrinsic synaptic mechanisms. These mechanisms are presently unknown.

C. Inhibition of the Postural Motor Systems

The tonic, postural system (Chapter 2) is too slow to act synergistically with the fast muscles. Instead, escape commands inhibit the postural system; and again we find that inhibition occurs at every level of the sensory-motor system that we have tested (Fig. 20E; Kuwada and Wine, 1979; Kuwada et al., 1980). In addition to driving the peripheral inhibitor to the tonic flexor muscles and causing postsynaptic inhibition of the tonic flexor motoneurons (Fig. 20), escape commands also block tonic flexor reflexes and prevent habituation of those reflexes (see below).

D. Heterogeneity of Inhibition of Sensory Interneurons and Flexor Motoneurons

Command-derived inhibition is surprisingly pervasive, sparing no level of the sensory-motor hierarchy investigated to date. However, not all elements within a level are inhibited. So far, two important omissions have been discovered for command-derived inhibition.

(1) Of the four identified sensory interneurons known to be presynaptic to LG, only one receives postsynaptic inhibition (Wine, 1977a). Other evidence suggests that command-derived inhibition of projecting sensory interneurons is, in general, rare. In separate studies of sensory interneurons in the sixth ganglion, Fricke et al. (1982) and Sigvardt et al. (1982) found only a single interneuron that was inhibited postsynaptically. (In the work by Sigvardt, 32 identified interneurons or about 60% of all projecting interneurons in the sixth ganglion have been studied to date.) The one interneuron so far found to receive postsynaptic inhibition, as first documented by Krasne and Bryan (1973), is a large interneuron that is excited by ipsilateral mechanoreceptor axons entering the sixth ganglion and whose axon position in the connectives is usually ventral and slightly lateral to the medial

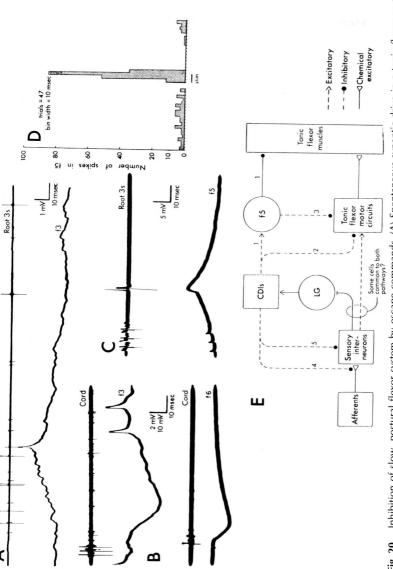

Fig. 20. Inhibition of slow, postural flexor system by escape commands. (A) Spontaneous synaptic drive in a tonic flexor motoneuron (f3) in second ganglion of isolated abdominal nerve cord. (B) Postsynaptic inhibition of tonic flexor motoneurons by LG impulses. Top two records show f3 and bottom two f6. (C) Excitation of peripheral inhibitor (f5 or fI) by a train of five LG impulses. (D) A histogram displaying driving of f1 by a triplet of LG impulses. (E) Summary diagram of relations between postural and escape circuits. Afferents and sensory interneurons presynaptic to LG's are inferred to activate tonic flexor circuits as well. Inhibition of tonic flexor reflexes occurs via pathways 4 and 5; pathways 2 and 3 inhibit tonic flexor activity (A–C: Kuwada et al., 1980; D and E: Kuwada and Wine, 1979; pathway 3 established by Tatton and Sokolove, 1975).

border of the LG. This neuron has been equated by various authors with Wiersma's A6 (Zucker, 1972a) or A3 (Krasne and Bryan, 1973; Bryan and Krasne, 1977a,b) and called interneuron A by Zucker (1972a), to which name we shall adhere here. Interneuron A is unusual in several other respects: its axon is the largest of any sensory interneuron in the abdomen, it produces the largest EPSP in the LG (Zucker, 1972a), and, when axotomized, its soma undergoes a time- and temperature-dependent progression of changes in membrane electrical properties not seen in many other interneurons but commonly seen in motoneurons (Kuwada and Wine, 1981).

(2) At the level of flexor motoneurons, we again find that command-derived inhibition singles out a neuron that is unusual in other respects. We emphasized earlier the distinctiveness of the motor giant (p. 258, Fig. 9, and Table III). Of the nine or ten flexor motoneurons in a hemiganglion, only the motor giant is inhibited.

At present we cannot explain these selective inhibitory actions. Indeed, they may turn out to be inexplicable in purely functional terms, requiring, in addition, a thorough knowledge of the animal's evolution and development.

E. Inhibitory Control of Habituation

The extensiveness of command-derived inhibition is at first sight surprising, but it can be shown that inhibition at each level has different consequences (Krasne et al., 1977). For example, presynaptic inhibition of afferent terminals lessens or entirely eliminates self-induced habituation (Krasne and Bryan, 1973; Bryan and Krasne, 1977a,b; see also Wine et al., 1975; Kuwada and Wine, 1979). When a crayfish escapes, its sensitive mechanosensory afferents are vigorously stimulated, but reafference and consequent self-induced habituation are prevented by command-derived presynaptic inhibition. Protection against habituation was demonstrated in the following way (Fig. 21). The rate and degree of habituation was first established by administering a stimulus repetitively (usually a root shock but taps have also been used) and measuring the response (usually an interneuron EPSP). Then, the LG or MG command cell was activated about 10–20 msec prior to each root shock except the last. As expected, the responses were all inhibited; but, on the last trial, the uninhibited response was significantly greater in amplitude than would be expected after a comparable number of uninhibited sensory volleys. Command cell impulses coming after the root stimulus offer no protection, which rules out sensitization as an explanation. Presynaptic inhibition may be widely used in nervous systems to prevent maladaptive plastic changes at synapses (Krasne, 1978).

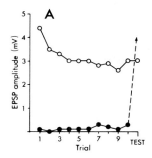
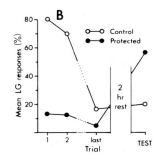

Fig. 21. Protection against habituation by command inhibition. (A) Protection against depression of EPSP in interneuron A6. Graph shows amplitude of compound EPSP's to a series of 11 root shocks at 4-sec intervals. Open circles show response to stimulus alone, closed circles show responses that followed a giant axon impulse by 20 msec. Each point is the mean of three runs. Eleventh stimulus was given alone for both conditions. (B) Comparable experiment using intact crayfish, with taps to abdomen as stimuli and LG-mediated tail flips as responses. Three blocks of ten trials each were given either alone (open circles) or following giant axon impulses, elicited via implanted electrodes (closed circles). As before, test consisted of stimulus alone for both conditions (A: from Bryan and Krasne, 1977a; B: from Wine et al., 1975).

VII. MODULATION OF ESCAPE TENDENCIES

So far, we have discussed escape as though it occurred automatically to the appropriate stimulus. It does not. The escape response is too costly to permit its indiscriminate use. It uses a good deal of energy, disrupts many other behaviors, and makes the crayfish highly visible to predators with motion-detecting systems. For these reasons, we expected internal controls for adjusting the tendency to escape. In fact, escape thresholds are highly variable, and in this section and the following one, some factors involved in the control of escape behavior are outlined, together with what is known about their neuronal substrates. Most of the results deal with the LG system and were discussed in a recent review (Krasne et al., 1977).

Tonic, descending inhibition is the most important factor we have found for controlling the LG escape response (Krasne and Wine, 1975). This can be appreciated by observing the dramatic swings in response threshold displayed by the LG system in an intact animal, followed by the immediate drop in threshold to a low and constant value following transection of the ventral nerve cord between thorax and abdomen (Fig. 22). Note that inhibition was graded over a wide range; but no sign of descending facilitation was observed, since the threshold in the intact animal never sank lower than that seen in neurally isolated abdomens. One would like to know the following about the inhibition: How is it turned on and off; where does it originate; over what pathways does it travel; and to which portions of the LG circuit

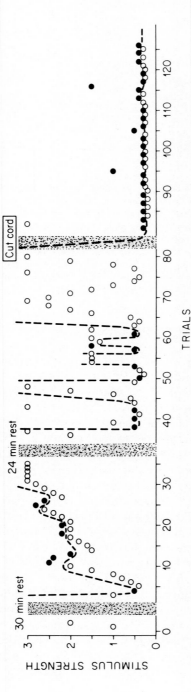

Fig. 22. LG response threshold during maintained restraint before and after cutting nerve cord. Single shocks to a second root of third abdominal ganglion were given at 30-sec intervals. Each point represents a single trial. Ordinate value gives the shock intensity; filled points indicate that LG's fired, while open points indicate that they did not. Dashed curve gives a subjective interpretation of moment-to-moment stimulus threshold for LG firing based upon data points. Note that upon cutting cord between thorax and abdomen, threshold dropped to slightly beneath lowest level seen while nerve cord was intact. (From Krasne and Wine, 1975.)

8. The Cellular Organization of Crayfish Escape Behavior

does it project? The answers to these questions for one kind of inhibition are summarized below.

A. Restraint-Induced Inhibition

The question of what turns inhibition on and off does not have a simple answer, since swings in excitability are often "spontaneous" in the sense that no specific stimulus can reliably be associated with either increases or decreases in threshold. However, the animal in Fig. 22 was restrained, and a consistent finding is that restraint of any kind causes strong inhibition of all three modes of escape. The ability to activate inhibition reliably permitted it to be studied. Lesions showed that most restraint-induced inhibition originates in the supra- and subesophageal ganglia, although some inhibition is also contributed by thoracic ganglia. Inhibition is exclusively unilateral, since transection of a hemiconnective causes a drop in threshold only for ipsilateral stimuli. This is an unusual finding, since most pathways in the escape systems are tightly coupled bilaterally, and it may mean that the animal can selectively inhibit one side.

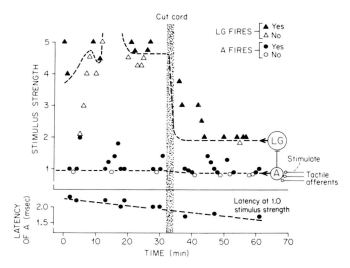

Fig. 23. Thresholds for LG and interneuron A (A6) firing in restrained crayfish before and after cutting nerve cord. Method was essentially the same as Fig. 22, except that LG and interneuron A were tested alternately, and stimuli were shocks to second and third roots of sixth abdominal ganglion. Measurements plotted in upper graph give a slight suggestion that interneuron A's threshold might have decreased very slightly at time of cord section, but measurements of interneuron A's firing latency at constant stimulus strength, plotted below, show that this decrease was manifestation of a gradual decline of interneuron A's threshold throughout experiment. (From Krasne and Wine, 1975).

Restraint-induced inhibition acts at the level of the command cell. Since a giant axon impulse always involves at least some motor response, uncoupling of the motor apparatus cannot be the primary mechanism of inhibition, although some motor inhibition may occur. Fig. 6 shows that transmission to the giants could be inhibited at any of four levels, and we saw earlier that command-derived inhibition acts at three of the four possible sites (Fig. 17). However, restraint-induced inhibition does not appear to influence transmission across the first central synapse (Fig. 23) and therefore presumably acts either directly upon the command neuron or upon the terminals innervating the command neuron. The functional significance of having inhibition at this site is that it is highly selective for escape, permitting sensory information to be transmitted to other neurons while the LG-mediated response is inhibited.

B. Other Modulating Factors

When large crayfish are disturbed, they often attack instead of fleeing, but animals that have lost their chelae appear timid. Only non-giant escape responsiveness is strongly enhanced following autotomy of the chelipeds (Krasne and Wine, 1975). One possible mechanism for this effect is reduced sensory influx from cheliped-associated receptors that normally inhibit responsiveness. This system may also contribute to increased thresholds for the escape response as the animals grow, since the claws grow with positive allometry (Lang et al., 1977; also Chapter 5 of Volume 3).

Mild inhibition also occurs when the animal is moving its legs or swimmerets (Fricke et al., 1982; F. B. Krasne, unpublished). Inhibition associated with limb movement arises, at least in part, from proprioceptive hairs associated with the legs, descends via a ventromedial pathway, and presynaptically inhibits transmission at the first central synapse (Fricke et al., 1982). Proprioceptive inhibition thus resembles command-derived inhibition and may use some of the same circuitry. Definite proof for convergence of proprioceptive and command drive onto a common inhibitory neuron has so far been obtained only for the flexor inhibitor, which receives both delayed central drive (Fig. 18) and proprioceptive drive from the stretch receptors (Fig. 16A). The stretch receptors also cause inhibition of transmission from mechanosensory afferents to interneuron A (F. B. Krasne, unpublished). It will be interesting to see if other instances of converging inhibition from these two sources can be found, since they are respectively commanding and detecting the fast flexion.

C. Functional Considerations

Unlike most circuits that we have discussed, those responsible for tonic modulation are known only by their effects—no modulating interneurons

8. The Cellular Organization of Crayfish Escape Behavior

have been identified, and nothing suggests that the search for them will be easy. Yet the demonstrated and conjectural roles for these circuits are of the highest interest, since they account for fluctuations in responsiveness usually described by terms such as state, mood, and drive. Maturational changes and perhaps associative learning may also be expressed via these circuits. Whereas these phenomena were formerly defined exclusively in behavioral terms, analysis has proceeded so that their effects can be studied physiologically in identified neurons. The remaining task is to trace the neural origins of the effects.

The three effects documented above are undoubtedly only a small portion of all the modulatory influences. Anecdotal evidence, experiments with other species, logical considerations, and preliminary experiments all suggest a plethora of effects. For example, when hungry crayfish detect food, they become highly aroused and explore vigorously. Escape thresholds are elevated, and the animals react aggressively when disturbed—until they seize the food and convey it to the maxillipeds. At that point, their behavior abruptly switches to a new mode. Rapid backward walking commences; the animal avoids others and tail flips vigorously at slight provocation (J. J. Wine, G. Hagiwara, and L. E. Lazar, unpublished results). Other examples are that females in berry seem to have raised thresholds for escape, while the timidity of freshly molted animals is striking (Huxley, 1880).

Modulation is certainly not the result of a unitary system. This is obvious when the effects are as discriminately different as are some effects documented above. But even when the effects appear similar, it is probably risky to infer common mechanisms. Wiersma (1970) and Rowell (1970) reviewed changes in responsiveness in Crustacea and insects, respectively, and related them mainly to general arousal. Arousal is a potent modulatory influence, but its correlation with escape thresholds is not simple (Krasne and Wine, 1975).

VIII. CONCLUDING REMARKS

It is now a cliche to say that simple systems are not so simple after all. Yet this need not be a lament, for in our view the most satisfying aspect of research on escape behavior has been the discovery at every juncture of cellular mechanisms and circuit designs of increasing subtlety. Perhaps much of the apparent baroqueness of the escape circuitry can be attributed to interplay between giant and non-giant systems. We presume that the giant axons evolved later, and, like the addition of new and powerful members to a governing committee, they required widespread adjustments and compromises to achieve optimal performance. Crustacea have had hundreds of millions of years to hone their nervous systems, and adaptations during the last half of this time must have been increasingly tuned to the rapidity and skill of myelin-equipped vertebrate predators. In this contest of ballistic movements, where life or death hinges on milliseconds and millimeters, we should not be surprised to find sophistication rather than simplicity.

ABBREVIATIONS

GX	Abdominal ganglion X
conn	Connective
TA	Tactile afferent sensory neuron
MRO	Muscle receptor organ (stretch receptor)
Acc	Accessory neuron to the muscle receptor organ
LG	Lateral giant command cell
MG	Medial giant command cell
I (INT)	Interneuron
SI	Sensory interneuron
T	Transient sensory interneuron
S	Sustained sensory interneuron
CDI	Corollary discharge interneuron
MoGI	Inhibitory interneuron to the motor giant
SG	Segmental giant neuron
E	Phasic extensor motoneuron
EI	Peripheral inhibitor to phasic extensor muscles
e	Tonic extensor motoneuron
eI	Peripheral inhibitor to tonic extensor muscles
FF	Phasic (fast) flexor motoneuron
MoG	Motor giant (largest phasic flexor motoneuron)
FAC	Phasic flexor motoneuron of the anterior contralateral group
FMC	Phasic flexor motoneuron of the medial contralateral group
FPI	Phasic flexor motoneuron of the posterior ipsilateral group
FI	Peripheral inhibitor to phasic flexor muscles
f	Tonic flexor motoneuron
fI (f5)	Peripheral inhibitor to tonic flexor muscles

ACKNOWLEDGMENTS

Many of our colleagues made their unpublished data available to us. For this we thank Gene Block, Russ Fricke, Grace Hagiwara, Donald Kennedy, Andy Kramer, Lee Miller, Gene Olson, Heinrich Reichert, Alan Roberts, Karen Sigvardt, and Konrad Wiese.

The draft was typed by Jan Ruby; Grace Hagiwara prepared the figures. Experiments from the authors' laboratories were supported by National Science Foundation grant number BNS 78-14179 and National Institutes of Health Biomedical Research Support grant (Jeffrey J. Wine), and by United States Public Health Service grant number NF 8108 (Frank B. Krasne).

8. The Cellular Organization of Crayfish Escape Behavior

REFERENCES

Atwood, H.L., and Wiersma, C.A.G. (1967). Command interneurons in the crayfish central nervous system. *J. Exp. Biol.* **46,** 249-261.

Barker, D.L., Herbert, E., Hildebrand, J.G., and Kravitz, E.A. (1972). Acetylcholine and lobster sensory neurones. *J. Physiol. (London)* **226,** 205-229.

Bowerman, R.F., and Larimer, J.L. (1974). Command fibres in the circumesophageal connectives of crayfish. I. Tonic fibres. *J. Exp. Biol.* **60,** 95-117.

Bruner, J., and Kennedy, D. (1970). Habituation: Occurrence at a neuromuscular junction. *Science* **169,** 92-94.

Bryan, J.S., and Krasne, F.B. (1977a). Protection from habituation of the crayfish lateral giant fibre escape response. *J. Physiol. (London)* **271,** 351-368.

Bryan, J.S., and Krasne, F.B. (1977b). Presynaptic inhibition: The mechanism of protection from habituation of the crayfish lateral giant escape response. *J. Physiol. (London)* **271,** 369-390.

Calabrese, R.L. (1976a). Crayfish mechanoreceptive interneurons. I. The nature of ipsilateral excitatory inputs. *J. Comp. Physiol.* **105,** 83-102.

Calabrese, R.L. (1976b). Crayfish mechanoreceptive interneurons. II. Bilateral interactions and inhibition. *J. Comp. Physiol.* **105,** 103-114.

Daniel, R.J. (1931). The abdominal muscular systems of *Homarus vulgaris* (L.) and *Palinurus vulgaris* (Latr.). *Proc. Trans. Liverpool Biol. Soc.* **45,** 3-49.

Fields, H.L. (1966). Proprioceptive control of posture in crayfish abdomen. *J. Exp. Biol.* **44,** 455-468.

Fields, H.L., and Kennedy, D. (1965). Functional role of muscle receptor organs in crayfish. *Nature (London)* **106,** 1232-1237.

Fricke, R.A., Block, G.D., and Kennedy, D. (1982). Inhibition of mechanosensory neurons in the crayfish. II. Inhibition of interneurons associated with proprioceptive feedback from locomotion. (Submitted for publication).

Furshpan, E.J., and Potter, D.D. (1959). Transmission at the giant motor synapses of the crayfish. *J. Physiol. (London)* **145,** 289-325.

Goldberg, R.A., Tillotson, C.L., and Wine, J.J. (1982). Habituation of two behavioral systems in the crayfish: Evidence for shared sensory pathways. In preparation.

Huxley, T.H. (1880). "An Introduction to the Study of Zoology, Illustrated by the Crayfish." Appleton, New York (reissued by MIT Press, 1974).

Johnson, G.E. (1924). Giant nerve fibers in crustaceans with special reference to Cambarus and Palaemonetes. *J. Comp. Neurol.* **36,** 323-373.

Johnson, G.E. (1926). Studies on the functions of the giant nerve fibers of crustaceans, with special reference to Cambarus and Palaemonetes. *J. Comp. Neurol.* **42,** 19-33.

Kandel, E.R. (1976). "The Cellular Basis of Behavior." Freeman, San Francisco, California.

Kandel, E.R., Krasne, F.B., Strumwasser, F., and Truman, J.W. (1979). Cellular mechanisms in the selection and modulation of behavior. *N.R.P. Bull.* **17,** 523-710.

Kao, C.Y. (1960). Postsynaptic electrogenesis in septate giant axons. II. Comparison of medial and lateral giant axons of crayfish. *J. Neurophysiol.* **23,** 618-636.

Kennedy, D., and Takeda, K. (1965a). Reflex control of abdominal flexor muscles in the crayfish. I. The twitch system. *J. Exp. Biol.* **43,** 211-227.

Kennedy, D., and Takeda, K. (1965b). Reflex control of abdominal flexor muscles in the crayfish. II. The tonic system. *J. Exp. Biol.* **43,** 229-246.

Kennedy, D., Evoy, W.H., and Fields, H.L. (1966). The unit basis of some crustacean reflexes. *Symp. Soc. Exp. Biol.* **20,** 75-109.

Kennedy, D., Calabrese, R.L., and Wine, J.J. (1974). Presynaptic inhibition: Primary afferent depolarization in crayfish neurons. *Science* **186,** 451–454.

Kramer, A.P. (1978). Some non-giant circuits mediating fast tail flexions in crayfish. Ph.D. Dissertation, University of California, Los Angeles.

Kramer, A.P., Krasne, F.B., and Wine, J.J. (1981). Interneurons between giant axons and motoneurons in the crayfish escape circuitry. *J. Neurophysiol.* **45,** 550–573.

Krasne, F.B. (1969). Excitation and habituation of the crayfish escape reflex: the depolarizing response in lateral giant fibres of the isolated abdomen. *J. Exp. Biol.* **50,** 29–46.

Krasne, F.B. (1976). Invertebrate systems as a means of gaining insight into the nature of learning and memory. *In* "Neural Mechanisms of Learning and Memory" (M.R. Rosenzweig and E.L. Bennett, eds.), pp. 401–429. MIT Press, Cambridge, Massachusetts.

Krasne, F.B. (1978). Extrinsic control of intrinsic neuronal plasticity: An hypothesis from work on simple systems. *Brain Res.* **140,** 197–216.

Krasne, F.B., and Bryan, J.S. (1973). Habituation: Regulation through presynaptic inhibition. *Science* **182,** 590–592.

Krasne, F.B., and Wine, J.J. (1975). Extrinsic modulation of crayfish escape behaviour. *J. Exp. Biol.* **63,** 433–450.

Krasne, F.B., and Woodsmall, K.S. (1969). Waning of the crayfish escape response as a result of repeated stimulation. *Anim. Behav.* **17,** 416–424.

Krasne, F.B., Wine, J.J., and Kramer, A. (1977). The control of crayfish escape behavior. *In* "Identified Neurons and Behavior of Arthropods" (G. Hoyle, ed.), pp. 275–292. Plenum, New York.

Kupfermann, I., and Weiss, K.R. (1978). The command neuron concept. *Behav. Brain Sci.* **1,** 3–39.

Kusano, K., and Grundfest, H. (1965). Circus reexcitation as a cause of repetitive activity in crayfish lateral giant axons. *J. Cell. Comp. Physiol.* **65,** 325–336.

Kuwada, J.Y., and Wine, J.J. (1979). Crayfish escape behaviour: Commands for fast movement inhibit postural tone and reflexes, and prevent habituation of slow reflexes. *J. Exp. Biol.* **79,** 205–224.

Kuwada, J.Y., and Wine, J.J. (1981). Transient, axotomy-induced changes in the membrane properties of crayfish central neurons. *J. Physiol. (London)* **317,** 435–461.

Kuwada, J.Y., Hagiwara, G., and Wine, J.J. (1980). Postsynaptic inhibition of crayfish tonic flexor motor neurones by escape commands. *J. Exp. Biol.* **85,** 344–347.

Lang, F., Govind, C.K., Costello, W.J., and Greene, S.I. (1977). Developmental neuroethology: Changes in escape and defensive behavior during growth of the lobster. *Science* **197,** 682–685.

Larimer, J.L., and Kennedy, D. (1969). Innervation patterns of fast and slow muscle in the uropods of crayfish. *J. Exp. Biol.* **51,** 119–133.

Larimer, J.L., Eggleston, A.C., Masukawa, L.M., and Kennedy, D. (1971). The different connections and motor outputs of lateral and medial giant fibres in the crayfish. *J. Exp. Biol.* **54,** 391–402.

Mellon, DeF., and Kaars, C. (1974). Role of regional cellular geometry in conduction of excitation along a sensory neuron. *J. Neurophysiol.* **37,** 1228–1238.

Mellon, DeF., and Kennedy, D. (1964). Impulse origin and propagation in a bipolar sensory neuron. *J. Gen. Physiol.* **47,** 487–499.

Miller, L., and Wine, J. J. (1982). Intersegmental differences in pathways between crayfish giant axons and fast flexor motor neurons. In preparation.

Mittenthal, J.E., and Wine, J.J. (1973). Connectivity patterns of crayfish giant interneurons: Visualization of synaptic regions with cobalt dye. *Science* **179,** 182–184.

Mittenthal, J.E., and Wine, J.J. (1978). Segmental homology and variation in flexor motoneurons of the crayfish abdomen. *J. Comp. Neurol.* **177**, 311-334.

Olson, G.C., and Krasne, F.B. (1978). Crayfish giant fibers are command and decision neurons. *Neurosci. Abstr.* **4**, 203 (Abstr. No. 629).

Otsuka, M., Kravitz, E.A., and Potter, D.D. (1967). Physiological and chemical architecture of a lobster ganglion with particular reference to gamma-aminobutyrate and glutamate. *J. Neurophysiol.* **30**, 725-752.

Paul, D.H. (1972). Decremental conduction over "giant" afferent processes in an arthropod. *Science* **176**, 680-682.

Pilgrim, R.L.C., and Wiersma, C.A.G. (1963). Observations on the skeleton and somatic musculature of the abdomen and thorax of *Procambarus clarkii* (Girard), with notes on the thorax of *Panulirus interruptus* (Randall) and *Astacus*. *J. Morphol.* **113**, 453-487.

Rayner, M.D., and Wiersma, C.A.G. (1964). Functional aspects of the anatomy of the main thoracic and abdominal flexor musculature of the crayfish *Procambarus clarkii* (Girard). *Am. Zool.* **4**, 285.

Rayner, M.D., and Wiersma, C.A.G. (1967). Mechanisms of crayfish tail flick. *Nature (London)* **213**, 1231-1232.

Reichert, H., Wine, J.J., and Hagiwara, G. (1981). Crayfish escape behavior: Neurobehavioral analysis of phasic extension reveals dual systems for motor control. *J. Comp. Physiol.* **142**, 281-294.

Roberts, A.M. (1968). Recurrent inhibition in the giant-fibre system of the crayfish and its effect on the excitability of the escape response. *J. Exp. Biol.* **48**, 545-567.

Roberts, A.M., Krasne, F.B., Hagiwara, G., Wine, J.J., and Kramer, A.P. (1982). The segmental giant: Evidence for a driver neuron interposed between command and motor neurons in the crayfish escape system. *J. Neurophysiol.* (In press).

Rowell, C.H.F. (1970). Incremental and decremental processes in the insect central nervous system. In "Short-term Changes in Neural Activity and Behaviour" (G. Horn and R.A. Hinde, eds.), pp. 237-280. Cambridge Univ. Press, London and New York.

Schrameck, J.E. (1970). Crayfish swimming: Alternating motor output and giant fiber activity. *Science* **169**, 698-700.

Selverston, A.I., and Remler, M.P. (1972). Neural geometry and activation of crayfish fast flexor motoneurons. *J. Neurophysiol.* **35**, 797-814.

Sigvardt, K.A., Hagiwara, G., and Wine, J.J. (1982). Mechanosensory integration in the crayfish abdominal nervous system. I. Structural and physiological differences between interneurons with single and multiple spike initiating sites. In preparation.

Stretton, A.O.W., and Kravitz, E.A. (1973). Intracellular dye injection: The selection of Procion yellow and its application in preliminary studies of neuronal geometry in the lobster nervous system. In "Intracellular Staining in Neurobiology" (S.B. Kater and C. Nicholson, eds.), pp. 21-40. Springer-Verlag, Berlin and New York.

Takeda, K., and Kennedy, D. (1964). Soma potentials and modes of activation of crayfish motoneurons. *J. Cell. Comp. Physiol.* **64**, 165-182.

Tatton, W.G., and Sokolove, P.G. (1975). Analysis of postural motoneuron activity in crayfish abdomen. II. Coordination by excitatory and inhibitory connections between motoneurons. *J. Neurophysiol.* **38**, 332-346.

Treistman, S.N., and Remler, M.P. (1975). Extensor motor neurons of the crayfish abdomen. *J. Comp. Physiol.* **100**, 85-100.

Uyama, C., and Matsuyama, T. (1980). Coordinated excitation of flexor inhibitors in the crayfish. *J. Exp. Biol.* **86**, 187-195.

Watanabe, A., and Grundfest, H. (1961). Impulse propagation at the septal and commissural junctions of crayfish lateral giant axons. *J. Gen. Physiol.* **45**, 267-308.

Wiersma, C.A.G. (1947). Giant nerve fiber system of the crayfish. A contribution to comparative physiology of synapse. *J. Neurophysiol.* **10,** 23-38.

Wiersma, C.A.G. (1949). Synaptic facilitation in the crayfish. *J. Neurophysiol.* **14,** 267-275.

Wiersma, C.A.G. (1967). Visual central processing in crustaceans. In "Invertebrate Nervous Systems" (C.A.G. Wiersma, ed.), pp. 269-288. Univ. of Chicago Press, Chicago, Illinois.

Wiersma, C.A.G. (1970). Reactivity changes in crustacean neural systems. In "Short-term Changes in Neural Activity and Behaviour" (G. Horn and R.A. Hinde, eds.), pp. 211-236. Cambridge Univ. Press, London and New York.

Wiersma, C.A.G., and Hughes, G.M. (1961). On the functional anatomy of neuronal units in the abdominal cord of the crayfish, *Procambarus clarkii* Girard. *J. Comp. Neurol.* **116,** 209-228.

Wiese, K. (1976). Mechanoreceptors for near-field water displacements in crayfish. *J. Neurophysiol.* **39,** 816-833.

Wiese, K., Calabrese, R.L., and Kennedy, D. (1976). Integration of directional mechanosensory input by crayfish interneurons. *J. Neurophysiol.* **39,** 834-843.

Wilkens, L.A., and Larimer, J.L. (1972). The CNS photoreceptor of crayfish: Morphology and synaptic activity. *J. Comp. Physiol.* **80,** 389-407.

Wine, J.J. (1971a). Hyperreflexia in the crayfish abdomen following denervation: Evidence for supersensitivity in an invertebrate central nervous system. Ph.D. Dissertation, University of California, Los Angeles.

Wine, J.J. (1971b). Escape reflex circuit in crayfish: Interganglionic interneurons activated by the giant command neurons. *Biol. Bull. (Woods Hole, Mass.)* **141,** 408.

Wine, J.J. (1975). Crayfish neurons with electrogenic cell bodies: Correlations with function and dendritic properties. *Brain Res.* **85,** 92-98.

Wine, J.J. (1977a). Neuronal organization of crayfish escape behavior: Inhibition of the giant motoneuron via a disynaptic pathway from other motoneurons. *J. Neurophysiol.* **40,** 1078-1097.

Wine, J.J. (1977b). Crayfish escape behavior. II. Command-derived inhibition of abdominal extension. *J. Comp. Physiol.* **121,** 173-186.

Wine, J.J. (1977c). Crayfish escape behavior. III. Monosynaptic and polysnaptic sensory pathways involved in phasic extension. *J. Comp. Physiol.* **121,** 187-203.

Wine, J.J., and Hagiwara, G. (1977). Crayfish escape behavior. I. The structure of efferent and afferent neurons involved in abdominal extension. *J. Comp. Physiol.* **121,** 145-172.

Wine, J.J., and Hagiwara, G. (1978). Durations of unitary synaptic potentials help time a behavioral sequence. *Science* **199,** 557-559.

Wine, J.J., and Krasne, F.B. (1972). The organization of escape behaviour in the crayfish. *J. Exp. Biol.* **56,** 1-18.

Wine, J.J., and Krasne, F.B. (1978). The cellular analysis of invertebrate learning. In "Brain and Learning" (T.J. Teyler, ed.), pp. 13-31. Greylock Publishers, Stamford, Connecticut.

Wine, J.J., and Mistick, D.C. (1977). Temporal organization of crayfish escape behavior: Delayed recruitment of peripheral inhibition. *J. Neurophysiol.* **40,** 904-925.

Wine, J.J., Kranse, F.B., and Chen, L. (1975). Habituation and inhibition of the crayfish lateral giant fibre escape response. *J. Exp. Biol.* **62,** 771-782.

Zucker, R.S. (1972a). Crayfish escape behavior and central synapses. I. Neural circuit exciting lateral giant fiber. *J. Neurophysiol.* **35,** 599-620.

Zucker, R.S. (1972b). Crayfish excape behavior and central synapses. II. Physiological mechanisms underlying behavioral habituation. *J. Neurophysiol.* **35,** 621-637.

Zucker, R.S., and Bruner, J. (1977). Long-lasting depression and the depletion hypothesis at crayfish neuromuscular junctions. *J. Comp. Physiol.* **121,** 223-240.

9

Views on the Nervous Control of Complex Behavior

PETER J. FRASER

I.	Introduction	293
II.	Studying Interneurons	294
	A. Ablation Experiments	294
	B. Sensory Interneurons	295
	C. Command Neurons	298
	D. Sensory Input to Command Neurons	302
	E. Integration in Interneurons	307
	F. Excited States and Complex Neuropils	308
III.	Multisegmental Interneurons and Behavior	309
	A. Behaviors Commanded and Numbers of Command Neurons Involved	309
	B. Correlations between Activity in Interneurons and Behavior	309
IV.	Perspectives	313
	References	314

I. INTRODUCTION

The intent of this chapter is to summarize broadly based studies on multisegmental interneurons aimed at a general understanding of behavior. Interneurons are classified by either their sensory input or their motor output. This leads to considerations of sensory coding, stimulus filtering, and line labelling, which fit with ethological ideas of feature detection, key stimuli, releasing factors, and orienting stimuli, or to considerations of motor control and central motor scores, which fit with the ethological idea of fixed action pattern.

In most cases it is not known how many interneurons are necessary or normal for the initiation, control, and fabrication of any behavior. We must discover what constitutes a functional group of cells sufficient for a behavior. Where the component interneurons and motor neurons can be identified, the problem becomes one of understanding the significance of connections and integrative mechanisms in behavior. One major weakness preventing an overall view is the paucity of investigated species. Our knowledge comes from crayfish, lobster, crab, and hermit crab. Little or nothing is known of the rest.

II. STUDYING INTERNEURONS

A. Ablation Experiments

Early workers in the field of nervous control of behavior employed a mixture of anatomical and physiological techniques. Bethe's work on the crab *Carcinus maenas* best exemplifies this approach (Bethe, 1895, 1897a,b). He first described the nervous system anatomically. Then he described in the normal crab a variety of reflexes, such as withdrawal of eyes, antennae, and mouthparts; closing and opening of the claws; reflexes of escape, defence, rearing, feeding, and righting; copulation; and finally locomotion. He compared normal behavior with behavior following inactivation of sense organs, such as compound eyes, statocysts, and antennulary chemoreceptors, or following surgical operations on parts of the brain or connectives. He was the first to show that the cell bodies of motor neurons were not required for the operation of the antennal withdrawal reflex (Bethe, 1897c).

Ablation techniques have been employed more recently to demonstrate the importance of the proximal eyestalk ganglia of crabs, lobsters, and hermit crabs in the performance of a variety of behavior patterns, including distance chemoreception, feeding behavior, general posture, and shell-entry behavior by hermit crabs (Hazlett, 1971). Results vary between species but implicate the terminal medulla of the eyestalk as a source of gain control over behavior. The demonstration of non-visual functions for the eyestalk ganglia [see also Maynard and Dingle (1963), Maynard and Sallee (1970), and Maynard and Yager (1968) for the involvement of the terminal medulla in chemically mediated and controlled feeding patterns in *Panulirus*] was predicted by work on single interneurons showing the flow of non-visual information from the brain toward the eyes (Waterman and Wiersma, 1963; Bush et al., 1964; Wiersma and Yamaguchi, 1967a; Chapter 1 of this volume). Ablation of eyestalks has also been shown to affect learning in the crayfish, *Procambarus clarkii* (Eisenstein and Mill, 1965).

Ablation experiments are not easily interpreted and are easily discredited. The structure of the central nervous system in decapods is such that it is impossible to cut certain pathways without interfering with others (see Chapter 1 of Volume 3). Furthermore, a cut anywhere will disrupt the extensive blood system and affect areas beyond the boundaries of the cut (Sandeman, 1966; Taylor, 1974). Nevertheless, where clear reproducible results follow surgical interference, any explanation of behavior must eventually account for such results. Of particular interest are cases in which the effects of an operation can be reversed. If one esophageal connective is cut, the crab *Cancer* walks in a circle (manégé, on circus movement: Fig. 1). Electrical stimulation with implanted electrodes of the posterior end of the cut connectives can counteract and eventually reverse the manégé. For more information, the reader is referred to Bethe's papers quoted above and to English summaries of his work and the work of others (Schöne, 1961; Wiersma, 1961a).

B. Sensory Interneurons

Pioneering work by Prosser on the crayfish showed that impulses could be recorded from whole esophageal connectives following stimulation of the eyes, caudal photoreceptors, antennae, antennules, uropods from both sides, and ipsilateral legs (Prosser, 1934a,b, 1935). Later, Wiersma and colleagues isolated single axons and stimulated them electrically, causing various complex body movements (Wiersma, 1938, 1947, 1952a,b). Emphasis shifted to mapping cells by means of their adequate sensory inputs in terms of modality, receptive fields, and response characteristics. This approach (Wiersma, 1958; Wiersma et al., 1955; Wiersma and Mill, 1965) allowed detailed mapping of about 130 cells [out of 3200 in each esophageal connective (Sutherland and Nunnemacher, 1968)], and over 100 cells in connectives between thoracic and abdominal ganglia (Hughes and Wiersma, 1960a; Wiersma and Hughes, 1961; Wiersma and Bush, 1963). Other cells were mapped in the optic tract between "brain" and eyestalk ganglia in crabs (Waterman and Wiersma, 1963; Bush et al., 1964; Waterman et al., 1964; Wiersma et al., 1964; Wiersma, 1966, 1970) and crayfish (Wiersma and Yamaguchi, 1966, 1967a,b). Some interneurons exhibited spontaneous discharges or could not be influenced by sensory stimulation. Following the original work by Wiersma, many other cells were classified and mapped in

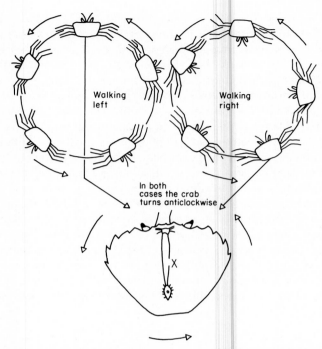

Fig. 1. Effect of cutting the right esophageal connective on locomotion in the crab, *Carcinus maenas*. Crab goes in a circle (manégé), as shown. Walking left is preferred. There are alterations in gait cycle following operation. Regardless of direction of locomotion, crab always turns counterclockwise. Information about counterclockwise turning is known to be carried in cell B in the right connective (Section II,D). (Redrawn from Bethe, 1897b.)

terms of their unique motor output. These cells were first termed "command fibres" by Wiersma (1962; Wiersma and Ikeda, 1964) and will be considered later (Section II,C). The studies outlined above established the uniqueness and constancy of crustacean interneurons and initiated the "identified cell" concept.

More recent studies have been made of selected interneurons. Many of these secondary studies have led to some generalities about behavior. Examples in crayfish are abdominal giant fibers (Chapter 8), abdominal command neurons (Chapter 2), and photoreceptors; and also visual interneurons (Chapter 1). Interneurons that integrate mechanoreceptor inputs have been studied in afferent pathways of giant fibers in crayfish (Zucker et al., 1971; Calabrese and Kennedy, 1974; Calabrese, 1976a,b). Multisegmental tactile interneurons of crayfish provide the clearest examples of multiple initiation zones for axonal spikes, both in different ganglia and within a single intraganglionic dendritic tree. These multiple sites of ex-

9. Views on the Nervous Control of Behavior

citation function in the integration of dynamic features of incoming stimuli (Kennedy and Mellon, 1964) and may make a given interneuron highly sensitive to directional stimuli, e.g., passage of water current in one direction (see Horridge, 1968).

Taylor has studied in crayfish the sensitivity of the large statocyst fibers, C4 and C87 (Fig. 2), to water-borne vibration, and he has examined C4 under a variety of conditions (Taylor, 1968, 1970). Wilkens and Larimer (1973) further investigated the input pathways and anatomy, as shown by intracellular injection of Procion Yellow dye of C4 (which they call the lateral hemigiant) and of another large cell, which they term the medial hemigiant. They equate the medial hemigiant with C87 studied by Taylor, although their diagram indicates a different location in the connective and input from antennae rather than statocyst (Fig. 2). There may be errors in identification.

The directionality of a large interneuron responding as a one-way movement detector in the Australian crayfish *Cherax destructor* has been studied (Fraser, 1977b), as have the responses of large interneurons that respond to

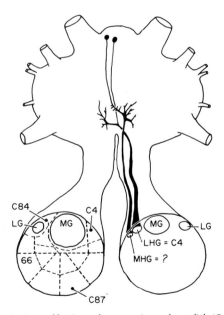

Fig. 2. Diagrammatic view of brain and connectives of crayfish (*Procambarus*). Right connective shows lateral (LG) and medial (MG) giant fibers and two large interneurons (LHG, MHG) as described by Wilkens and Larimer (1973). Left connective shows the standard cross section used by Wiersma and colleagues to map positions of interneurons in the connectives. Each location received a number (e.g., 66) and each cell a letter followed by a number. Statocyst interneurons C4, C84, and C87 are shown.

stimulation of the statocysts (homologous to those described in *Procambarus*) during tilting of the whole animal. In a rested crayfish, head-ascending tilt evokes a large response from C4 (unpublished observation). Glantz has performed detailed analyses of various visual interneurons believed to be involved in the defense reflex (Section III,B; Glantz, 1977). Horche (1971) described responses in the esophageal connectives of *Ocypode* to airborne sounds. Roye (1972) described several units in connectives of *Callinectes sapidus* with phasic statocyst input. These interneurons have not yet been identified anatomically, although the cells described by Roye are clearly homologous to cells discovered at the same time in *Carcinus* (Fraser, 1974b).

Interneurons in the esophageal connectives of *Carcinus* and *Scylla* have been studied by techniques of intracellular recording and injection of dye (Fraser, 1973, 1974a,b, 1975a,b, 1977b; Fraser and Sandeman, 1975). This has permitted identification of interneurons physiologically and anatomically, in terms of not only their position in the connectives, but also their branching patterns in the brain. Recently, anatomical studies have been extended to thoracic ganglia (P. J. Fraser and E. A. Campbell, unpublished; Chapter 1 of Volume 3).

It is obvious that studies on sensory input to interneurons demand greater knowledge of the sensory system. This is being supplied, and many previously investigated cells now require more detailed reinvestigation in the light of new knowledge.

C. Command Neurons

The idea of command neurons dates back to the finding that electrical stimulation of single nerve cells elicits recognizable behavior (Wiersma, 1938; Wiersma and Novitski, 1942). The term command was first applied to single cells by Wiersma [1961b (quoted in Wiersma, 1978), 1962], but it gained most acceptance when applied to single nerve cells capable of eliciting the coordinated movements of swimmerets in crayfish (*Procambarus clarkii*) (Wiersma and Ikeda, 1964).

The finding that repeatable behavior could be evoked on electrical stimulation of the nervous system was not new. In 1897, Bethe put platinum rings around the eyestalks of crabs (*Carcinus maenas*) and formed an electrical circuit via a metal plate on which the crab stood. Electrical stimulation of both eyes elicited a symmetrical rearing reflex (Aufbaumreflex), without flight, but with repeated opening and closing of pincers, leaping, and hitting together of the pincers. A weak stimulus gave only a general convulsion of the legs, with feeble raising of the pincers (Bethe, 1897a). Stimulation of one eye led to asymmetrical rearing and flight to the unstimulated side. There are

9. Views on the Nervous Control of Behavior

many explanations possible for these results, but it is not unlikely that command neurons or pathways in the optic tract were stimulated electrically. Many behavioral effects that in crabs can be elicited by stimulating esophageal connectives can also be elicited by stimulating the optic tract (P. J. Fraser, personal observation).

The concept of the command neuron has been of great importance in relating nerve cells to behavior, and it has been extended to mollusks, insects, and vertebrates. There has recently been an extensive open-forum debate, running to about 25,000 words, regarding definition and classification of command cells and the meaning of the concept in understanding behavior (Kupfermann and Weiss, 1978). Command cells have been reviewed in general terms (Davis, 1976, 1977; Kennedy and Davis, 1977) and specifically in relation to Crustacea (Bowerman and Larimer, 1976; Larimer, 1976; Larimer and Gordon, 1977). None of these reviews includes work on command neurons of crab, which illustrate aspects of the input organization of command neurons not readily seen in cells of crayfish (Fraser, 1973, 1974a, 1975b, 1978a,b, Section II,D,2).

The main concern of early studies on command cells was to determine whether effects seen on electrical stimulation come about through activation of a single cell or of more than one cell. By recording and stimulating from small bundles at different locations in the nervous system of crayfish (*Procambarus clarkii*), Kennedy et al. (1966) claimed to have shown that only a single cell was responsible for the evoked behavior. However, the fact that axons of crustacean interneurons are known to run together for long distances discounts to some extent their claim. The main evidence for command effects by a single cell comes from intracellular stimulation of individual lateral and medial giant fibers of *Procambarus* (Larimer et al., 1971), and from the volume of results showing repeatable discrete motor patterns evoked from small bundles of axons.

It is now fully accepted that single cells can produce complete complex cyclical or non-cyclical behaviors.

The range of behavior patterns evoked is considerable and, as for sensory neurons, there is great difficulty in summarizing the behavior (as might be expected, since the cells were identified in the first place by means of their unique behavioral outputs). In brief summary, the range in crayfish includes:

(1) Positioning and moving the abdomen (Chapter 2; see references in Larimer, 1976), including rapid abdominal flexion and extension [escape responses, mediated by giant and non-giant cells; and swimming (Chapter 8; Schrameck, 1970; Wine and Krasne, 1972; Bowerman and Larimer, 1974b)].

(2) Various movements of the appendages, such as opening and closing

the dactylopodite of the claw (Smith, 1974), displacement of several limbs (Atwood and Wiersma, 1967; Bowerman and Larimer, 1974a,b), and straight or cyclic movements on both sides (Larimer and Kennedy, 1969a,b; Kovac, 1974a,b).

(3) Metachronal beating and movements of swimmerets (also in lobster) (Hughes and Wiersma, 1960b; Wiersma and Ikeda, 1964; Atwood and Wiersma, 1967; Stein, 1971; Davis and Kennedy, 1972a,b,c).

(4) Whole-body movement involving angular displacement in various planes.

(5) Forward and backward walking (Atwood and Wiersma, 1967; Bowerman and Larimer, 1974a,b).

(6) Defense reflex (Wiersma, 1952b; Atwood and Wiersma, 1967; Bowerman and Larimer, 1974a).

(7) Freezing of positions of body and limbs (Bowerman and Larimer, 1974a).

In crabs, a variety of behavioral sequences is generated by stimulation of small bundles isolated from esophageal connectives or by local stimulation of whole connectives through metal or glass microelectrodes. These behavioral sequences include swimming, involving cyclical beating of fifth legs and characteristic movements of other legs; jumping; extension of all legs; sequential flexion and extension of legs on one side; extension and raising upward of legs on one side; and cyclical flexion of legs on the other side (Fraser, 1973, 1974a,b, 1975b; P. Fraser and E. A. Campbell, unpublished).

In a variety of Crustacea, command fibers have been shown to influence rate of heartbeat (Wiersma and Novitski, 1942; Wilkens et al., 1974; Field and Larimer, 1975), rate of beating of scaphognathite and other appendages involved in respiration (Mendelson, 1971; Wilkens et al., 1974), and movements of the stomach, hindgut, and anus (Wolfe and Larimer, 1971; Dando and Selverston, 1972; Winlow and Laverack, 1972; Muramoto, 1977).

There is no general agreement regarding definition of command interneurons. This author prefers the strictly operational definition that they are neurons that evoke some behavioral response when stimulated by the experimenter (see Kupfermann and Weiss, 1978, and commentaries). Some common features of command neurons are accepted:

(1) Command neurons control a sequence of behavior, and hence they cause widespread, complex effects involving different motor neurons. For example, some command neurons that control abdominal flexion also affect uropods and swimmerets and hence influence as many as 300 motor

neurons distributed over many segments (Kennedy et al., 1966). Other command cells may influence several thousand.

(2) In terms of our usual descriptions of behavior, more than one command cell can initiate a particular behavior. Thus, any one of five neurons can elicit beating of swimmerets in crayfish (Wiersma and Ikeda, 1964).

(3) The detailed motor output of each command cell differs in most cases, although some bilaterally homologous command neurons isolated in the esophageal connectives are thought to evoke identical outputs (Atwood and Wiersma, 1967).

(4) The quantitative relationship between behavioral output and frequency of firing of a command cell element is often complex. Increasing frequency of firing often allows recruitment of more and more segments; furthermore, simple displacement of a limb can alter as more muscles are recruited, and can turn into oscillatory beating. Although command neurons are conveniently stimulated with a train of pulses of constant frequency, they can, when stimulated in short phasic bursts, still elicit organized behavior (Atwood and Wiersma, 1967; Davis and Kennedy, 1972a,b).

(5) Although, in certain cases, sensory input to command cells is known, or can be inferred, most command cells have a very high threshold to sensory activation, and simultaneous stimulation of several different sensory inputs may be required to make them respond (Atwood and Wiersma, 1967).

(6) Certain interneurons (termed "suppression neurons") inhibit the actions of command cells (see Bowerman and Larimer, 1974a).

(7) Repeated stimulation of command cells causes habituation, as shown by decreased frequency of beating of appendages or gradual loss of motor units from the overall output.

The swimmeret system in the lobster (*Homarus americanus*) illustrates many properties of command cells (Davis and Kennedy, 1972a,b). Stimulation of command neurons causes organized beating of swimmerets, and this can conveniently be monitored by recording from the two nerves in each segment containing motor neurons and inhibitor neurons that, respectively, innervate power-stroke and return-stroke muscles. Figure 3A shows a record obtained in this way following electrical stimulation of a small (less than 20 μm diameter) strand isolated from an abdominal connective. Tonic stimulation produces cyclical behavior. Furthermore, on occasions, spikes can be evoked from the strand following tactile stimulation. Such spikes correlate with the behavior (Fig. 3B), and all recordings are consistent with the idea that the spikes are causal to the behavior. Similar recordings can be made following total deafferentation. Burst period was shown to be inversely pro-

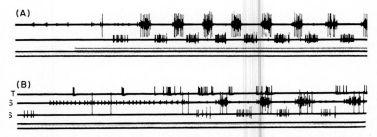

Fig. 3. (A) Recordings from power-stroke (PS) and return-stroke (RS) motor neurons during electrical stimulation (monitored by the third oscilloscope trace) of a small bundle of axons. (B) Recording from the same bundle of axons during beating of swimmerets evoked by tactile stimulation. (For more details, see Davis and Kennedy, 1971; reproduced with permission.)

portional to the intensity of motor output. Apparently not all command neurons can evoke the full frequency range of beating. Stimulating two cells together causes a summation of effects on a common oscillator, as shown by decreased burst period and increased number of motor neuron spikes. Associated with a decrease in length of the burst period is an increase in frequency of motor neuron spikes within the burst and recruitment of motor neurons, usually in ascending order, from smallest to largest (fitting the rough rule of thumb for neurons: small are tonic, large are phasic). Comparisons of command-induced behavior and that evoked by natural stimulation (e.g., tilting the lobster) indicate that to explain the natural behavior, several command cells must be firing. Inhibitory command cells that are capable of increasing the burst period are also found.

Most command neurons of swimmerets produce an asymmetrical motor pattern, strongest ipsilaterally in all recorded segments; but a few produce nearly symmetrical beating. Bilateral homologues produce different directions of force and torque. Some cells produce phase-locking of motor neuron spikes with stimulating pulses. In such cases, the latency of the phase-locking response relative to the stimulus pulse is shorter in more anterior segments. This shows that the spread of excitation is from anterior to posterior. However, the most posterior appendages are usually excited most strongly and are the last to habituate following repetitive stimulation.

D. Sensory Input to Command Neurons

1. INPUT TO COMMAND NEURONS OF CRAYFISH AND LOBSTER

In general, command neurons have complex inputs that are not easily related to sensory pathways. The techniques involved often do not allow simultaneous measurements of input and output. Most command cells have been studied in semi-isolated preparations, which would be unlikely to reveal normal inputs. Some of the input organization of lateral giant fibers of crayfish is well known. Some sensory inputs to other command neurons controlling abdominal position have also been described (Chapter 2). Tactile stimulation of the abdomen evokes activity in bundles of axons containing command neurons for movements of swimmerets in lobsters (Davis and Kennedy, 1972a; Fig. 3).

2. EQUILIBRIUM INTERNEURONS OF CRABS

Apart from the above examples, the clearest known input to command cells is found in equilibrium interneurons of crabs (*Carcinus maenas, Scylla serrata*); these interneurons are command fibers for righting and swimming (Fraser, 1974b, 1975b, 1978a). In these crabs, eight large interneurons have been identified as equilibrium interneurons (Fig. 4). The cells have three

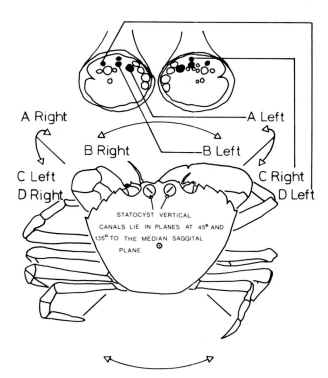

Fig. 4. Location of axons of eight equilibrium interneurons in the crab, *Carcinus*, showing axes and directions of rotation that optimally excite each interneuron. Cells A, C, and D each receive input from the vertical canal of one statocyst, which lies in the vertical plane at 45° to pitch and roll planes. Cell B receives input from horizontal canal of one statocyst, which lies in the horizontal plane.

main inputs; i.e., from statocyst, proprioceptors of the leg joints, and central neuropils (Fraser, 1975b).

a. *Input*

i. INPUT FROM THE STATOCYST. Brachyura have a complicated statocyst, with infoldings forming two fluid-filled toroids similar in some respects to vertebrate semicircular canals. Fluid movements in these statocyst canals are monitored by two types of receptor hair, thread hairs and free hook hairs. Each equilibrium interneuron receives input from thread hair and free hook hair receptors in one statocyst. It is difficult to describe coding in the cells simply because thread hair receptors are remarkably position-sensitive, and they respond to low frequency (0.1-10.0 Hz) oscillations of fairly large amplitude and high frequency oscillations (10-100 Hz) of small amplitude (Silvey *et al.*, 1976; Fraser, 1982; P. J. Fraser and E. A. Campbell, unpublished observations). Free hook hair receptors exhibit marked phase-locking of spikes to high frequency oscillations and are a significant input channel to the interneurons only following stimuli that produce this phase-locking (hence optimizing spatial summation). This sensory input is, hence, multicomponent, yielding complicated gain and phase plots (Fraser, 1975a, 1980). At times, the interneurons code for angular position; at other times, for angular velocity or angular acceleration (Fig. 5). Although the coding properties are complicated, each cell clearly codes for one direction of the component of rotation in one plane. The set of cells together can code for all directions and planes in three dimensions in terms of their relative signals (Fraser, 1974b, 1975a, 1977a, 1979). The only redundancy of information is found in cells C and D, which send their information down different connectives.

ii. PROPRIOCEPTIVE INPUT FROM LEGS. Cells A, C, and D are known to have a major input from proprioceptors monitoring at least three joints of all walking legs. This input has not been extensively analyzed, and the exact identity of the proprioceptors involved has not yet been worked out. Nevertheless, the functional significance of this input can be inferred from responses in the interneurons following manipulation of leg joints. Consider cell C. This cell fires in response to manipulation of thoracic-coxopodite (M-C) joints, as shown in Fig. 6. Interestingly, this cell responds to forced extension of ipsilateral M-C joints and to flexion of contralateral M-C joints. Although it is difficult to measure the directionality of the responses because the crab quickly ceases to oppose rotation of the body relative to the legs, all observations are consistent with the idea that the leg-joint proprioceptive

9. Views on the Nervous Control of Behavior

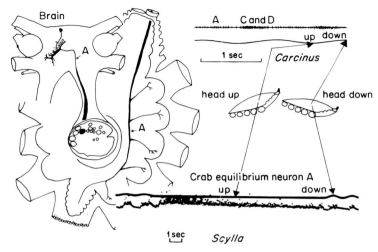

Fig. 5. Anatomy of right equilibrium neuron A in brain and thoracic ganglion of the crab, *Carcinus*. Typical extracellular recordings from *Carcinus* (showing spikes from neurons A, C, and D) and *Scylla serrata* (showing instantaneous frequency of spikes from neuron A). Note response to maintained displacement (position coding) compared to the usual coding for angular velocity seen during sinusoidal oscillation (peak response has a c90° phase lead over peak displacement of the crab). At high frequencies (not shown), the cell codes for angular acceleration. Position traces are indicated by arrows. This cell receives input from the vertical canal of the left statocyst and is most sensitive to rotation in plane indicated by Fig. 4.

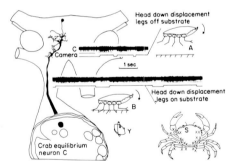

Fig. 6. Left equilibrium neuron C in *Carcinus*, showing response to a 15° rotation in the plane of vertical canal of the left statocyst (inset Y) with the crab clear of the substrate (A) and with its legs in contact with the substrate (B). Leg-joint proprioceptive inputs (X) reinforce input from the statocysts (S). Neuron C responds to retraction of thoracic-coxopodite joints and elevation of coxopodite-basi-ischiopodite joints; also to flexion of contralateral meropodite-carpopodite joints and extension of ipsilateral meropodite-carpopodite joints. Widely distributed leg propriocetive input can be most simply described in terms of properties of the single interneuron, i.e., responding best to forced displacement in plane of vertical canal of one statocyst.

input reinforces input from the statocysts for forced rotations. The leg-joint input to any one cell hence codes for a single direction of rotation in one plane in a homogeneous medium (Fraser, 1974b, 1975b; unpublished observations).

iii. NONSPECIFIC INPUT. A third main input to these equilibrium cells is manifest as a sustained tonic discharge, which can occur either following a general stimulus, such as touch anywhere on the body, or spontaneously in a restrained crab. This discharge is correlated with movements of the legs but, unlike many activity cells (Wiersma and Oberjat, 1968), the tonic activity precedes and outlasts the movements of the leg (Fraser, 1975b). The non-specific input is abolished by cutting both optic tracts, although pathways from the statocysts and proprioceptors of the legs are still functional after this operation.

b. Output. Cell A has been proved a command neuron. The best proof is that sustained depolarization of the axon by low current through an intracellular microelectrode [which elicits a train of action potentials (Fraser, 1973)] causes organized leg movements. The same sequence of leg movements can be elicited by rotating the crab in the plane of one vertical canal (Fraser, 1974b, 1975b) and occurs spontaneously when the crab is held on its back. Preceding and during the "spontaneous" motor activity, cell A fires tonically (Fraser, 1975b). The behavioral sequence evoked on stimulation of neuron A has been termed the righting reflex (Fraser, 1974b, 1975b). Rotation of the crab in the plane of a single vertical canal, which will uniquely stimulate one equilibrium cell, elicits the righting reflex asymmetrically. Thus, in Fig. 5, the right neuron A will be excited, and the right fifth leg will beat with greater amplitude and frequency than the left fifth leg. The frequency and phase of oscillation of legs 2–4 resemble those during walking. If the fifth leg encounters a support during its oscillation, or if it can be laid behind the crab, which is on its back, and gain purchase on the ground, then it acts to lever the crab upward in a head-descending direction.

The other equilibrium interneurons are also command fibers. Head-down angular acceleration evokes swimming in crabs (Fraser, 1974b, 1975b). Swimming can also be evoked by low-current electrical stimulation of cells C and D (Fraser, 1973, 1974a).

The important point is that swimming (which is performed by all crabs, Hartnoll, 1971) and the righting reflex may be considered simply as directional components of a general equilibrium reflex. Thus, the fractionation of the set of equilibrium interneurons into cells, each coding for one direction of rotation in one of three orthogonal planes, is preserved at the behavioral

9. Views on the Nervous Control of Behavior

level, where different behavioral outputs are produced by these separate cells.

The complicated outputs of these cells may be simplified by considering the output of each of the cells as mainly producing torque (in a uniform medium) in the same plane and in the opposite direction to the angular acceleration exciting a particular cell via the statocyst. This is hard to measure experimentally, because the righting reflex and swimming reflex are labile under conditions of restraint and the reflexes are altered by local proprioceptor reflexes caused by contact of the limbs with the substrate. Indirect confirmation is provided by the asymmetric output of single cells, and by the observation that *Carcinus,* when dropping through the water, will make swimming movements that will prevent it from turning onto its back.

E. Integration in Interneurons

The need to understand integrative properties of cells has long been accepted. Maynard (1965) wrote:

> The laws governing biologically meaningful neural interaction require statements about both the functional connectivity patterns *and* the permissable modes of action of the participating elements. The problem of mechanisms becomes one of describing those properties of a neuron critical for normal function as an element in a specific integrating system and of determining how those particular properties are selected from the potential range of capabilities of the individual cell.

Maynard's approach was, first, to sample the properties of cells in the lobster brain and start a component's list for the integrating brain; second, to study so called "crystals of neural activity": systems of small numbers of neurons, such as cardiac and stomatogastric ganglia, which permit exhaustive description (Chapter 6). He described three types of integrative action in spiny lobster:

(1) The motor unit with a limited dendritic field that appeared to sum all inputs electronically, producing spikes at a single locus.

(2) Interneurons with widespread aborizations (which correspond to many of the descending interneurons in esophageal connectives). They communicate intracellularly with propagated spikes. In the presence of one input, other inputs may be ineffective.

(3) Interneurons of olfactory and accessory lobes, which produce temporal rather than spatial integrative biases (Maynard, 1965).

Integration in interneurons and motor neurons in abdominal ganglia of crayfish has been investigated by Kennedy and co-workers, and the study of crayfish lateral giant neurons, which was carried out in Kennedy's laboratory, gives us our most complete knowledge of integration in crustacean nerve cells involved in behavior (Preston and Kennedy, 1960, 1962; Zucker et al., 1971).

Much of the work on integration in crustacean ganglia concerns motor neurons rather than interneurons. Networks of motor neurons controlling movements of eyes in crabs have been extensively analyzed in terms of input arrangement, integrative properties, and mutual interactions (Sandeman, 1977; Chapter 5). Much of the integration in this system is done at the level of the motor neurons. Thus, care must be taken in attempts to correlate properties of interneurons with behavior. The more integration that is done at the level of motor neurons, the less behaviorally relevant will the signal in the interneurons appear.

In general, although knowledge of integration will be essential for any understanding of behavior, at present the problem is largely that of identifying cells or groups of cells participating in a pattern of behavior.

F. Excited States and Complex Neuropils

During recordings from interneurons in partially tethered decapod Crustacea, it is common for the animal to produce forceful movements. During this struggling, which has been called the "excited state" (Wiersma and Yamaguchi, 1967a), thresholds and activity levels in many interneurons are altered, revealing the presence of central pathways. The sensitivity of C4 in crayfish to water-borne vibration is altered during movements of appendages, struggling, tail flips, and most kinds of visual stimuli (Taylor, 1968). A multimodal interneuron (giant cell 1) in the crab *Carcinus* shows altered response to a repeated "off" of light during the "excited" state (Fraser, 1974a).

The lateral giant fiber of crayfish species can show greatly altered excitability as judged by response threshold. Transection of the nerve cord above the abdomen abolishes all variability, and excitability rises to a level close to the maximum displayed in the intact animal (Krasne and Wine, 1977; Chapter 8). This pathway (modulatory pathway III: Krasne and Wine, 1977) is the largest source of variability in the reflex. Not all tail-flip behavior is suppressed along with suppression of the lateral giant fibers. Escape responses mediated by non-giant fibers still occur, although these in turn can be inhibited centrally, e.g., during defensive behavior.

Activity fibers, which give a sustained discharge during the excited state, are best known in the visual system (Chapter 1). Equilibrium interneurons of crabs show many features in common with activity fibers (Fraser, 1974b, 1975b; Section II,D). The excited state is not a unitary phenomenon. There is little evidence concerning the origin and maintenance of these modulating effects. The "non-specific" input to equilibrium interneurons of crabs may be dependent on pathways from the hemiellipsoid bodies in the eyestalks, because cutting optic tracts abolishes this input (Fraser, 1975b). The idea that activity effects and "excited state" pathways originate from these bodies is consistent with the behavioral evidence (Section I,A) that they seem to have a role in controlling the time-course of behavior.

III. MULTISEGMENTAL INTERNEURONS AND BEHAVIOR

A. Behaviors Commanded and Numbers of Command Neurons Involved

Patterns of behavior evoked by stimulating command neurons can be broadly classified as defense, locomotor, respiratory, or circulatory behaviors. All of these are high on any hierarchy of behavior patterns. Davis and Kennedy (1972a) found it necessary to stimulate several command neurons of the swimmerets to reproduce the full range of swimmeret-beating found in an intact lobster. Davis (1976, 1977) and Larimer (1976), reviewing the range of behaviors evoked by command neurons, concluded that more than one command neuron is necessary for most normal behavior. In the crab, more than one command neuron is active during unrestrained locomotion (see Section II,A). A consideration of vector coding in the nervous system predicts that more than one command neuron should be involved in most behaviors (Fraser, 1978a).

B. Correlations between Activity in Interneurons and Behavior

1. NON-IDENTIFIED CELLS

There have been several recent attempts to investigate the role of command neurons in normal behavior, by monitoring the activity of various command cells during behavior in tethered or free-walking Crustacea. In lobsters, Ayers and Davis (1977) have recorded activity from neurons discharging before and during optokinetically induced locomotion. Electrical stimulation often had no effect, causing coordinated walking in only 2 of 31 animals "perhaps simply because normal proprioceptive cues were not available to the lobsters." Their results are not conclusive, but they suggest that neurons categorized as command elements on the basis of electrical stimulation may be active during voluntary execution of the same behavior (Davis, 1977).

Larimer and Gordon (1977) cut one esophageal connective of the crayfish (*Procambarus*) for ease of recording and tried to correlate activity in the cut connective with behavior. Of course, with the connective cut, none of the recorded activity can be causal to the behavior. They justified this experimental approach by assuming that the two connectives are neurally equivalent and that one is sufficient for the transmission of at least some behaviors. This assumption is perhaps valid in crayfish for giant fibers and various abdominal command fibers that produce symmetrical output; but it is unjustified at least in the case of the crab (see later), and there is no strong reason for justifying it generally in crayfish or in any other arthropod.

In crabs (*Carcinus maenas*), it has proved possible to record from whole esophageal connectiones in unrestrained animals (Fraser, 1973, 1974a, 1978a; P. J. Fraser, unpublished). Quiescent behavior correlates with no activity in large interneurons. Inputs to identified giant cells are fully functional, as in the restrained crab. Active behavior is preceded and accompanied by activity in several interneurons. No obvious correlation of particular cells and particular behaviors emerge in the sense of one-to-one mapping of cells onto behavior, with the exception that in several crabs, which had obviously suffered brain damage during implantation of electrodes, single spikes in a large interneuron occurred spontaneously and were correlated with jumping. This type of jumping is a component of the rearing reflex and occurs occasionally as an aberrant spontaneous behavior in crabs kept in aquaria. Many cells, such as equilbrium interneurons, are active during many different patterns of behavior.

2. IDENTIFIED COMMAND NEURONS IN CRABS

In the crabs *Carcinus maenas* and *Scylla serrata*, cells A, B, C, and D (Section II,A,4) are active before and during righting, swimming, and walking (Fraser, 1974b, 1978a,b). It has proved possible, to some extent, to correlate activity in these cells with gaits and body position (Figs. 7 and 8). In the resting crab, cells A, B, C, and D and most other large cells are silent. There is always activity in one or more of the cells preceding and during locomotion. There is no evidence for a consistent involvement of any other command cells during locomotion.

Consideration of these results poses the question as to whether the set of cells A, B, C, and D is sufficient to control locomotion. Central driving of one or more equilibrium cells will produce torque, depending on the pattern of fulcra and forces produced by the legs on the ground. The resulting displacement will recruit equilibrium cells to compensate. Integrated over time, torque will cancel out, but linear forces will sum to give a net locomotive force. Plastic changes in gait due to removal of legs can be explained by the altered forces and fulcra leading to a changed pattern of recruitment in cells A, B, C, and D. The above hypothesis is compatible with Bethe's results from ablation experiments (Bethe, 1897a,b). Normal sideways walking is driven

9. Views on the Nervous Control of Behavior

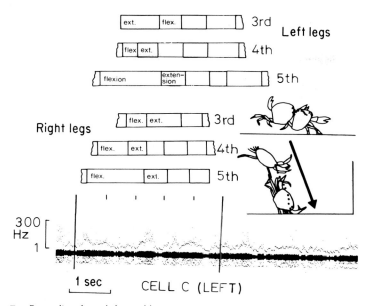

Fig. 7. Recording from left equilibrium neuron C in *Carcinus* with implanted electrodes (middle trace) during a short walking sequence (to the crab's left), showing relationship between gait and instantaneous frequency of interneuron (top trace). Behavior was monitored by cinecamera (60 frames/sec) with a mirror arranged to give side and top views (insets). Although not shown in this sequence, activity in neuron C preceded the start of walking. Dots show location of carapace edge every 15 frames.

largely by the legs on the trailing side, and the trailing-side esophageal connective and leading-side statocyst have to be intact for approximately normal walking. When one connective is cut, walking is largely in the direction that places the intact connective on the trailing side. Furthermore, the crab always circles in a direction that can be predicted from the particular cell B that is severed. Hence in Fig. 1, the crab always goes anticlockwise from the observer's point of view, which, in the normal crab, would be corrected by activity in the right cell B.

3. IDENTIFIED SENSORY INTERNEURONS

Recently there have been attempts to correlate activity in sensory interneurons with behavior. Glantz (1977) correlates activity in visual interneurons with claw raising, which is part of the defense reflex. Jittery-movement neurons habituate over the same time-course as does the defense reflex, whereas sustaining neurons and dimming neurons do not (see Chapter 1 of this volume). Levator muscles of the cheliped (claws) show a decline in latency of firing that depends on the velocity of a visual target over a cer-

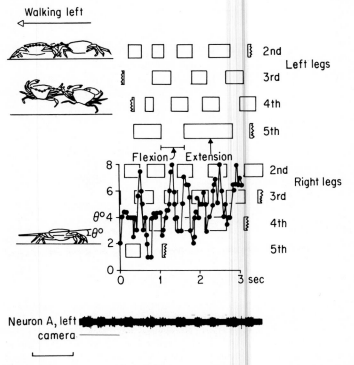

Fig. 8. Extracellular record from neuron A in the left connective during crab's walking to the left. In this case, interneuron activity did not precede locomotion but is clearly related to oscillation of body position (increasing $\Theta°$) during gait cycle. This is consistent with ablation evidence suggesting that the main drive for locomotion is from the contralateral connective, i.e., from the right connective when the crab is going left, as in this case. Dots show carapace edge every 15 frames, as in Fig. 7.

tain range. Jittery-movement detector neurons show a linear relationship with target velocity over this range. No such relationship is apparent for other visual interneurons.

Glantz was also able to isolate single, large interneurons (including C99: Wiersma, 1958) sensitive to motion; these neurons could then be stimulated electrically as well as visually. It was possible to elicit an entire reflex with any one of three separate cells, but motor latencies following electrical stimulation were always in excess of those following visual stimulation.

In crabs (*Carcinus maenas*), giant neuron 1 resembles C99 in crayfish (Fraser, 1974a). Recordings from this cell show that often, but not invariably, there is activity preceding and during the initial part of the rearing reflex. Equilibrium interneurons A respond during the rearing-up part of the reflex, as may be expected from their input from the statocyst. If the pathways from

9. Views on the Nervous Control of Behavior 313

the statocyst are inactivated by ablation, the crab falls backward after rearing up (Bethe, 1897a). Part of the observed behavior in crab is hence attributable to non-visual equilibrium pathways.

Observation of many identified sensory neurons in freely moving crabs has led to the conclusion that a given interneuron is involved in many patterns of behavior and that subdivisions produced by consideration of behavior of the whole animal are not related in a simple way to the subdivisions of the nervous system into interneurons (Fraser, 1974a,b, 1975b, 1978a). It is obvious that if any such simple relationship existed, it would have been noticed either in the experiments above or in others by Taylor (1968, 1970) or Schrameck (1970) involving recording from crayfish with implanted electrodes.

IV. PERSPECTIVES

There are many large gaps in our understanding of the mechanisms and components involved in the neural control of complex behavior. More detailed mapping of properties and relevant firing patterns of interneurons is required to even out the bias toward large cells, to understand levels and strategies of information-processing in specific sensory channels, and especially to extend our knowledge to Crustacea other than a few large decapods. Possible and desirable approaches include detailed study of behavior where ablation experiments point to input pathways. For example, behavioral effects attributable to displacement-sensitive pathways from statoliths are known (Chapter 5), but no statolith interneurons are known to code for angular displacement. In Crustacea, no attempts have been made to isolate and examine neurons tuned to behaviorally relevant stimuli, such as those involved in communication. Thus, although Horche (1971) has found auditory interneurons presumably involved in auditory communication in crabs (*Ocypode*), these await identification and detailed study. Decapod Crustacea are known to respond to particular colored spots or patches on claws or carapace (Hazlett, 1971), but no visual interneurons tuned to these features have been described.

Crustacean multimodal interneurons offer a unique insight into the ways in which widely separated sense organs are linked or excluded. Identified cells allow comparative and developmental studies, in which changing behavioral capabilities in both a phylogenetic and an ontogenetic sense may be correlated with changing properties or features of the cell. Thus, the crayfish *Procambarus clarkii* has numerous command neurons evoking abdominal extension, and this behavior is easily evoked by removing support from the legs. *Orconectes*, however, has few such command cells, and these cells produce weak effects. Removal of support elicits only slight extension behavior (see Chapter 2).

Monitoring identified cells during unrestrained behavior is necessary to define normal involvement of such cells in behavior. The requirement for simultaneous detailed monitoring of behavior makes meaningful experiments technically difficult, but not impossible. Once relevant groups of cells are identified, it will be desirable that detailed studies of connection and integration be made. More work is needed on the origins of the various modulatory pathways to interneurons that may represent activity in complex neuropils, such as hemiellipsoid bodies. There are difficulties regarding classification of identified cells and measurements of behavior. Where a behavior involves vectorial output in more than one dimension, or where feedback loops are involved, then the output measured by displacement of appendages or by muscle activity is no longer an adequate measure of the behavior. It is important to try to classify cells in a way that will facilitate the comparative approach and to adopt measurements of behavior that will fit with the properties and organization of the central cells. Hence, in terms of understanding the underlying neural activity, locomotion in crabs may be better described in terms of components of force and torque in the planes used by the equilibrium cells rather than in terms of the gaits employed. A wide variety of approaches should be employed until sufficient examples are known to define a strict framework regarding the neural control of complex behavior.

REFERENCES

Atwood, H. C., and Wiersma, C. A. G. (1967). Command interneurons in the crayfish central nervous system. *J. Exp. Biol.* **46,** 249-261.
Ayers, J., and Davis, W. J. (1977). Neuronal control of locomotion in the lobster, *Homarus americanus*. I. Motor programs for forward and backward walking. *J. Comp. Physiol. A* **115A,** 1-27.
Bethe, A. (1895). Studien uber das Centralnervensystem von *Carcinus maenas* hebst Angaben über ein neues Verfahren der Methylenblau fixation. *Arch. Mikrosk. Anat. Entwicklungsmech.* **44,** 579-622.
Bethe, A. (1897a). Das Nervensystem von *Carcinus maenas*. Ein anatomisch-physiologischer Versuch. I. Theil, I. Mittheil. *Arch. Mikrosk. Anat. Entwicklungsmech.* **50,** 460-546.
Bethe, A. (1897b). Das Nervensystem von *Carcinus maenas*. Ein anatomisch-physiologischer Versuch. I. Theil, II. Mittheil. *Arch. Mikrosk. Anat. Entwicklungsmech.* **50,** 589-639.
Bethe, A. (1897c). Vergleichende Untersuchungen uber die Funktionen des Centralnervensystem der Arthropoden. *Arch. Gesamte Physiol. Menschen Tiere* **68,** 449-545.
Bowerman, R. F., and Larimer, J. L. (1974a). Command fibres in the circumoesophageal connectives of crayfish. I. Tonic fibres. *J. Exp. Biol.* **60,** 95-117.
Bowerman, R. F., and Larimer, J. L. (1974b). Command fibres in the circumoesophageal connectives of crayfish. II. Phasic fibres. *J. Exp. Biol.* **60,** 119-134.

9. Views on the Nervous Control of Behavior

Bowerman, R. F., and Larimer, J. L. (1976). Command neurons in crustaceans. *Comp. Biochem. Physiol. A* **54A**, 1-5.

Bush, B. M. H., Wiersma, C. A. G., and Waterman, T. H. (1964). Efferent mechanoreceptive responses in the optic nerve of the crab *Podophthalmus*. *J. Cell. Comp. Physiol.* **64**, 327-346.

Calabrese, R. L. (1976a). Crayfish mechanoreceptive interneurons 1. Nature of ipsilateral excitatory inputs. *J. Comp. Physiol. A* **A105**, 83-102.

Calabrese, R. L. (1976b). Crayfish mechanoreceptive interneurons 2. Bilateral interactions and inhibition. *J. Comp. Physiol. A* **A105**, 103-114.

Calabrese, R. L., and Kennedy, D. (1974). Multiple sites of spike initiation in a single dendritic system. *Brain Res.* **82**, 316-321.

Dando, M. R., and Selverston, A. I. (1972). Command fibres from the supraoesophageal ganglion to the stomatogastric ganglion in *Panulirus argus*. *J. Comp. Physiol.* **78**, 138-175.

Davis, W. J. (1976). Organizational concepts in the central motor networks of invertebrates. In "Neural Control of Locomotion" (R. M. Herman, S. Grillner, P. S. G. Stein, and D. G. Stuart, eds.), pp. 265-292. Plenum, New York.

Davis, W. J. (1977). The command neuron. In "Identified Neurons and Behavior of Arthropods" (G. Hoyle, ed.), pp. 293-305. Plenum, New York.

Davis, W. J., and Kennedy, D. (1972a). Command interneurons controlling swimmeret movements in the lobster. I. Types of effects on motor neurons. *J. Neurophysiol.* **35**, 1-12.

Davis, W. J., and Kennedy, D. (1972b). Command interneurons controlling swimmeret movements in the lobster. II. Interactions of effects on motor neurons. *J. Neurophysiol.* **35**, 13-19.

Davis, W. J., and Kennedy, D. (1972c). Command interneurons controlling swimmeret movements in the lobster. III. Temporal relationships among bursts in different motor neurons. *J. Neurophysiol.* **35**, 20-29.

Eisenstein, E. M., and Mill, P. J. (1965). Role of the optic ganglia in learning in the crayfish *Procambarus clarkii* (Girard), *Anim. Behav.* **13**, 561-565.

Field, L. H., and Larimer, J. L. (1975). The cardioregulatory system of crayfish: The role of circumoesophageal interneurons. *J. Exp. Biol.* **62**, 531-543.

Fraser, P. J. (1973). Giant fibres and directional statocyst fibres: A study of interneurons between the brain and thoracic nervous system in the shore crab *Carcinus maenas* (L.). Ph.D. Thesis, Aberdeen University, Scotland.

Fraser, P. J. (1974a). Interneurons in crab connectives (*Carcinus maenas* (L.)): Giant fibres. *J. Exp. Biol.* **61**, 593-613.

Fraser, P. J. (1974b). Interneurons in crab connectives (*Carcinus maenas* (L.)): Directional statocyst fibres. *J. Exp. Biol.* **61**, 615-628.

Fraser, P. J. (1975a). Free hook hair and thread hair input to fibre 5 in the mud crab, *Scylla serrata*, during antennule rotation. *J. Comp. Physiol.* **103**, 291-313.

Fraser, P. J. (1975b). Three classes of input to a semicircular canal interneuron in the crab, *Scylla serrata*, and a possible output. *J. Comp. Physiol.* **104**, 261-271.

Fraser, P. J. (1977a). How morphology of semicircular canals affects transduction, as shown by response characteristics of statocyst interneurons in the crab *Carcinus maenas* (L.). *J. Comp. Physiol.* **115**, 135-145.

Fraser, P. J. (1977b). Directionality of a one way movement detector in the crayfish *Cherax destructor*. *J. Comp. Physiol.* **118**, 187-193.

Fraser, P. J. (1977c). Cercal ablation modifies tethered flight behaviour of cockroach. *Nature (London)* **268**, 523-524.

Fraser, P. J. (1978a). Vector coding and command fibres. *Behav. Brain Sci.* **1**, 22-23.

Fraser, P. J. (1978b). Equilibrium interneurons and locomotion of arthropods. *Neurosci. Lett., Suppl.* **1**, 594.

Fraser, P. J. (1982). Semicircular canal morphology and function in crabs. In "Vestibular Function and Morphology" (T. Gualtierotti, ed.). Springer-Verlag, Berlin and New York (In press).

Fraser, P. J., and Sandeman, D. C. (1975). Effects of angular and linear accelerations on semicircular canal interneurons of the crab *Scylla serrata*. *J. Comp. Physiol.* **96**, 205–221.

Glantz, R. M. (1977). Visual input and motor output of command interneurons of the defense reflex pathway in the crayfish. In "Identified Neurons and Behavior of Arthropods" (G. Hoyle, ed.), pp. 259–274. Plenum, New York.

Hartnoll, R. G. (1971). The occurrence, methods and significance of swimming in the Brachyura. *Anim. Behav.* **19**, 34–50.

Hazlett, B. A. (1971). Non visual function of crustacean eyestalk ganglia. *Z. Vergl. Physiol.* **71**, 1–13.

Horche, K. (1971). An organ for hearing and vibration senses in the ghost crab *Ocypode*. *Z. Vergl. Physiol.* **73**, 1–21.

Horridge, G. A. (1968). "Interneurons." Freeman, San Francisco, California.

Hughes, G. M., and Wiersma, C. A. G. (1960a). Neuronal pathways and synaptic connections in the abdominal cord in the crayfish. *J. Exp. Biol.* **37**, 291–307.

Hughes, G. M., and Wiersma, C. A. G. (1960b). The coordination of swimmeret movements in the crayfish, *Procambarus clarkii* (Girard). *J. Exp. Biol.* **37**, 657–670.

Kennedy, D., and Davis, W. J. (1977). The organization of invertebrate motor systems. In "Handbook of Physiology" (E. R. Kandel, ed.), 2nd ed., Sect. 1, Vol. 2, pp. 1023–1087. Am. Physiol. Soc., Bethesda, Maryland.

Kennedy, D., and Mellon, De F. (1964). Synaptic activation and receptive fields in crayfish interneurons. *Comp. Biochem. Physiol.* **13**, 275–300.

Kennedy, D., Evoy, W. H., and Hanawalt, J. T. (1966). Release of coordinated behavior in crayfish by single central neurons. *Science* **154**, 917–919.

Kovac, M. (1974a). Abdominal movements during backward walking in crayfish. I. Properties of the motor program. *J. Comp. Physiol.* **95**, 61–78.

Kovac, M. (1974b). Abdominal movements during backward walking in crayfish. II. The neuronal basis. *J. Comp. Physiol.* **95**, 79–94.

Krasne, F. B., and Wine, J. J. (1977). Control of crayfish escape behavior. In "Identified Neurons and Behavior of Arthropods" (G. Hoyle, ed.), pp. 275–292. Plenum, New York.

Kupfermann, I., and Weiss, K. R. (1978). The command neuron concept. *Behav. Brain Sci.* **1**, 3–39.

Larimer, J. L. (1976). Command interneurons and locomotor behavior in crustaceans. In "Neural Control of Locomotion" (R. H. Herman, S. Grillner, P. S. G. Stein, and D. G. Stuart, eds.), pp. 293–326. Plenum, New York.

Larimer, J. L., and Gordon, W. H. (1977). Circumoesophageal interneurons and behavior in crayfish. In "Identified Neurons and Behavior of Arthropods" (G. Hoyle, ed.), pp. 243–258. Plenum, New York.

Larimer, J. L., and Kennedy, D. (1969a). Innervation patterns of fast and slow muscles in the uropods of crayfish. *J. Exp. Biol.* **51**, 119–133.

Larimer, J. L., and Kennedy, D. (1969b). The central nervous control of complex movements in the uropods of crayfish. *J. Exp. Biol.* **51**, 135–150.

Larimer, J. L., Eggleston, A. C., Masukawa, L. M., and Kennedy, D. (1971). The different connections and motor outputs of lateral and medial giant fibres in the crayfish. *J. Exp. Biol.* **54**, 391–402.

Maynard, D. M. (1965). Integration in crustacean ganglia. *Symp. Soc. Exp. Biol.* **20**, 111–149.

Maynard, D. M., and Dingle, H. (1963). An effect of eyestalk ablation on antennulary function in the spiny lobster, *Panulirus argus*. *Z. Vergl. Physiol.* **46,** 515-540.

Maynard, D. M., and Sallee, A. (1970). Disturbance of feeding behavior in the spiny lobster, *Panulirus argus*, following bilateral ablation of the medulla terminalis. *Z. Vergl. Physiol.* **66,** 123-140.

Maynard, D. M., and Yager, J. G. (1968). Function of an eyestalk ganglion, the medulla terminalis in olfactory integration in the lobster, *Panulirus argus*. *Z. Vergl. Physiol.* **59,** 241-249.

Mellon, De F. (1977). Central and peripheral features of crayfish oculomotor organization. In "Identified Neurons and Behavior of Arthropods" (G. Hoyle, ed.), pp. 149-166. Plenum, New York.

Mendelson, M. (1971). Oscillator neurons in crustacean ganglia. *Science* **171,** 1170-1173.

Müller, J. (1835). On the peculiar properties of nerves. In "Handbuch der Physiologie des Menshen für Vorlesungen," 2nd ed., Vol. 1, Book 3, Sect. 4. Coblene.

Muramoto, A. (1977). Neural control of rhythmic anal contraction in the crayfish. *Comp. Biochem. Physiol. A* **56A,** 551-557.

Preston, J. B., and Kennedy, D. (1960). Integrative synaptic mechanisms in the caudal ganglion of the crayfish. *J. Gen. Physiol.* **43,** 671-681.

Preston, J. B., and Kennedy, D. (1962). Spontaneous activity in crustacean neurons. *J. Gen. Physiol.* **45,** 821-836.

Prosser, C. L. (1934a). Action potentials in the nervous system of the crayfish. I. Spontaneous impulses. *J. Cell. Comp. Physiol.* **4,** 185-209.

Prosser, C. L. (1934b). Action potentials in the nervous system of crayfish. II. Responses to illumination of the eye and caudal ganglion. *J. Cell. Comp. Physiol.* **4,** 363-377.

Prosser, C. L. (1935). Action potential in the nervous system of the crayfish. III. Central responses to proprioceptive and tactile stimulation. *J. Comp. Neurol.* **62,** 495-505.

Roye, D. B. (1972). Evoked activity in the nervous system of *Callinectes sapidus* following phasic excitation of statocysts. *Experientia* **28,** 1307-1309.

Sandeman, D. C. (1966). The vascular circulation in the brain, optic lobes and thoracic ganglia of the crab *Carcinus*. *Proc. R. Soc. London, Ser. B* **168,** 82-90.

Sandeman, D. C. (1977). Compensatory eye movements in crabs. In "Identified Neurons and Behavior of Arthropods" (G. Hoyle, ed.), pp. 131-148. Plenum, New York.

Schöne, H. (1961). Complex behaviour. In "The Physiology of Crustacea" (T. H. Waterman, ed.), Vol. 2, pp. 465-520. Academic Press, New York.

Schrameck, J. E. (1970). Crayfish swimming: Alternating motor output and giant fibre activity. *Science* **169,** 698-700.

Silvey, G. E., Dunn, P. A., and Sandeman, D. C. (1976). Integration between statocyst sensory neurons and oculomotor neurons in the crab *Scylla serrata*. II. The thread hair sensory receptors. *J. Comp. Physiol.* **108,** 45-52.

Smith, D. O. (1974). Central nervous control of excitatory and inhibitory neurons of opener muscle of the crayfish claw. *J. Neurophysiol.* **37,** 108-118.

Stein, P. S. G. (1971). Intersegmental coordination of swimmeret motor neuron activity in crayfish. *J. Neurophysiol.* **34,** 310-318.

Sutherland, R. M., and Nunnemacher, R. F. (1968). Microanatomy of crayfish thoracic cord and roots. *J. Comp. Neurol.* **132,** 499-518.

Taylor, R. C. (1968). Water-vibration reception: A neurophysiological study in unrestrained crayfish. *Comp. Biochem. Physiol.* **27,** 795-805.

Taylor, R. C. (1970). Environmental factors which control the sensitivity of a single crayfish interneuron. *Comp. Biochem. Physiol.* **33,** 911-921.

Taylor, R. C. (1974). A saline transfusion technique for crayfish C.N.S. studies. *Comp. Biochem. Physiol. A* **A47**, 1185-1190.

Waterman, T. H., and Wiersma, C. A. G. (1963). Electrical responses in decapod crustacean visual systems. *J. Cell. Comp. Physiol.* **61**, 1-16.

Waterman, T. H., Wiersma, C. A. G., and Bush, B. M. H. (1964). Afferent visual responses in the optic nerve of the crab, *Podophthalmus. J. Cell. Comp. Physiol.* **63**, 135-155.

Wiersma, C. A. G. (1938). Function of the giant fibres of the central nervous system of the crayfish. *Proc. Soc. Exp. Biol. Med.* **38**, 661-662.

Wiersma, C. A. G. (1947). Giant nerve fibre system of the crayfish: A contribution to comparative physiology of synapse. *J. Neurophysiol.* **10**, 23-38.

Wiersma, C. A. G. (1952a). The neuron soma: Neurons of arthropods. *Cold Spring Harbor Symp. Quant. Biol.* **17**, 155-163.

Wiersma, C. A. G. (1952b). Repetitive discharges of motor fibres caused by a single impulse in giant fibres of the crayfish. *J. Cell. Comp. Physiol.* **40**, 399-419.

Wiersma, C. A. G. (1958). On the functional connections of single units in the central nervous system of the crayfish, *Procambarus clarkii* Girard. *J. Comp. Neurol.* **110**, 421-472.

Wiersma, C. A. G. (1961a). Reflexes and the central nervous system. *In* "The Physiology of Crustacea" (T. H. Waterman, ed.), Vol. 2, pp. 241-279. Academic Press, New York.

Wiersma, C. A. G. (1961b). Comparative neurophysiology, its aims and its present status. *Festschr. Kh. Koshtoyants* pp. 82-88 (in Russian).

Wiersma, C. A. G. (1962). The organization of the arthropod central nervous system. *Am. Zool.* **2**, 67-78.

Wiersma, C. A. G. (1966). Integration in the visual pathway of Crustacea. *Symp. Soc. Exp. Biol.* **20**, 151-178.

Wiersma, C. A. G. (1970). Neuronal components of the optic nerve of the crab, *Carcinus maenas. Proc. K. Ned. Akad. Wet.* **73**, 25-34.

Wiersma, C. A. G. (1978). The original definition of command neuron. *Behav. Brain Sci.* **1**, 34-35.

Wiersma, C. A. G., and Bush, B. M. H. (1963). Functional neural connections between the thoracic and abdominal cords of the crayfish, *Procambarus clarkii* (Girard). *J. Comp. Neurol.* **121**, 207-235.

Wiersma, C. A. G., and Hughes, G. M. (1961). On the functional anatomy of neuronal units in the abdominal cord of the crayfish, *Procambarus clarkii* (Girard). *J. Comp. Neurol.* **116**, 209-228.

Wiersma, C. A. G., and Ikeda, K. (1964). Interneurons commanding swimmeret movements in the crayfish, *Procambarus clarkii* (Girard). *Comp. Biochem. Physiol.* **12**, 509-525.

Wiersma, C. A. G., and Mill, P. J. (1965). Descending neuronal units in the commissure of the crayfish central nervous system; and the integration of visual tactile and proprioceptive stimuli. *J. Comp. Neurol.* **125**, 67-94.

Wiersma, C. A. G., and Novitski, E. (1942). The mechanisms of the nervous regulation of the crayfish heart. *J. Exp. Biol.* **19**, 225-265.

Wiersma, C. A. G., and Oberjat, T. (1968). The selective responsiveness of various crayfish oculomotor fibres to sensory stimuli. *Comp. Biochem. Physiol.* **26**, 1-16.

Wiersma, C. A. G., and Yamaguchi, T. (1966). The neuronal components of the optic nerve of the crayfish as studied by single unit analysis. *J. Comp. Neurol.* **128**, 333-358.

Wiersma, C. A. G., and Yamaguchi, T. (1967a). Integration of visual stimuli by the crayfish central nervous system. *J. Exp. Biol.* **47**, 409-43.

Wiersma, C. A. G., and Yamaguchi, T. (1967b). The integration of visual stimuli in the rock lobster. *Vision Res.* **7**, 197-204.

Wiersma, C. A. G., Ripley, S. H., and Christensen, E. (1955). The central representation of sensory stimulation in the crayfish. *J. Cell. Comp. Physiol.* **46,** 307-326.

Wiersma, C. A. G., Bush, B. M. H., and Waterman, T. H. (1964). Efferent visual responses of contralateral origin in the optic nerve of the crab, *Podophthalmus. J. Cell. Comp. Physiol.* **64,** 309-326.

Wilkens, J. L., Wilkens, L. A., and McMahon, B. R. (1974). Central control of cardiac and scaphognathite pacemakers in the crab, *Cancer magister. J. Comp. Physiol.* **90,** 89-104.

Wilkens, L. A., and Larimer, J. L. (1973). Sensory interneurons: Some observations concerning the physiology and related structural significance of two cells in the crayfish brain. *Tissue Cell* **5,** 393-400.

Wine, J. J., and Krasne, F. B. (1972). The organization of escape behavior in the crayfish. *J. Exp. Biol.* **56,** 1-18.

Winlow, W., and Laverack, M. S. (1972). The control of hind gut motility in the lobster *Homarus gammarus* (L.). 3. Structure of the sixth ganglion (6 A.G.) and associated ablation and microelectrode studies. *Mar. Behav. Physiol.* **1,** 93-121.

Wolfe, G. E., and Larimer, J. L. (1971). The intestinal control system in the crayfish *P. clarkii. Am. Zool.* **11,** 666.

Zucker, R. S., Kennedy, D., and Selverston, A. I. (1971). Neuronal circuit mediating escape responses in crayfish. *Science* **173,** 645-650.

Systematic Index*

Alpheus armillatus H. Milne Edwards [Crangon armillatus], Alpheoides, 23, 24
Alpheus californiensis Holmes, 75
Alpheus heterochelis Say, 75
Aplysia, Gastropoda, 223, 226
Astacus, Astacoidea, 34, 145, 172
Astacus astacus (Linnaeus) [Astacus fluviatilis], 72
[Astacus fluviatilis] see Astacus astacus
Astacus leptodactylus Eschscholtz, 86, 87, 90, 92, 94, 96, 147
[Astacus pallipes] see Austropotamobius pallipes
Austropotamobius pallipes Lereboullet [Astacus pallipes], Astacoidea, 169

Caenorhabditis elegans, Nematoda, 221
Calappa hepatica (Linnaeus), Oxystomata, 135
Callinectes, Brachyrhyncha, 34, 62, 64, 74
Callinectes sapidus Rathbun, 74, 84, 117, 185, 298
[Cambarus affinis] see Orconectes limosus
[Cambarus ayersii] see Cambarus setosus
[Cambarus clarkii] see Procambarus clarkii
Cambarus hubrichti Hobbs, Astacoidea, 24
[Cambarus limosus] see Orconectes limosus
Cambarus setosus Faxon [Cambarus ayersii], 24
[Cambarus virilis] see Orconectes virilis
Cancer, Brachyrhyncha, 34, 111, 295
Cancer anthonyi Rathbun, 212

Cancer magister Dana, 37, 95, 185
Cancer pagurus Linnaeus, 91, 112, 205
Cancer productus Randall, 185
Carcinus, Brachyrhyncha, 4, 10, 13, 20, 34, 43, 70, 111, 137, 138, 139, 140, 141, 143, 146, 298, 303, 305, 307, 308, 311
Carcinus maenas (Linnaeus), 3, 4, 6, 43, 76, 84, 88, 89, 91, 92, 94, 109, 110, 111, 112, 113, 117, 118, 121, 122, 123, 124, 125, 126, 127, 128, 129, 135, 136, 215, 218, 294, 296, 298, 310, 312
Carcinus mediterraneus Czerniavsky, 76, 94
Cardisoma, Brachyrhyncha, 70, 121, 123
Cardisoma guanhumi Latreille, 43, 86, 90, 95, 115, 117, 118, 119, 121, 122, 125, 126
Cherax destructor, Parastacoidea, 297
Chionoecetes opilio (Fabricius), Oxyrhyncha, 112
Crangon, Crangonoidea, 24, 145
[Crangon armillatus] see Alpheus armillatus

Emerita, Hippoidea, 64, 73, 77, 88
Emerita analoga (Stimpson), 73
Eriocheir japonicus De Haan, Brachyrhyncha, 212
[Eupagurus bernhardus] see Pagurus bernhardus

Gecarcinus lateralis (Fréminville), Brachyrhyncha, 129

*Names that have been superseded appear in brackets. Parentheses around name of author of scientific name indicates currently assigned genus is not the original one.

Gecarcinus quadratus Saussure, 112

Hemigrapsus oregonensis (Dana), Brachyrhyncha, 110
Hemisquilla ensigera (Owen), Stomatopoda, 75
Homarus, Nephropoidea, 34, 64, 66, 72, 86, 88, 111, 149, 152, 172, 173, 175, 185, 206, 207, 209, 214, 221
Homarus americanus H. Milne Edwards, 23, 43, 65, 70, 72, 76, 84, 85, 91, 93, 94, 95, 97, 180, 185, 273, 310
Hommarus gammarus (Linnaeus) [*Hommarus vulgaris*], 36, 108, 109, 124, 125, 167, 168, 169, 172, 186, 188
[*Hommarus vulgaris*] see *Hommarus gammarus*
Homola barbata (Fabricius), Archaeobrachyura, 74

Jasus lalandii (H. Milne Edwards), Palinuroidea, 95

[*Leander serratus*] see *Palaemon serratus*
Leptograpsus, Brachyrhyncha, 143
Leptograpsus variegatus (Fabricius), 139
Limulus, Xiphosura, 225

Maja squinado (Herbst), Oxyrhyncha, 118

Nephrops, Nephropoidea, 149, 152

Ocypode, Brachyrhyncha, 62, 70, 298, 313
Ocypode ceratophthalma (Pallas), 69, 77
Oratosquilla oratoria (De Haan) [*Squilla oratoria*], Stomatopoda, 4, 213
Orconectes, Astacoidea, 313
Orconectes limosus (Rafinesque) [*Cambarus affinis, C. limosus*], 23
Orconectes virilis (Hagen) [*Cambarus virilis*], 23, 50, 76

Pachygrapsus, Brachyrhyncha, 4, 13, 18, 19
Pachygrapsus crassipes Randall, 4, 7, 143
Pachygrapsus transversus (Gibbes), 70
Pagurus [*Eupagurus*], Paguroidea, 17, 34, 52, 111
Pagurus bernhardus (Linnaeus) [*Eupagurus bernhardus*], 111, 113, 117, 118, 121, 123, 124, 125, 126, 127, 129

Pagurus ochotensis Brandt, 74, 86, 87, 91, 93, 97
Pagurus pollicaris Say, 46, 72
Palaemon serratus (Pennant) [*Leander serratus*], Palaemonoidea, 111, 128
[*Pagurus varians*] see *Palaemonetes varians*
Palaemonetes, Palaemonoidea, 145, 146
Palaemonetes varians (Leach) [*Palaemon varians*], 146
Palinurus, Palinuroidea, 40, 41, 92, 152, 172
Palinurus elephas (Fabricius) [*Palinurus vulgaris, Panulirus vulgaris*], 64, 88, 89, 92, 94, 95, 152, 156, 172, 175, 176
(*Palinurus interruptus*) see *Panulirus interruptus*
(*Palinurus vulgaris*) see *Palinurus elephas*
Panulirus, Palinuroidea, 18, 141, 172, 194, 195, 206, 207, 208, 209, 210
Panulirus argus (Latreille), 4, 13, 23, 147, 148, 171, 172, 176, 196
Panulirus gracilis Streets, 4
Panulirus guttatus (Latreille), 23
Panulirus interruptus (Randall) [*Palinurus interruptus*], 3, 4, 5, 13, 19, 139, 152, 175, 176, 180, 208
Panulirus japonicus (von Siebold), 211
[*Panulirus vulgaris*] see *Palinurus elephas*
Paralithodes camtschatica (Tilesius), Paguroidea, 112
Penaeus, Penaeoidea, 72
Penaeus monodon Fabricius, 113
Penaeus setiferus (Linnaeus), 72
Podophthalmus, Brachyrhyncha, 13, 15
Podophthalmus vigil (Fabricius), 136
Porcellana platycheles (Pennant), Galatheoidea, 111
Portunus sanguinolentus (Herbst), Brachyrhyncha, 74, 210, 212
Potamocarcinus richmondi, Brachyrhyncha, 111
Praunus, Mysida, 146
Procambarus, Astacoidea, 13, 16, 18, 19, 34, 48, 139, 141, 172, 297, 298, 299, 309
Procambarus blandingii (Harlan), 76
Procambarus clarkii (Girard) [*Cambarus clarkii*], 3, 4, 10, 11, 13, 14, 20, 23, 24, 41, 42, 50, 72, 73, 76, 81, 84, 88, 90, 91, 92, 95, 96, 136, 137, 138, 155, 156, 172, 186, 216, 243, 295, 298, 299, 313

Systematic Index

Scylla, Brachyrhyncha, 215, 298
Scylla serrata (Forskål), 12, 215, 305, 310
Squilla, Stomatopoda, 62, 206, 209, 214
Squilla mantis Latreille, 169
[*Squilla oratoria*] see *Oratosquilla oratoria*

Tritonia, Nudibranchia, 227

Uca, Brachyrhyncha, 62
Uca pugnax (S. I. Smith), 76, 95
Uca rapax (S. I. Smith), 111
Uca thayeri Rathbun, 111

Subject Index

A

Abdominal musculature, decapods, 44, 46
Abdominal proprioceptors, 47
Abdominal reflexes, 47
Ablation experiments, 294
Anus, 186
Anal nerves, 186
Anal proprioceptors, 188
Apodeme tension receptors, 40
Autotomizer muscles, 118
Autotomy, 107
  behavioral role, 128
  consequences, 109, 112
  forced, 111
  mechanism, 123
  motor activity, 122
  natural incidence, 109
  physical factors, 112
  role of cuticular stress detector, 127
  sensory influences, 126

B

Basi-ischiopodite, crabs, 117
  tendons, cuticular structures, 117
  muscles, 120, 121, 124
Breakage plane, crustacean limb, 108, 117
Breaking force, in autotomy, 115

C

Cardiac ganglion, 205
  activity patterns, 207
  anatomy, 206
  cardioregulatory innervation, 214
  driver potentials, 212
  endogenous properties, 210
  intrinsic reflexes, 213
  neural organization, 208
  synaptic interactions, 208, 209
Cardiac sac, 170, 195
  musculature, 175
Caudal photoreceptor, 23
  anatomy 24
  physiology, 24
Chordotonal organs, 37
Command fibers, 41, 49
Command interneurons, 288
  abdominal, 296
  activity and correlated behavior, 309
  identified crab, 311
  sensory input to, 302
Command systems, 79
Commissural ganglion, 170, 172, 179, 183
Compensatory eye movements, 133
Complex neuropils, 308
Control of limb movements, 61
Coordinating neurons, 80
Crayfish claw
  central inputs, 218
  claw motoneurons, 219
  control of closing, 216
  interactions of motoneurons, 218
  neurons, 217
  proprioceptive reflexes, 216
  tactile reflexes, 217

Subject Index

Crayfish escape behavior, 241
 corollary discharge, 267
 escape response, 244
 inhibition, 275, 278, 280, 282, 285
 lateral giant mediated, 246
 medial giant system, 254
 modulation, 283, 286
 non-giant system, 254
 uropod muscles, 266
Crayfish optic nerve, 2
Cuticular strain detector, 115
Cuticular stress detectors, 39

D

Defecation, 186
Dimming fibers, 12
Distributed reflexes, 40

E

Efferent copy, 80
Efferent sensory control, 42
Equilibrium interneurons, crabs, 303
 anatomy, 305
 input from leg proprioceptors, 304
 input from statocyst, 304
 nonspecific inputs, 306
Equilibrium reflexes, 144
Esophageal chemoreceptors, 185
Esophageal ganglion, 172, 175, 178, 183
Esophageal movements, 173
Esophageal nerves, 179
Esophageal oscillators, 175
Esophageal sensors, 173
Excited states, 308
Eye control system, 147
Eye cup
 anatomy, 134
 motor neurons, 136
 muscles, 135, 137
 withdrawal, crab, 214
Eye muscles, 136
Eye withdrawal motoneurons, 215

F

Fast phase generator, 142
Fracture plane, crustacean limb, 113

Foregut, control, 170
Foregut neurons, 174
Foregut rhythm, 184

G

Gastric mill, 179, 195
 chemoreceptors, 184
 mechanoreceptors, 184
 motoneurons, 180, 182, 195
 neural circuit, 196
 teeth, 180, 182, 183
Gastric rhythm, 180, 184
Gastropyloric movements, 185
Gravity responses, 97, 152

H

Hepatopancreas, 177
Hindgut movements, 186

I

Illumination field effect, 144
Inferior ventricular nerve, 178
Interganglionic efferent interneurons, 187
Integration in interneurons, 307
Integrating segment, 215
Inter-limb coordination, 75
Interneurons, crayfish
 descending motion sensitive, 22
 mechanoreceptive, 19
 multimodal, 20
Intersegmental control systems, 79
Intrasegmental motor programs, 64

J

Jittery movement detectors, 312
Jittery movement fibers, 4, 13

L

Labrum, 168
Lateral giant neuron, crayfish, 246, 297
 afferent inputs, 247, 248
 physiological response, 249
 habituation, 251
 sensory interneurons, 250

Lateral hemi-giant neuron, crayfish, 297
Leg muscles, lobster, 64
Load compensation, 51, 91
Locomotion, 61

M

Mandibles, 166, 167
 muscle receptor organs, 168
 proprioceptors, 167
 rhythms, 167
Medial giant neuron, crayfish, 254, 297
Medial hemi-giant neuron, crayfish, 297
Metachronal rhythm, 75
Midgut control, 186
Motoneurons
 inhibition in small systems, 224
 interactions with interneurons, 223
 interactions with motoneurons, 224
 intrinsic properties, 220
 oscillatory properties, 221
 sensory functions, 221
 sensory inputs, 221
 size principle, 221
Mouthpart receptors, 169
Movement fibers, 13, 15, 17
Multimodal interaction, 154
Multimodal interneurons, 313
Multisegmental neurons, 309
Muscle receptor organs, 47
Myochordotonal organ, 38, 92

N

Network rhythm generator, 228
Nerve cord stretch receptors, 48
Neuronal integration, 155
Neuronal oscillators, 81
Nystagmus, 137

O

Oculomotor neurons
 crabs, 149
 crayfish, 138
Optokinetic memory, 143
Optokinetic nystagmus, 138
Optomotor fibers, 148
Optomotor responses, 96, 138

P

Paragnatha, 166, 169
Posterior intestinal nerves, 186
Posterior stomach nerve, 184
Postural control, 33
 abdomen, 44, 49
 thoracic legs, 35, 41
Proprioceptive feedback, 141
Proprioceptive reflexes, 39
Pyloric cycle, 179
Pyloric filter, 176
Pyloric motoneurons, 195
Pyloric neural circuit, 196
Pyloric neurons, 176
Pyloric press, 176
Pyloric rhythm, 195

R

Reflex behavior, 154
Reflexes, 83
 apodeme tension receptor, 91
 cuticular stress detector, 91
 distributed, 93
 exteroceptive, 96
 interactions between, 94
 intersegmental, 94
 muscle receptor organ, 91
 negative feedback, 84, 87
 pathways in swimmeret system, 85
 positive feedback, 88
 proprioceptive, 83
 resistance, 84
 sensory hair, 92
 variable, 88
Resistance reflexes, 36, 43
Rotational nystagmus, 149
Rhythmicity
 cardiac ganglion, 230
 from network connectivity, 227
 gastric mill, 229
 intrinsic to single neurons, 225
 mechanisms, 225
 pyloric, 228
Rhythmic limb movements, 77

S

Seeing fibers, 17
Segmental oscillators, 77
Sensory interneurons, 295

Subject Index

Sensory modulation, of motor programs, 83
Space constant fibers, 16
Statocyst canals, 149
Statocyst responses, 144
Statolith hairs, 145
Statoreceptors, 147
Stepping cycle, 65, 70
Stomatogastric ganglion, 172, 175, 179, 180
    activity patterns, 194
    endogenous rythmicity, 204
    neural organization, 196
    neuropil, 199
    non-spike mediated inhibition, 200
    plateau potentials, 203
    synaptic morphology, 199
    transmitter chemistry, 199
Stomatogastric nerve, 178
Substrate responses, 152
Sustaining fibers
    circadian rhythm, 9
    polarization sensitivity, 12
    receptive fields, 3, 4, 6, 7
    shadow effect, 10
    spectral sensitivity, 12
Swimmerets, 70, 79, 82, 301
Swimming, 63, 71
    exopodite, 72
    lobsters and crayfish, 71
    uropod, 73, 88
    walking legs, 75

T

Thoracicocoxal muscle receptors, 38

Thread hairs, 150
Tilt compensation, 145

V

Ventral mechanoreceptors, 49
Vestibular eye reflex, 151
Visual feedback, 142
Visual interneurons, crayfish
    bilateral, 18
    circumesophageal commissure, 20
    heterolateral, 17
    homolateral, 3
    optic peduncle, 2
    synaptic interactions, 21
Visual movement fibers
    crayfish, 5
    receptive fields, 3, 6, 7

W

Walking
    brachyuran, 68, 76
    command systems, 80
    gaits, 75
    joint movements, 67
    lateral, 69
    macruran, 64, 76
    motor output, 68
    motor programs, 68
    stepping cycle, 70
Weighted neuronal integration, 155